SOUL AND BODY IN HUSSERLIAN PHENOMENOLOGY

ANALECTA HUSSERLIANA

THE YEARBOOK OF PHENOMENOLOGICAL RESEARCH

VOLUME XVI

Editor-in-Chief:

ANNA-TERESA TYMIENIECKA

*The World Institute for Advanced Phenomenological Research and Learning
Belmont, Massachusetts*

SOUL AND BODY IN HUSSERLIAN PHENOMENOLOGY

Man and Nature

Edited by

ANNA-TERESA TYMIENIECKA

The World Institute for Advanced Phenomenological Research and Learning

Published under the auspices of *The World Institute for Advanced Phenomenological Research and Learning*, A-T. Tymieniecka, President

D. REIDEL PUBLISHING COMPANY

A MEMBER OF THE KLUWER ACADEMIC PUBLISHERS GROUP

DORDRECHT / BOSTON / LANCASTER

Library of Congress Cataloging in Publication Data

Main entry under title:

Soul and body in Husserlian phenomenology.

(Analecta Husserliana ; v. 16)
"Published under the auspices of the World Institute for Advanced
Phenomenological Research and Learning."
Includes index.
1.   Husserl, Edmund, 1859–1938—Congresses.
2.   Phenomenology—Congresses.   3.   Mind and body—Congresses.
I.   Tymieniecka, Anna Teresa.   II.   World Institute for Advanced
Phenomenological Research and Learning.   III.   Series.
B3279.H94A129   vol. 16   142'.7s   [128'.1]   83–3218
ISBN 90–277–1518–1

Published by D. Reidel Publishing Company,
P.O. Box 17, 3300 AA Dordrecht, Holland.

Sold and distributed in the U.S.A. and Canada
by Kluwer Academic Publishers,
190 Old Derby Street, Hingham, MA 02043, U.S.A.

In all other countries, sold and distributed
by Kluwer Academic Publishers Group
P.O. Box 322, 3300 AH Dordrecht, Holland

TABLE OF CONTENTS

PART VI

THE HORIZON OF NATURE AND BEING

PART VII

HUSSERL AND THE HISTORY OF PHILOSOPHY

ANNEX

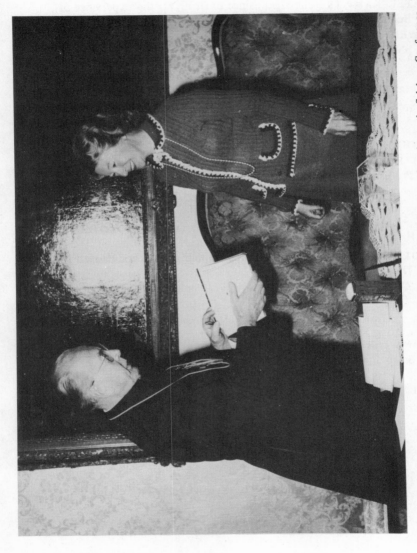

H. E. Dr. Franz König, Cardinal of Vienna, in a discussion with A-T. Tymieniecka prior to the Salzburg Conference

ACKNOWLEDGMENTS

The present volume contains a distinctive segment of the research program on 'Nature and Man' conducted by *The World Institute for Advanced Phenomenological Research and Learning* and its three international phenomenology societies. In the main it has been presented at the Institute's *VIIIth International Phenomenology Congress in Salzburg*, August 15–21, 1980, *The American Phenomenology Seminar* in North Pomfret, Vt., June 10–25, 1980, and at *The Boston Forum For The Interdisciplinary Phenomenology of Man* for the year 1980–81. One of the essays was read at the Institute's *First Oriental Phenomenology Conference*, which took place in Hong Kong in November 1980.

The Salzburg Conference was held under the auspices of Dr. Rudolf Kirchschläger, President of the Federal Republic of Austria.

Acknowledgments are due to Professors Karl Ulmer, Peter Kampits, and Hans Köchler of the *Oesterreichische Gesellschaft für Phänomenologie* — born on the occasion of the conference — for acting as co-sponsors of this event and introducing us to the Austrian authorities. We owe our gratitude to pesonalities from Austrian academic and public sectors for their endorsement as well as concrete help in carrying out the conference. Our thanks are due to the Austrian Preparatory Committee, to The Honorary Committee of the Conference, and especially to H. E. Dr. Franz König, Cardinal of Vienna.

I express, in the first place, however, my appreciation and thanks to H. E. D. Dr. Karl Berg, Archbishop of Salzburg, Messr. Dr. Herbert Batliner, Austrian Consul General in Liechtenstein, Komm. Rat Dr. Hans Leobacher, and Dr. Franz Ruedl, Director der Salzburger Sparkasse for their generous contribution to our enterprise.

Professor Dr. Georg Pfligersdorffer, *President des Internationalen Forschungszentrum für Grundfragen der Wissenschaften*, Salzburg, has been the host of the conference held in Mönchsberg and Dr. Wielfried Haslauer, Landeshauptman of Salzburg offered to the participants a reception in the Residenz Palast. Their understanding of the intellectual and social significance of our work and their generosity has not only given us encouragement but also made this conference possible. Dr. Blaikler, *Obersekretär des Internationalen Forschungszentrum*, Herr Hans Hochgeratener, and Frau Hilde Gruber, who did not spare themselves any effort to make our stay in Salzburg an unforgettable occasion, cannot be forgotten.

I want to thank personally Dr. Hans Rudnick for his invaluable assistance in running the conference.

And last, but not least, appreciation and thanks are due to the numerous phenomenological societies and centers as well as to the distinguished personalities from the philosophical sector, who have endorsed this important occasion for the meeting of minds.

Dr. Kadria Ismail, research fellow in the WPI, has helped with the editorial work of this volume.

A-T. T.

INAUGURAL ESSAY

A group of the participants at the Salzburg Congress.

ANNA-TERESA TYMIENIECKA

FROM HUSSERL'S FORMULATION OF THE SOUL–BODY ISSUE TO A NEW DIFFERENTIATION OF HUMAN FACULTIES

1. INTRODUCTION

Our choice of the soul–body problem in Husserl's phenomenology as the theme of this congress is not an arbitrary one. On the contrary: this problem presents itself to us inevitably as we carry out the research objective of our symposia and congresses, which is to explore, step-by-step, the fundamental questions that phenomenology poses, by means of criticism and then refashioning. The soul–body problem reveals itself, astonishingly, as Husserl's deepest concern and thus as an integral part of the foundation of phenomenology as such. In the course of my Husserl/Ingarden critique the problem presented itself to me compellingly because Ingarden was quite apparently stumped by it. At the end of his monumental opus, Ingarden, who adopted all of Husserl's acquisitions in this area, simply stopped short at the soul–body problem and could not get beyond it. We could then say with T. S. Eliot: "And the end of all our exploring/ Will be to arrive where we started/ And know the place for the first time." We have met here to look at this place together "for the first time."

In his *Cartesian Meditations*, Husserl relates his work specifically to Descartes, to whom the dichotomizing, the origin of the soul–body problem in modern philosophy, is ascribed. It was the overcoming of the idealism/ realism antithesis arising from this position that constituted the basic intent of transcendental phenomenology, as has so often been pointed out. Husserl expressly recognizes and affirms Descartes' "cogito ergo sum," which seeks the certainty of cognition in the awareness, of having made the breakthrough to immanent consciousness despite the necessity for the completion of cognition by pure transcendental consciousness to guarantee this assurance of cognition. Husserl devoted his studies to this purpose, especially in the second volume of his *Logical Investigations*[1] and in *Ideas I*.[2]

Pure consciousness, which is the domain of the certainty of cognition, correlates with the assumption of the intentionality of consciousness which has, in Husserl's view, the essential task of sense-bestowing, establishing the priority of the intellect.

But for Husserl it was not enough to pursue the Cartesian concern for the

3

A-T. Tymieniecka (ed.), Analecta Husserliana, Vol. XVI, 3–10.

certainty of cognition. With a grand sweep he undertakes to bridge the gap between intellect and empirical psyche and to make good the loss of actuality/real world that resulted from the Cartesian undertaking. *Ideas II* can be regarded as the attempt to regain reality by catching it up in the same net with pure consciousness. Examined more closely, the investigations contained in this work constitute a coming-to-grips with Descartes. Point by point Husserl works out, in his own way, the intuitions that are contained in the Cartesian "cogito" or that are merely suggested by it. He demonstrates, in this fundamental essay, that the full development of the system of immanent consciousness — which Descartes lacked — extends to indubitable cognition by way of the connexion to the empirical and to reality altogether, he charts a new course. Yet, insofar as he is attempting to regain the organism and reality that eluded Descartes' immanent consciousness, it is no longer preeminently a matter of the manner of cognition, but of the *existential connections* between the human functions that can be differentiated by the phenomenologically objectivating consciousness.

The continuous existential unity (in which the body/organism plays out its existential autonomy) of the functional provinces would also include the human "I can" in all its fullness.

In our present conference program Husserl's efforts to set forth the unity of the human being will be treated from many standpoints in serious explications and discussions.

Husserl's differentiation (which has not been given sufficient attention) of functional realms in the human being ("domains," as I have called them in my Ingarden/Husserl critique), i.e., body/organism, organism/soul, soul/mind, constitutes an undertaking that I consider to be most important and interesting. I shall also begin with the *distribution of the functions* and not with the customary constitutive/cognitive orientation. But I maintain that, in spite of Husserl's frequent treatment of this differentiation, his way is not the way to achieve full access to the faculties underlying the functions — the "I can" in all its manifestations: I think, I value, I know, I will, I do. But now it seems that Husserl's overall conception is to be found not only in his analyses that are oriented to research of cognition, i.e., of the objectivating subject, but must be sought first in his explication of the *faculties* that participate in the ethical, i.e., practical subject — the design of the collective functions which, in his view, makes up the unity of consciousness.

In Husserl's view — as appears already in his lectures, before the first volume of the *Ideas* was published — the acting subject extends not only from the cognitive, i.e. logical functions of reason with which the feeling functions of the affective mind are tied but also to the valuing, willing and doing which stem from both functions. Nevertheless, following Husserl one finds that it is only in the intentional founding of objectivating, valuing and willing, whose unity is accomplished as, and comes to an apex in, the unity of the "total consciousness" in human action, that one could find an essential correlation between actuality and present consciousness as a presupposition for an absolute cognition.

It is universally assumed that Husserl identifies consciousness with intentionality. De Waelhens emphasizes this in 'L'idée phénoménologique de l'intentionalité,' but asks whether Husserl understood the presuppositions of his own theory. In his opinion, intentionality, on which the whole structure of consciousness rests, instead of being "un point de départ facile et évident," must itself be grounded "dans le cadre d'une théorie complège touchante les rapports de la conscience et du réel." Now it is in the idea of intentionality that the attempt to overcome the alternatives of realism/idealism rests (also in the differentiation of the functional domains). Does intentionality really go to the root of the interrelations of consciousness even if the latter were grasped in the whole breadth of its functional operating, and does it really go to the root of the real? As Levinas emphasizes, intentionality has nothing to do with the relations between real objectivities; it is essentially " . . . l'acte de préter un sens" – *Sinngebung* – sense-bestowing. But I wish to emphasize that sense-bestowing should not be limited only to the functions of the intellect.

In Husserl himself one does not see the relationship of consciousness to the real treated only in the intellective, fully constitutive or primordially constitutive perspectives. The question is, however, whether the unity of "total consciousness" in Husserl does not ultimately rest on intentionality thus hindering the explanation of the emergence of the other normative functions.

Here I shall undertake a short presentation of Husserl's conception of the human faculties that are said to make up the "total consciousness," and then I shall proceed to my own conception of sense-bestowing which I call "cyphering," in its imaginative form-giving.

2. TOTAL CONSCIOUSNESS AND THE DIVISION OF FACULTIES IN HUSSERL

Husserl proceeds from the basic faculty of the subject "I can" and ascribes to reason all the functions of consciousness which in their unity activate the human faculties. There is a fundamental distinction made between the objectivating, the valuing and the practical consciousness (or attitudes of consciousness). Corresponding to these distinctions there are, respectively: the object of judgment, the object of value and the object of will. But this differentiation refers (as Diemer also shows) to fundamentally different faculties of reason: understanding, which is altogether responsible for the logical structures of cognition; and affective reason, which is on the one hand the source of affective valuing and value-measuring and on the other hand, however, is the source of all affective conscious processes. This means that in every conscious process, in every really immanent act, we "are-there-too-with the object" ("mit dem Gegenstand dabei sind") and that we "live" in the act of experiencing.[3] This also implies the corresponding division of kinds of objects according to their predicates, relations and states of affairs (e.g., in the objectivating orientation of consciousness the objects are constituted by means of logically predelineated, objectively valid predicates). Yet the value-objects and practical objectivities which arise in the affective mind and are supplied with predicates of the affective mind (referring to purely subjectively-valuing acts of the subject) are half subjective.

Although there are extensive interweavings between the objectivating understanding, which is directed toward rational structures, and affective reason — interweavings which are constantly changing in the changing life of consciousness — and although values stemming from affective valuation are supported by rudimentary objective references, the two faculties coincide, in Husserl's view, most sharply not in the higher or originally intellective spheres but in practical consciousness.

Decision-making, which, as is known, distinguishes the intellect as an autonomous domain of functioning from the psyche, is to be related to the intellect, i.e., understanding, only in its objective aspect; but it derives, as willing decision-making, from the active volitional orientation which, proceeding from valuing done by the affective mind, actualises itself in willing and doing. Although it is the objective valuing, i.e., the valuing explicated by the understanding, and the volition that radiates from value that constitute the ethical Subject, yet it is in the feeling-faculty proper, of the affective mind, which enjoys, and accordingly brings forth valuing and feeling, that the unique subjectivity comes about through the purely subjective predicates of the object that carries value. These predicates signify a unique *subjective sense-bestowing*, which takes place above all in the psychic-mental realm.

The life-world as a "harmonious social living," which Husserl ascribes especially to the effect of the unity of the functional domains of body/ organism, organism/psyche, psyche/mind, reaching their apex in the mind, is to be traced — on the part of the faculties — to the ethical Subject, i.e., to the valuing and acting Subject. As Husserl himself expresses it:

Gesetze des Handelns in einer sozialen Gemeinschaft von Personen, die auf eine gemeinsame Welt von möglichen Werten bezogen sind, die sie realisieren, in deren Realisierung sie sich kennen und fordern können ... sind in der Übereinstimmung von den aus subjektivbedingten Lebenssituationen hervorquellenden Wertungs-und Willensakten der einzelnen singulären ethischen Subjekte untereinander in den durch die Vernunft-Einsichtigkeit-Objektivation intersubjektiv ....[4]

Ultimately this unity of reason — i.e., of understanding and of the affective faculty — operates in a combined way in the personal Ego, which promotes the whole range of the "I can" in the human being. As Husserl expresses it: "das Ich ist der identische Brennpunkt aller Aktionen und Affektionen, das Ich des 'Ich kann,' 'ich tue', 'Ich denke', 'ich werte', 'Ich will'.[5]

I have tried to show how the functional domains of the human being — which in *Ideen II* are differentiated only from the standpoint of the objectivating consciousness, in their existential links — in the perspective of Husserl's total thinking, first of all derive from the division of faculties. Second, I have tried to show how by the major weight being shifted to practical consciousness, i.e., to acting and doing, and proceeding from this standpoint, there exists still another kind of existential interweaving of the functional totality: body-organism-soul-spirit. In fact, it is apparent that without taking these into account one cannot expect to find a solution to the difficulties that Husserl's view of the unity of the soul/body entails.

Without occupying myself further with these perplexities of Husserl's, I turn now to the constructive treatment of the problems — first by proceeding from what we have gained from Husserl's view and continuing, then, with my own research.

3. EXISTENTIAL SENSE-BESTOWING AND THE NEW DIVISION OF THE FACULTIES

At the outset I would like to throw open two questions. If one takes some distance from the objectivating direction of the immanent processes and turns to conscious living itself, the life of consciousness appears to us in a unique unity and fullness of conscious living and of living-through. The question arises: "is this fullness and unity a matter of the affective mind alone?" As a matter of fact, if one turns to the social living of the life-world where we see, with Husserl and also beyond him, how the individual acting subject effectuates itself, this self-effectuating being not merely self-enclosed but also extending over the individual span of time, the question arises as to the possibility of continuity of social living. If it were *only* a matter of sense-bestowing by the understanding and the affective mind linked together, how could one account for the undeniable continuity of living through one's life within the intricacies of inter-subjective relations?

(1) *Enjoyment and Creative Orchestration*

As far as the first question is concerned, I wish to assert that the unique fullness of the mental life process in which the Ego is rooted is by no means only the fullness of feeling of the acts of mental life, whether of knowing or of valuing. Nor is it a result of all the constitutive and preconstitutive acts that bring about the "identification" of the conscious I. On the contrary, the fullness of the mental life process is the basis of this constitutive process of identification. It is a uniquely peculiar fullness of conscious life which not only fills out everything but also encompasses everything and which leaves no gap between the self-singularizing mental life processes; it is, rather, a *self-renewing* fullness of mental life processes which, living itself out in all acts, always means more than these; the living subject lives and rests in it even though it also contains it. It does not allow itself to be taken up in the cognitive, valuing and acting consciousness alone. In it the consciousness of the child, not yet broadly developed, finds an incomparably rich fullness of the life-world that by far exceeds that of the fully constituted consciousness of the adult. While the objectivating reason aiming at a logical order and objective distinctions, sets up rifts and cleavages; while the affective mind produces valuing distinctions and the will produces action, causing sharp breaches in the continuity of the individual's life-world constitution – all directed to objectivating distinctions – what else but the imagination[6] could bring about this unity which is so fundamentally essential to individual mental living? It seems that what I call "imagination" is a particular faculty which has the unique role between the cognitive, objectivating reason and the affective, valuing sphere of emotion and volition; the role of filling in, from within, the emptied life of consciousness which is continually impoverished by the differentiating functions, taken apart and weakened, and filling it in with affective phantasma that hover freshly before it. In this way it enriches what has become impoverished and transforms what has become unenjoyable with a new enchantment. In this way (1) imagination would prevent the constitutive composite image of the conscious functions from being torn apart by producing the fullness of immanent life and providing a bridge of invention; moreover (2) the possibilities, on the part of conscious-ness, for coalescence and adaptation, which form the basis for its alterations, and otherwise remain incomprehensible would be ascribable to imagination; (3) when imagination enters the scene in a completely unfolded *creative context of the total functions* which unfolds together with the imagination, the possibility of making a moral decision would be ascribable to imagination as a *morally unconditional decision*. (As we remind ourselves, it is in the moral decision that Husserl sees the autonomous standing-out of the spirit from the soul and the ground of social community life.)

(2) *Orchestration of the Existential Sense-bestowing*

With this we come to the second question thrown open above. How does sense-bestowing occur in the social life-world? It is there, in the fully devel-oped acting and doing that employs all the human faculties, where full *sense-bestowing*, i.e., preconstitutive and supraconstitutive activity of all the human functions, taking shape in man's inwardness, is exercised. It is there, also, where we can observe all the faculties that activate these functions, since here it is no longer a matter of a conscious identification on the part of the conscious subject, but rather one of *existential* "self-individualization" of the human monads in action and work. The question as to the *unity* of the functions is sharpened here to that of the *continuity of the process of the existence of the human being that is being carried on by means of the constitution of the life-world*. The being of the lower animal is rigidly bound to its surrounding life-world, whose vital, meaningful net has been cast forth by the animal, with the result that when the sense links in its environment are broken, or changed by external circumstances, its sense-bestowing organs of sensation are thus also prevented from functioning. The human being, on the other hand, when he meets up with the inadequacy of the domain of his life-world, will immediately search out a way and *undertake* to accommodate the latter to himself once more. He develops inventive powers for reshaping this meaningful network of life, or even for casting forth a new one. This is the distinguishing mark of the human mind and of the human decision. Thus human existence stands out as *historical*.

When Eugen Fink says that "the constituting life of transcendental subjec-tivity is in itself historical," one can see in that much more than the continuity of the course of the processes of consciousness cast forth logically and as an object, and of the universal life-world constituting itself in them. What we mean by human "history" is precisely the continuity of the sense-bestowing human existence which must continually be cast forth anew, through and above the drastic reshapings of his life-world that the human being must undertake again and again from one epoch to another, from one revolution to the next. The fact that there is a continuing weaving of the connexions in spite of the continual emergence of the new and in the midst of radically opposed tendencies and currents, indicates that there are connexions that cannot ultimately be explored. It is this which brings about the sense of wonderment and pondering about history and which gives human life deep and unexplorable roots in the common fate of mankind. But this connexion is shaped through interweaving by inventively and enterprisingly oriented sense-bestowing on the part of the existentially self-individualizing human monad. This brings to light the fundamental role of the imagination, which — as a third faculty together with the intellect and the affective mind — sets human sense-bestowing, from its origins, on a unique plane. It is only on this plane that one can see how the functional unity of the human faculties, which come into play together, brings about a specifically human totality (i.e., of body/organism, organism/soul, soul/spirit) in a venturesome, active nexus of meaning that carries out its effect on human existence.

Just as reason is oriented to cognitional identification and the affective mind is oriented to affective valuing and willing, imagination is oriented to action. It is only when it emerges an an autonomous faculty of the human being that one can grasp the psyche-organism relation by its roots.

This is the image that I propose, on the basis of my attempts to look at these things in a new way.

The World Institute for Advanced Phenomenological Research and Learning

Translated from the German original by Dr. Barbara Haupt Mohr

NOTES

1 Edmund Husserl, *Logische Untersuchungen, Untersuchungen zur Phänomenologie und Theorie der Erkenntnis*, II/1, fifth ed., Tübingen: Max Niemeyer Verlag, 1968.
2 *Ideen zu einer reinen Phänomenologie und phänomenologischen Philosophie; erstes Buch*, The Hague: Martinus Nijhoff, 1971.
3 Cf. *Ideen II*, p. 8.
4 *Die Ethischen MS.* F121, p. 4, quoted according to Roth, cf. Alois Roth, *Edmund Husserl's ethische Untersuchungen*, The Hague: Martinus Nijhoff, 1960.
5 Op. cit., F128, p. 247.
6 'Imaginatio Creatrix,' in *Analecta Husserliana*, Vol. III.

PART I

THE PROBLEM OF EMBODIMENT
AT THE HEART OF PHENOMENOLOGY

In the grounds of Mönchsberg: Professors Yoshihiro Nitta, Hirotaka Tatematsu, Keiichi Noe, Mrs Noe, Ichiro Yamaguchi, Mamoru Takayama, Evelyn M. Barker, Marie-Rose Barral, Stephen F. Barker, Anna-Teresa Tymieniecka.

YOSHIHIRO NITTA

THE SINGULARITY AND PLURALITY OF THE VIEWPOINT IN HUSSERL'S TRANSCENDENTAL PHENOMENOLOGY

In his later work, *The Crises of European Sciences and the Transcendental Phenomenology*, Husserl designates the indivisible bond among science, life, and philosophy as the conceptual unity of the intellectual culture of modern Europe. According to Husserl, the crisis of European culture and science is rooted in the dissolution of this conceptual unity. Therefore, this crisis can be overcome only if the authentic bond of these three moments is reestablished; and to Husserl, this is precisely the task of transcendental phenomenology.

The triadic idea of *The Crises* corresponds structurally to the unity of the teleology in immanent consciousness which is formed from three moments: "the external conceptualization (*Limesidee*), the horizon, and the transcendental reflection; namely, the adequate evidence, the process as motivation, and the apodeictic evidence." Based on the results of his inquiries in genetic phenomenology, Husserl was able to have the triad of the teleological structure of reason revealed concretely as the dimension of the transcendental historicity. According to Husserl, this triad constitutes the basic structure of the modern way of knowing (*neuzeitliches Wissen*); and only from this triad can various misconceptions about science such as objectivism be properly criticized, as they appear in modern philosophy of science.

If we construe Husserl's attempt as a new justification for the theory of cognition being essentially perspective, this attempt purports to view both the world of the senses given in the perspective and the supposedly de-perspectivized scientific world altogether as being possible only in a perspective (horizon). It means further to dispose even the transcendental reflection itself in the primordial perspective of life.

To be sure, recent investigations into the perspectivity of cognition deal mainly with the domain of the sensory cognition. If here the concept of the sensory be construed within the Platonistic frame of reference of reason and senses, then the perspective structure may be limited merely to the sensory world. However, it appears appropriate to extend this perspectivity above and beyond the sphere of the world of senses as the only way of knowing. As Heidegger points out about Nietzsche's perspectivism, it is not only of utmost significance to get rid of the Platonic conception of the senses, but also to open up a new path in which the sensory as well as the nonsensory world of

13

A-T. Tymieniecka (ed.), Analecta Husserliana, Vol. XVI, 13–17.
Copyright © *1983 by D. Reidel Publishing Company.*

the spirit may be asserted. Such a new scheme of order, designed to provide the theory for the perspectival structure of cognition, cannot be accomplished by the return from the idealized world of sciences to the world of senses, but by revealing a new dimension of philosophical reflection, which enables us to rectify and reconstrue the de-perspectified world of science as a possible perspective, i.e., as a mode of appearance of the world.

As a doctrine of appearance phenomenology primarily procures the meaning of a theory of the perspectivity. For the concept of appearance characterizes no other than the problem of identity and difference between what appears and its appearance. Therefore, phenomenology may be reformulated as the analysis of the "mediated immediacy in the mode of givenness of being." This phenomenon of the cognition's being primarily mediated may be here understood as the perspective structure of cognition. In the following discussion we extract the most important problems of perspectivity from Husserl's phenomenology, insofar as it is to be regarded as a doctrine of appearance.

(1) First of all, and in the most primary sense, the phenomenon of perspective reveals itself in the perspectivistic *Abschattung*, namely, that a thing is never given on all sides, but only from one side. Husserl's analysis of perception shows the role of the body as the function of locus as well as the mutual conditional relationship of aspect-datum and locus-datum. Thus, the concept *Hylē* is investigated in connection with the kinesthetic function, for the appearances of what appears are kinesthetically motivated. "The appearances constitute dependent systems. Only as the dependent (elements) of kinesthesis can they merge continuationally into each other and constitute a unity of a sense" (*Husserliana*, 11: 14). In looking at the analysis of corporeality and of the kinesthetic function in regard to the problem of perspectivity, one may notice two important problem areas, namely, first, the function of the primary viewpoint which is given with the body as an orientation-zero-point, and second, the problem of the primary distance.

(a) The body as a zero-point of the appearance (of phenomenon) is given with all appearances (of phenomena). "My body is always in its (being) there. It does not require any phenomenal reference back to the body" (*Husserliana*, 15: 304). As "a moment of the primary experience of the world" (*Husserliana*, 15: 304), the appresentation of the body is un-thematically and anonymously present simultaneously. The primacy of a viewpoint for the phenomenon of perspectivity lies in the fact that the un-thematic being-there simultaneously, together with the consciousness of the "absolute being here," functions as the condition for all phenomena.

(b) Between the body's being un-thematically there simultaneously as a moment of constitution and of the perceived environment, the lived distance always thus plays a role. This distance, which is the condition of the constitution of space, is rooted in the final analysis in the double phenomenality of the body, namely, in the body-corporeal unity-difference (*Husserliana*, 15: 304). On the one hand, the body belongs to the environment as a corporeal unity; on the other hand, it is the un-thematic being-present-to which cannot be objectified. One can find the primary form of this double phenomenality of the body in the "functional togetherness of the touching and the touched sense-organ" (*Husserliana*, 15: 298), or in the double sensation of touching oneself (*Husserliana*, 15: 302); in other words, in the (mutual) reversion between the body and the corporeal unity. So far as the interplay of the body and the corporeal unity is basically the transcendental condition of all lived distance between the point of view and the field of view, it forms a moment of constitution of perspectivity.

(2) Through the thematic ascertainment of the object, there now results the phenomenon of the horizon. If in the horizon, whether it be an inner or an outer horizon, a sensory frame of reference, in other words, a definable indefiniteness, is pre-projected, then this means that the process of being mediated in the immediate givenness of an object takes place. In the continual process of experience, which contains horizons of more and more distant, more and more vague universality, the anticipated universality is gradually particularized. If experience shows itself as the process of identification and differentiation of what appears and its appearance, then it is, in the final analysis, the self-unfolding process of the structure of cognition-being known as knowledge. Since the horizon arises at the same time with the ascertainment of the object, the incompleteness of the horizons points to the inexhaustibility of cognition.

(3) The endless process of sensory experience is promoted by the tendency to take complete possession of the object. Such a complete acquisition remains, however, an unrealizable goal, an infinitely distant telos which is, in principle, unattainable. It is a system of endless processes of appearing (of an object) which is to be absolutely ascertainable by its essential "type" (*Husserliana*, 3: 352). The complete givenness (of an object) is therefore rather an external conceptualization which is indispensible as the regulative principle for the process of appearance. So far as experience is the process of endless approximation to the external conceptualization, it can be interpreted as the dynamic interplay of perspectivation and de-perspectivation.

(4) Husserl understands scientific knowledge as idealization. However, the

method of idealization, according to Husserl, transcends the perspectivistic world and objectivizes and theoreticizes it, but it is only possible in the continuous actualization of the apperceptive horizon. For idealization means a change (shift) in the appearance of the world. If the transcendental reflection can attain that dimension in which all changes in the appearance of the world become transparent, it becomes possible to regard the sensory world and the scientific world in unity only as a perspectivization of the world.

The world never appears in itself thematically, but it gives itself to each subject in a special and different manner. Such a biased one-sidedness of point of view belongs to the nature of the appearance of the world, and yet the world, in accordance with its nature, must at the same time always be one; the concept "world" refers to a singular entity. Yet how is the consciousness of the one world possible, when the world appears in a different manner for each subject?

Since the world is always given to us already as a ground, the world precedes all objectivization. The passive, primordial doxa affects, together with the appresentation of being-present-to (*Dabeisein*) of the body, already before the objectivizing act. In the passive, primordial doxa, however, the world is understood un-thematically as a singular world. Therefore, there originates, in opposition to the apparitional manifoldness of the world, a universal, teleological movement which aims at the one world as the final representation of world. "The world itself in its truth is the constantly intended world in this life of consciousness, and it is a real and possible telos of a verifying knowledge as the ideal" (*Husserliana*, 15: 548). Within the endless movement toward the ideal of the one world, the objective worlds constitute themselves, namely, the particular worlds which are distinguishable according to their specific purposes. Intersubjectivity thus functions first of all as the condition of the individual particular appearance of the world and then as the executor of the world-constitution; the intersubjectivity as the executor belongs to the process of this universal movement. The problem of the relationship between the primacy and plurality of the point of view is therefore very closely connected to the question of interpersonality.

(5) The primacy which is peculiar to the point of view lies in its singularity. Husserl attributes such singularity to the world, to the body, and to the primordial ego. In this connection, by "singular" he does not mean simple numerical unity, in other words, a mere singular, but it means the liveliness in which the subject lives through itself. Thus, Husserl speaks in *The Crises* of a unique ego, namely, the primordial ego, the "ego of my epoché which can never lose its singularity and personal invariableness" (*Husserliana*,

6: 188). Perhaps the liveliness and the concept of the "living present" is to be understood as the functioning of the ego in its singularity. Only on the basis of such originality which is lived by me and which is therefore single is understandable the functioning of the point of view with its perspectivity. Only in such primacy which is lived by me and therefore single is the functioning of the point of view with its perspectivity founded. If, however, the point of view can function at the same time only in relation to the plurality of other subjects, how then does the singularity of the point of view and its plurality relate to each other? Do they relate to each other? If one understands the pluralization as egocentric, then the point of view loses it primacy and can no longer be a living point of view. If, on the other hand, the singularity of the point of view is understood as a numerical unity, then this again excludes the other points of view.

In a manuscript of Husserl's, the following statement stands. "I cannot exceed my factual being and in it I cannot exceed the intentionally determined being-together of others, etc., in other words, the absolute reality." In this connection, in other words, my factual being is a nucleus of primordial accidentality which is carried or borne within me. Only in such an irreducible absolute factuality is the living being-together of others included, for the single primordial ego functions precisely as always together with others as a living fact. To be sure, Husserl did not succeed in thematisizing such a fundamentally formulated being-together in the frame of problems connected with intersubjectivity. But in the last sketch of Husserl's thought, namely, in the teleology of history, an indication of the equal primacy of being-together with the factual being of the primordial ego can be found. "The whole personal responsibility for our own truthful being as a philosopher in our interpersonal appeal bears at the same time in itself the responsibility for the truthful being of mankind, which exists only as being directed to a telos . . . " (*Husserliana*, 6: 15). If the self-responsibility of the philosopher is at the same time the responsibility for mankind for the totality of monads, then the living relationship of the single ego to the other egos can be realized only in a practice which is the primary execution of transcendental reflection itself. In this sense, transcendental phenomenology as a theory of perspectivity is capable of solving the *aporia* of the point of view, namely, of reconciling the conflict between its singularity and its plurality.

From the left: Professors Yoshihiro Nitta, Hirotaka Tatematsu, Karl Ulmer, A-T. Tymieniecka, H. E. Dr. Karl Berg (Archbishop of Salzburg), Angela Ales Bello, Hans Beilner (Rektor of the University of Salzburg), Erling Eng, Stephan Strasser, and Georg Pfligersdorffer (President of *Das Internationale Forschungszentrum*, Salzburg).

STEPHAN STRASSER

DAS PROBLEM DER LEIBLICHKEIT IN DER PHÄNOMENOLOGISCHEN BEWEGUNG

Leiblichkeit ist eines der groszen Themen der abendländischen Philosophie. Die Denker des Okzidents entdecken in der Natur, in dem organischen Leben und vor allem im Menschen selbst einen Kontrast, ein Spannungsverhältnis, einen Bruch, und sie versuchen, ihn hermeneutisch auszulegen. Ihre Deutungen divergieren in hohem Masze. Es ist gewisz nicht gleichgültig, ob sie "sarx" und "pneuma", erste Materie und beseelender Akt, "extensio" und "cogitatio", "Natur" und "Geist" als die für die philosophische Menschenkunde entscheidenden Kategorien betrachten; der Unterschied zwischen monistischer und dualistischer Sichtweise ist grosz. Doch auch die Monisten werden mit dem anfänglich angedeuteten Rätsel konfrontiert. Wenn sie — etwa wie die griechischen Atomisten — einerseits von plumpen, schweren, eckigen Leibes-, andererseits von leichten, beweglichen, feurigen Seelenatomen sprechen; wenn sie wie Leibniz spiritualistisch schlafende von vollbewuszten, selbstbewuszten, erinnerungsfähigen Monaden unterscheiden, dan zeigt es sich, dasz sie denselben Urgegensatz zu erklären suchen, allerdings in einer sehr verschiedenen philosophischen Sprache.

Meine These lautet dahin, *dasz die phänomenologische Bewegung einen bedeutsamen Beitrag zur Hermeneutik der menschlichen Leiblichkeit geliefert hat*. Dies möchte ich im Folgenden zeigen, und zwar an der Hand der Werke dreier groszer Phänomenologen: *Husserl, Sartre* und *Merleau-Ponty*. Uebrigens spreche ich bewuszt von dem Problem der "Leiblichkeit" und nicht von dem der "Verleiblichung". Wenn Richard M. Zaner den Begriff des "embodiment" als Ausgangspunkt wählt;[1] dann erleichtert er das Verständnis einiger phänomenologischer Denker, erschwert jedoch den Zugang zu anderen.

Von "embodiment" ist allerdings bei dem Initiator der phänomenologischen Bewegung, bei *Edmund Husserl*, die Rede. Denn, man mag es drehen und wenden, wie man will: die Philosophie des historischen Husserl ist ursprünglich eine Bewusztseinsphilosophie. (Ich betone Letzteres mit Absicht. Denn so, wie man seinerzeit manchen Bewunderern der deutschen Klassik vorgeworfen hat, dasz sie von Goethe zu *"ihrem* Goethe" flüchten, so will es mir heute scheinen, dasz manche Phänomenologen "ihren Husserl" gegen Husserl auszuspielen geneigt sind.) Der bewusztseinsphilosophische Charakter seiner Philosophie tritt über deutlich im Ersten Buch der "Ideen" hervor, in dem der

19

A-T. Tymieniecka (ed.), Analecta Husserliana, Vol. XVI, 19–36.
Copyright © 1983 *by D. Reidel Publishing Company*.

Gründer der Phänomenologie zum ersten Mal seine Methode der Reduktion systematisch rechtfertigt. Ausdrücklich erklärt er in diesem Zusammenhang, dasz "ein leibloses und, so paradox es klingt, wohl auch ein seelenloses, nicht menschliche Leiblichkeit beseelendes Bewusztsein denkbar [ist], d.h. ein Erlebnisstrom, in dem sich nicht die intentionalen Erfahrungseinheiten Leib, Seele, empirisches Ich konstituierten".[2] M.a.W. in der Einstellung der Reduktion sind Leib sowohl als Seele — Seele als Trägerin von Erlebnissen animalischer Wesen — transzendente, mundane Realitäten; sie stehen im Gegensatz zu dem reinen Bewusztsein und seiner transzendentalen Absolutheit. Ichsubjekte mit Leib und Seele, mit Erlebnissen im psychologischen Sinn werden zwar faktisch konstituiert, aber nicht notwendigerweise. Der Husserl des Jahres 1913 insistiert darauf, dasz ein transzendentaler Bewusztseinsstrom, in dem nichts Derartiges in Erscheinung tritt, "nicht eine metaphysische Konstruktion [ist], sondern durch entsprechende Einstellungsänderung in seiner Absolutheit zweifellos Aufweisbares, in direkter Anschauung zu Gebendes . . . ".[3]

Hier regen sich Zweifel. Ist solch ein "reiner" Bewusztseinsstrom wirklich zu erschauen? Kommt Husserl mit diesem Denkmotiv nicht in die Nähe der transzendentalen Apperzeption Kants oder des "Bewusztsein überhaupt" der Neukantianer? Husserl hat jene Positionen niemals ausdrücklich widerrufen. Sie müssen aber als Bestandteile einer programmatischen "Fundamentalbetrachtung" gelesen werden. Und es ist bei jedem Denker angezeigt, nicht nur auf das zu achten, was er als sein Absicht *ankündigt*, sondern auch auf das, was er konkret philosophierend *tut*. Zwischen beidem herrscht nicht immer völlige Uebereinstimmung.

Gehen wir etwa Husserls Beschreibungen der "Konstitution der materiellen Natur" nach, die unmittelbar nach dem ersten Buch der "Ideen" geschrieben wurden. Wir stellen dann fest, dasz dort von Anfang an, auch da, wo Husserl bewuszt abstraktiv an der rein theoretischen Einstellung festhält, die Rede von "Empfindungen", von "in der Sinnessphäre konstituierten Einheiten", von [sinnlicher] "Rezeption" ist.[4] Kurzum, auch bei der Konstitution der Natur, die, wie Husserl versichert, eine "Sphäre bloszer Sachen ist",[5] gilt, was er später betont, "dasz bei aller Erfahrung von raumdinglichen Objekten der *Leib* als Wahrnehmungsorgan des erfahrenden Subjekts mit dabei ist".[6] Dazu kommt eine These, die die Erkenntnis fremder Subjektivitäten betrifft: "Die äuszerlich mir gegenüberstehenden Leiber erfahre ich wie andere Dinge in Urpräsenz, die Innerlichkeit des Seelischen durch Appräsenz".[7]

Die Weise, wie Husserl diesen Grundgedanken ausgearbeitet und ihm in der fünften Cartesianischen Meditation eine relativ endgültige Form verliehen hat,

kann hier nicht im einzelnen besprochen werden. Nur eines darf nicht übergangen werden: Husserl lädt uns hier ein, das Denkexperiment einer dem
"solus ipse" erscheinenden oder, wie er es ausdrückt: einer "eigenheitlichen"
Natur mitzumachen. Aber auch in dieser eigenheitlichen Sphäre, die aus
lauter Körpern besteht, ist *ein* Körper in besonderer Weise ausgezeichnet: der
Körper, der empfindet, der Körper, der in einem typischen Stil mit anderen
Körpern umgeht, und schlieszlich der Körper, der in eigenartiger Weise auf
sich selbst zurückbezogen ist, Dadurch wird dieser ausgezeichnete Körper zu
meinem Leib. [8]

Dazu kommt noch die wichtige Erkenntnis Husserls, dasz sich mein Leib
nicht in dem idealen Raum befindet, den die Geometrie beschreibt. Wo ich
leiblich bin, ist *hier*, alles andere ist *dort*. Dank meinem Leibe gibt es ein
rechts und *links*, ein *oben* und *unten*, ein *vorne* und *hinten*. Wenn ich nun
im Modus des "dort" einen Körper wahrnehme, der ebenso wie mein Leib
fungiert, dann musz ich eine "apperzptive Uebertragung von meinem Leibe
her" vornehmen. Es kommt zu einer analogisierenden Auffassung des fremden
Leibes und zu der bereits erwähnten Appräsentation von dessen seelischer
Innerlichkeit. Der Andere ist "jetzt mitdaseiendes ego im Modus Dort (*wie
wenn ich dort wäre*)", versichert Husserl. [9] Dabei musz das Begreifen des
räumlich entfernten Anderen in prototypischem Sinne verstanden werden:
es bildet die wahrnehmungsmäszige Grundlage für alles Verstehen fremder
Menschheiten, auch der sich in sozialer, historischer, kultureller Hinsicht
"dort" befindlichen. Dank der Leiblichkeit ist *"Einfühlung"* möglich; Husserl
übernimmt diesen Begriff von Theodor Lipps, verleiht ihm jedoch transzendentale Bedeutung.

Wer den Stand der philosophischen Anthropologie und Psychologie des
beginnenden 20. Jahrhunderts kennt, wird zugeben müssen, dasz Husserl in
vielfacher Hinsicht einen entscheidenden Durchbruch erzielt hat. Vor allem
hebt er die alte cartesianische Unterscheidung zwischen der "res cogitans"
und der "res extensa" auf. In Husserls reduktiver Sicht bin ich in erster
Instanz ein "cogito" ohne jegliche "res", d.h. ich bin nichts als ein transzendentaler Bewusztseinsstrom. In diesem meinem Bewusztsein konstituiert sich
allererst alles, was irgendwie als "res", als Substanz, als Ich-einheit aufgefaszt
werden kann. Andererseits ist mein Leib keine "res extensa", die sich in der
geometrischen "extensio" befindet. Mein Leib gliedert, gestaltet, strukturiert
den Raum auf eine typische Weise. Diese Gliederung ermöglicht die Konstitution einer materiellen Natur, einer animalischen Natur und einer sozialen Welt.

Unsere Bewunderung darf uns nicht blind für die Tatsache machen, dasz
Husserl in anderer Hinsicht der Vergangenheit verhaftet bleibt, und zwar einer

idealistischen Vergangenheit. Dies geht u.a. aus seiner Auffassung vom "ego" des "ego cogito" hervor. Ich als meditierendes Ich bin ein transzendentales Bewusztsein und als solches "*Geltungsgrund* für alle objektiven Geltungen und Gründe".[10] Alles andere, was mir als seiend gilt, ist konstituierte Wirklichtkeit, wenn auch in sehr verschiedener Weise konstituierte. Husserl spricht in diesem Zusammenhang von "ontologischen Regionen". Auch der Eigenleib gehört einer solchen Region an, ist ein Phänomen, das eine bestimmte Regelstruktur aufweist.[11] So ist es erklärlich dasz Husserl von dem Leib spricht, "*in* dem ich unmittelbar *schalte* und *walte*".[12] Dieses "in" ist bedeutungsvoll. Es beweist, dasz ich in Husserls Sicht nicht Leib bin, auch nicht Seele im psychologischen Sinn, nicht psychophysischer Organismus, nicht unmittelbar eingegliedert in einen Kosmos, in eine soziale und historische Welt. Ich bin ein konstituierendes Bewusztsein. Infolge bestimmter konstitutiver Leistungen verleibliche ich mich; ich *bin* aber nicht urspränglich Leib.

Bei Jean-Paul Sartre verhalten sich die Dinge bereits ganz anders. Ich denke dabei an den jängeren Sartre, den Verfasser von *L'Être et le Néant*.[13] Auf ihn möchte ich mich beschränken, denn nur in diesen Betrachtungen über *Das Sein und das Nichts* stellt Sartre das Problem der Leiblichkeit ausdrücklich, behandelt er es ausführlich und bietet er eine originelle Lösung.

Das soeben genannte Werk trägt den Untertitel "Versuch einer phänomenologischen Ontologie". Damit ist bereits gesagt, dass man Sartre's Philosophie der Leiblichkeit nicht versteht, wenn man sie von seiner ontologischen Positionen loslöst. Die groszen Thesen seiner Seinslehre darf ich als bekannt voraussetzen. Sartre stellt zwei grosze Bereiche einander gegenüber, die einen schroffen Gegensatz bilden: das inhaltlose, grundlose, beziehunglose An-sich-sein, von dem man nur sagen kann, dasz es faktisch besteht, einerseits; und das Bewusztsein, das sich von dieser massiven Sinnlosigkeit belagert weisz, andererseits. Das Bewusztsein konstituiert demnach nicht das Sein; (Sartre kehrt dem phänomenologischen Idealismus entschlossen den Rücken zu). Das Bewusztsein verhält sich zum Sein ähnlich wie ein Bildhauer, der eine formlose Steinmasse bearbeitet. Der Stein nimmt nur dadurch Gestalt an, dasz der Künstler Stücke wegmeiszelt, Vertiefungen schafft, Furchen anbringt. Auf analoge Weise enthüllt das Bewusztsein eine Welt durch seine "nichtenden" Akte ("néantisations"): durch seine Auffassungen als "nicht dies", "nicht-hier", "nicht-jetzt", durch sein "vielleicht, aber nicht gewisz", "möglich, aber nicht wirklich", "wertvoll, aber nicht real". Kurzum, das Bewusztsein ist dasjenige, was das "nicht" in das Sein trägt; seine Nichtungen gleichen den Schlägen, die der Bildhauer mit seinem Meiszel führt und durch die er dem Formlosen Gestalt verleiht.

Dies ist jedoch nicht alles. Das Bewusztsein findet das Sein nicht nur *um* sich her, es entdeckt es auch *in* sich selbst. Das Bewusztsein existiert ja gleichfalls an sich, es koïnzidiert mit sich selbst in einer sinnlosen Faktizität; und auch dieses Ur-Faktum musz es verneinen. Darum ist das Bewusztsein in Sartre's Sicht nicht Spiegelung einer Auszenwelt, sondern das, was sich gegen die Erstarrung und Versteinerung im eigenen An-sich-sein zur Wehr setzt. So erklärt sich die (an Heidegger gemahnende) Umschreibung: "Das Bewusztsein ist ein Seiendes, für das es in seinem Sein das Bewusztsein des Nichts seines Seins gibt".[14] Mit anderen Worten: das Bewusztsein ist nicht ein identischer "Ich-Pol", von dem man nur sagen kann, dasz er faktisch besteht. Es ist vielmehr in seinem tiefsten Wesen Widerstand gegen jene Faktizität, ein Widerstand, der sein eigentliches "Ek-sistieren" ausmacht. In diesem Sinne kann Sartre die Behauptung wagen, dasz das Bewusztsein ist, was es nicht ist, und dasz es nicht ist, was es ist. Sartre faszt also das Bewusztsein nicht als ein Wissen auf, sondern als immerwährende Selbstüberschreitung, als rastloses Tranzendieren, als grenzenlose Freiheit.

Auf jene Dialektik zwischen Faktizität und Transzendenz möchte ich nicht eingehen, sondern mich dem Problem der Leiblichkeit zuwenden. — Befände sich das Bewusztsein in einem einsamen vis-à-vis mit dem An-sich-sein, dann würde sich das Problem des Eigenleibes nicht stellen. Das Bewusztsein würde dann eine um das eigene ego zentrierte Welt entwerfen, die *seine* Welt wäre, und bei deren Enthüllung es nicht gestört würde. Der Philosoph könnte es dann immer noch als ein transzendentales Bewusztsein betrachten, auch wenn seine konstitutiven Leistungen in nichtenden Akten bestünden. Diese solipsistische Sicht entspricht jedoch nicht der wirklichen Situation des Bewusztseins. Sartre betrachtet den Solipsismus als eine Klippe für die philosophische Anthropologie, eine Klippe, an der viele berühmte Denker nur scheinbar nicht gestrandet sind. Wie kann man ihr wirklich entgehen?

Gleich anderen Philosophen der Existenz geht Sartre von einer fundamentalen Erfahrung aus: von der der Scham; genauer gesagt, von der des *Beschämtwerdens*. Ich schäme mich *vor jemandem*. Ein Anderer ertappt mich bei einer schändlichen Tat. Er mustert mich, er betrachtet mich, er sieht mich, wie ich bin. Der Blick ist in Sartre's Anthropologie ein transzendierender Akt: er schafft Entfernungen um mich hin, Räume, in denen sich Dinge befinden, Objekte, die identische Pole für mich sind und für niemanden anderen. Nun ist aber die Rede von einem *Betrachtet-werden*; als Ertappter und Beschämter bin ich mir dessen intensiv bewusst. In dieser Situation bin ich nicht mehr Transzendenz; ich bin vielmehr eine transzendierte Transzendenz. Aus einem Subjekt bin ich in ein Objekt für den Anderen verwandelt. Der Andere sieht mich, wie ich bin, d.h. er nagelt mich durch seinen Blick auf eine bestimmte Faktizität fest. In seiner Welt bin ich nicht ein freies Wesen, das unübersehbare Möglichkeiten besitzt; in seiner Welt erscheine ich als ein Ding unter Dingen. So entdeckt mein Bewusztsein, dasz es eine Auszenseite hat. Es ist endlich, es ist faktisch, und zwar dadurch, dasz es nicht nur für ist, sondern auch für Andere.

Die "Auszenseite meines Bewusztseins" ist noch nicht mein Leib. Sie ist mein cogito, sofern es von dem Anderen objektiviert, identifiziert, verdinglicht werden kann. Die "Auszenseite" ist meine unaustilgbare Faktizität, die mir durch das Auftreten des Anderen enthüllt wird. Der Eigenleib ist etwas noch Konkreteres. Er ist "die *zufällige Form . . . , welche die Notwendigkeit meiner Zufälligkeit annimmt*".[15] Mein Sein ist kontingentes Sein, so würde man in der Sprache der Scholastik sagen; aber mein Leib ist die konkrete Form, in der meine Kontingenz erscheint. Infolge meines Leibes bin ich, dieses nicht-notwendige Wesen, zufälligerweise "hier" und nicht "dort", "jetzt" und nicht "dann", "Mann" und nicht "Frau" usw. Und zwar bin ich dies alles in erster Linie *für mich*. Der Leib ist keineswegs ein An-sich *im* Für-sich, versichert Sartre, sonst würden wir wiederum einem extremen Dualismus à la Descartes verfallen. Der Leib ist ursprünglich auf mich bezogen. Das bedeutet in negativer Hinsicht, dasz er mir ursprünglich nicht erscheint, anonym bleibt, sich primär nicht als Objekt für mich darstellt.

Illustrieren wir letzteres an der Hand eines Beispiels, das dem wahrnehmenden Leben entnommen ist. Wenn ich Peter gegenüber an einem Tisch sitze, dann befindet sich dieses Glas hier links vom Krug und etwas hinter ihm; für Peter ist es rechts und ein wenig vor dem Krug. Jedes Bewusztsein hat infolge seines Leibes ein bestimmtes Hier, es entwirft seine Welt von einem bestimmten Standpunkt aus. Wäre dem nicht so, so würde *meine* Welt und die Peters in einer chaotischen Unterschiedslosigkeit untergehen. Den alles überschwebenden Blick, den standpunktlosen Standpunkt der Rationalisten und Idealisten gibt es nicht. Mein Bewusztsein ist notwendigerweise situiert, und aus seiner Situation erwachsen ihm Regelstrukturen. Der Umstand, dasz mein Bewusztsein situiert ist, ist meinem Leib zu verdanken, der zwar selbst nicht als Objekt erscheint, aber alle Objekte in einer bestimmten Ordnung erscheinen läszt.

Wir haben von den bekannten Schwierigkeiten gesprochen, die aus der Cartesianischen Leib-Seelen-Lehre erwachsen. Niemand, der das Geistesleben der neueren Zeit kennt, wird den nachhaltigen Einflusz dieser Doktrine unterschätzen. Sartre geht ihr von Anfang an zu Leibe, aber auf Grund von anderen Argumenten als Husserl. Worauf beruht dieser Dualismus, fragt er sich. Offenbar darauf, dasz ich den Leib als ein Ding beschreibe, das u.a. ein Groszhirn, endokrine Drüsen, Verdauungsorgane usw. aufweist, und dann dieses so beschaffene Ding mit meinem Bewusztsein in Verbindung zu bringen suche, das ich als meine absolute Innerlichkeit auffasse. Der Leib, den ich auf diese Weise beschrieben habe, ist jedoch nicht *mein* Leib. Ich habe ja niemals mein Hirn oder meine endokrinen Drüsen gesehen, betastet, gemessen. Die Deskriptionen der Physiologen, Anatomen, Mediziner betreffen jeweils den *Leib des Anderen* und nicht die von mir erlebte Leiblichkeit. Mein Leib — der Leib für mich — ist dasjenige, wodurch ich zur Welt bin, wodurch ich situiert bin; er ist somit kein Bestandteil der Welt, kein Teil einer mundanen Situation. — Dies gilt in analoger Weise auch für leibliche Organe, etwa Sinnesorgane. Durch irgendwelche Geräte — etwa Spiegel — ist es mir möglich, mein Auge zu sehen, ich kann es aber nicht sehen sehen. "Entweder ist es Ding unter Dingen, oder aber es ist das, wodurch sich mir die Dinge entdecken. Aber beides gleichzeitig kann es nicht sein", ruft Sartre aus.[16] Durch die Einführung dieser Alternative enthüllt er den Grundirrtum der Cartesianischen Dichotomie.

Sartre ist daher konsequent, wenn er in seiner ausführlichen Abhandlung in erster Linie zwei Aspekte der Leiblichkeit unterscheidet. Der Leib ist zunächst *für sich*. Was bedeutet dies für die menschliche Praxis? Sartre bezieht sich hier auf Heideggers bekannte Analysen von dem zuhandenen Zeug und den Zeugganzen. Es ist wahr, meint Sartre: der Nagel verweist auf den Hammer, der Hammer auf den Balken usw.; aber diese Beziehungen sind nur scheinbar gegenständlicher Art. In Wirklichtkeit weisen sie den Charakter lateinischer Gerundiva auf: der Nagel ist einzuschlagen, der Hammer am Stiel anzufassen, der Balken auf der Schulter zu tragen usw. Kurzum, alle zuhandenen Zeuge verweisen letztlich nach einem Ur-Werkzeug, das seinerseits nur mehr auf sich selbst weist. Infolgedessen ist mein Leib vergleichbar mit einem Schlüssel, der zahlreiche Schlösser aufschlieszt; der Schlüssel ist mir aber nie in der Weise eines Dinges "gegeben", sondern immer nur indirekt von zuhandenem Zeug angezeigt. Denn wenn ich etwa mit der Hand den Hammer schwinge, so erscheint mir meine Hand nicht als ein Objekt oder Instrument; sie ist dann für mich nichts anderes als die Handhabung des Hammers.

"Ich lebe bei meinen Objekten"; diese Wendung kommt wiederholt in unveröffentlichten Manuskripten Husserls vor. Sie könnte ebenso von Sartre stammen. Wahrnehmend befinde ich mich "dort" bei den Dingen, die ich betrachte; handelnd nehme ich das Ziel vorweg, das ich zu erreichen suche. Im Zuge meiner intentionalen Hinwendung fasse ich meinen Leib nie als einen Projektionsschirm, niemals als ein Werkzeug auf. In dieser Situation kann mein Leib "immer nur das *Ueberschrittene* sein", stellt Sartre fest.[17] Dieser Umstand hat jedoch für ihn auch ontologische Bedeutung. Wenn es wahr ist, dasz mein Leib die zufällige Form meiner Faktizität ist, dann ist mein leibliches Wahrnehmen und Handeln jeweils ein Transzendieren dieser Faktizität, ist es Entwurf, Wahl, Entscheidung. In diesem Sinne ist mein Leib die notwendige Bedingung meines Ek-sistierens.

Mein Leib ist aber auch *Leib für den Anderen*; darauf gründet sich die zweite Dimension der Leiblichkeit. Der Leib, der für mich kein Objekt ist, hat demnach lateral die Eigenschaften eines Objekts: er läszt sich betrachten, beschreiben, lokalisieren, beeinflussen, benützen. Diese zweite Perspektive ergänzt die erste. Sie erklärt die Möglichkeit einer objektivierenden und universalisierenden Auffassung. In dieser Sicht erscheint mir mein Leib als "ein Standpunkt, demgegenüber ich einen Standpunkt einnehme, als ein Werkzeug, das ich mittels anderer Werkzeuge gebrauchen kann".[18] Die zuhandenen Zeuge weisen dann nicht unmittelbar auf mich als unentbehrliches Urinstrument, sie zeigen indirekt ein zweites pragmatisches Beziehungszentrum an, das der Andere ist. Die Faktizität des Anderen enthüllt auf doppelte Weise die Kontingenz meines eigenen leiblichen Seins: theoretisch die Zufälligkeit meines Standpunkts, praktisch die Zufälligkeit meines Ausgangspunkts.

Dasz es noch eine dritte Dimension der Leiblichkeit gibt, ging bereits aus unserer Analyse des Beschämt-werdens hervor; sie betrifft *meinen Leib für den Anderen*. Insoferne ich für den Anderen bin, gilt mir der Andere als das Subjekt, für das ich Objekt bin. Ich bin dann eine transzendierte Transzendenz — und das ist bedeutungsvoll für mein Selbstverständnis; denn dann existiere ich für mich als ein vom Anderen Erkannter, und zwar gerade in seiner Faktizität Erkannter.

Aus dieser Dialektik des Erkennes, des Erkanntwerdens und des erkannten Erkanntwerdens, des Gebrauchens, des Gebrauchtwerdens und des gebrauchten Gebrauchtwerdens ergeben sich dann meine konkreten Verhaltensweisen zum Anderen. Sartre entwickelt unter dem Titel "Les relations concrètes avec autrui"[19] eine sehr originelle Sozialphilosophie, die gleichzeitig eine Philosophie des menschlichen Trieblebens, der Sexualität und Emotionalität darstellt. Hierauf einzugehen, ist nicht meine Aufgabe.

Nur die letzten Folgerungen jener Analysen kann ich nicht unerwähnt lassen. In Sartre's Sicht verfolge ich in all meinen konkreten, auch-leiblichen Verhaltensweisen zum Anderen ein geheimes Ziel, ein Ziel, das ich nur in der Sprache der Metaphysik andeuten kann: Ich möchte existierend die in sich ruhende Substanzialität des An-sich-seins mit der freien Transzendenz des Für-sich vereinigen. M.a.W. ich möchte durch den Anderen werden wie

Gott — wie der Gott der Metaphysik: ein "ens causa sui", eine Substanz, die an und für sich besteht. Doch alle meine verschiedenartigen Versuche, dies mittels des Anderen zu bewerkstelligen — die Liebe, der Hasz, der Masochismus, der Sadismus, die Gleichgültigkeit, das Verlangen — sind zum Scheitern verurteilt. Sie erreichen ihr metaphysisches Ziel nicht. Sie schlagen fehl — und zwar nicht etwa infolge der Endlichkeit, der Materialität, der Verderbtheit des Leibes, sondern weil ihr "telos": das Sein an und für sich zu verwirklichen, einen Widerspruch in sich schliesst. Der Schluszakkord von "Das Sein und das Nichts" darf nicht, wie dies vielfach geschehen ist, als "Absurdismus" karikiert werden; er klingt eher tragisch. Die Menschheit in ihrem sozialen, leiblich-emotionalen Verhalten verzehrt sich vergeblich. "Der Mensch ist eine nutzlose Leidenschaft", stellt Sartre fest.[20]

Sehen wir von jener Metaphysik ab und fragen wir rückblickend, inwiefern sich Sartre's "approach" unseres Problems von dem Husserlschen unterscheidet und inwiefern er ihm ähnelt. Beschränken wir uns auf die essentiellen Positionen beider Denker, dann fällt vor allem Sartre's Absage an den phänomenologischen Idealismus auf, zumal an den Idealismus der konstitutiven Leistungen. Ich konstituiere nicht meinen Leib, er ist vielmehr die faktische Form meiner Faktizität. An die Stelle des transzendentalen Bewusztseins tritt somit die Kategorie der Faktizität. (Dasz hier ein Einflusz des jüngeren Heidegger vorliegt, kann kaum bezweifelt werden.)

Ferner führt der Weg zur Erfassung des alter ego nicht über das Wissen um die eigene Leiblichkeit. Sartre schlägt geradezu den umgekehrten Weg ein. Dadurch, dasz mir ein anderes Bewusztsein, eine andere Freiheit gegenübertritt, entdecke ich allererst, dasz ich erscheine, und zwar als Leib erscheine. Bei Sartre wird die Sozialität nicht erst in einer letzten Meditation Gegenstand philosophischer Besinnung; sie ist von Anfang an eine Dimension meiner Existenz, auch meiner Existenz als leibliches Wesen.

Bricht Sartre gänzlich mit der groszen Tradition, die von Descartes zu Husserl führt? In *einer* wichtigen Hinsicht ist dies nicht der Fall. Auch Sartre's Philosophie wird von der Alternative beherrscht: Bewusztsein oder Ding, Subjekt oder Objekt, Urquellpunkt von Intentionen oder identischer Gegenstand von Intentionen. Und gerade hier könnte eine Kritik an Sartre's Philosophie der Leiblichkeit ansetzen. Ist es wahr, so kann man sich fragen, dasz der Eigenleib gänzlich dem Bereich des Für-sich angehört, dasz jegliche Objektivierung ihn in den Leib für den Anderen verwandelt? Ist diese Alternative zwingend? Weist die Erfahrung des Leibes nicht einen schillernden Charakter auf?

Dies sind Fragen, mit denen sich *Maurice Merleau-Ponty* sein Leben lang beschäftigt hat. Seine Positionen in Kürze wiederzugeben, ist aus zwei Gründen schwierig. Zum ersten hängt ein groszer Teil seines Oeuvres direkt oder indirekt mit seiner Philosophie der Leiblichkeit zusammen; von Vollständigkeit der Wiedergabe kann schon darum keine Rede sein. Zweitens hat der Denker Merleau-Ponty im Laufe seiner Entwicklung seine Sichtweise so stark geändert, dasz man − ähnlich wie bei Wittgenstein − beinahe von zwei Philosophien sprechen könnte. Die erste liegt, deutlich und sorgfältig ausgearbeitet, in zahlreichen Publikationen vor, vor allem in den beiden Hauptwerken *Structure du comportement*[21] und *Phènoménologie de la perception.*[22] Durch; seinen frühzeitigen Tod wurde Merleau-Ponty daran gehindert, seiner zweiten Philosophie endgultige Form zu verleihen. Nur auf Grund einer Interpretation seiner Spätschriften − u.a. seiner Abhandlung *Le Visible et l'invisible*[23] und seines Artikels *L'Oeil et l'Esprit*[24] kann man die wichtigsten Motive seines neuen Denkens rekonstruieren. Da dieses Denken jedoch gleichfalls um das Thema der Leiblichkeit kreist, kann es hier nicht gänzlich übergangen werden.

Der Löwensche Historiker Alphons de Waelhens charakterisiert Merleau-Ponty's Position als "Eine Philosophie der Ambiguität".[25] Damit gibt de Waelhens zugleich die Motive an, die Merleau-Ponty veranlaszt haben, sich von Sartre zu distanzieren. Wenn man nämlich den Menschen auf Grund seines Zur-Welt-Seins beschreiben will, dann musz man die Alternative "Ding oder Bewusztsein" fahren lassen. Denn die Welt ist nicht in den Dingen, sie ist der Horizont der Dinge. Andererseits würde ein reines Bewusztsein die Dinge ohne Perspektiven, Verdeckungen, Hindernisse überblicken Wie soll man nun die menschliche Existenz nennen, soferne sie in dem mundanen Horizont gänzlich befangen ist, sofern sie mit jenen Perspektiven, Verdeckungen, Hindernissen zu rechnen und zu kämpfen hat? Merleau-Ponty nennt sie das "natürliche", das "anonyme", das "vor-persönliche" Subjekt; man könnte auch von einem "Leibsubjekt" sprechen. Als Leibsubjekt ist der Mensch unmittelbar zur Welt, und zwar infolge seiner leiblichen Intentionalität, seiner wahrnehmenden, motorischen, sexuellen, expressiven Intentionen. − Man begreift, dasz hier ein Gegensatz zur spiritualistischen und idealistischen Tradition hervortritt. Bei Merleau-Ponty ist keine Rede mehr von einem reinen Ich, das "in" seinem Leib waltet. Das Walten des Subjektes hat zunächst den Charakter eines leiblichen Fusz-fassens in der Welt. Aus dieser elementaren, anonymen Hinwendung zur Welt erhebt sich allerdings so etwas wie ein persönliches ego. Aber die menschliche Person vermag niemals, die Spuren ihres leiblich-anonymen Ursprungs abzustreifen. Auch die Reflexion verwandelt sie nicht in ein völlig durchsichtiges Bewusztsein. Die Reflexion kann niemals total werden; sie setzt ja jeweils etwas Unreflektiertes voraus. Auch die kristallklaren Denkgebilde ruhen letztlich auf dem dunklen Untergrund einer vitalen naturhaften Verwurzelung in der Welt. Aus dieser Ursituation erklärt sich die essentielle Ambiguität der menschlichen Existenz.

Dies sind kühne Thesen, wird man bemerken; wie beweist sie unser Philosoph? Das elementare Zur-Welt-sein des Mensch ist ja niemals als solches gegeben. Wie kann es erfaszt, phänomenologisch beschrieben und analysiert werden? – Merleau-Ponty macht sich die Sache nicht leicht. Er vertieft sich in die Forschungsergebnisse gewisser Erfahrungswissenshcaften, namentlich in die der Physiologie, der Psychologie und der Psychiatrie. Es ist wahr, dasz das elementare Zur-Welt-sein im Bereich der menschlichen Normalität nicht in Erscheinung tritt. Was nicht direkt Phänomen wird, läszt sich jedoch auf Umwegen erschlieszen. Merleau-Ponty bedient sich dabei zweier Hilfswissenschaften: einerseits der Patho-psychologie, die, ausgehend von krankhaften Störungen der elementaren Hinwendung zur Welt, vorsichtige Schlüsse hinsichtlich der normalen subjektiven Leistungen zieht; andererseits der vergleichenden Psychologie, die das Funktionieren der anonymen Leiblichkeit auf dem Niveau des animalischen und frühkindlichen Lebens beobachtet und systematisch mit dem des erwachsenen Menschen vergleicht. – Dies bietet Merleau-Ponty gleichzeitig Gelegenheit zu einer Kritik gewisser Humanwissenschaften. Er beweist, dasz ihre Ergebnisse nicht immer mit den stillschweigenden philosophischen Voraussetzungen der Forscher übereinstimmen; ihre Resultate sind weder auf naturalistisch-empiristischer, noch auf intellektualistisch-idealistischer Basis erklärlich. Die "philosophie des savants" läszt vielfach zu wünschen übrig; die Erfahrungswissenschaftler sollten ohne Vorurteile zu Werk gehen. Kein Wunder, dasz gewisse Reformbestrebungen innerhalb der empirischen Psychologie sich sehr ausdrücklich auf Merleau-Ponty berufen und nicht so sehr auf Husserl.[26]

Merleau-Ponty's Analysen sind zahlreich und beziehen sich auf vielerlei Phänomene des normalen und abnormalen Verhaltens. Ich musz mich darauf beschränken, den Stil seines Philosophierens an der Hand einiger Beispiele zu illustrieren. – Was bedeutet etwa Merleau-Ponty's Wort: "Der Leib *bewohnt* den Raum"? Wir haben von leiblichen Intentionen gesprochen und als erste die wahrnehmenden und motorischen genannt. Diese beiden Arten von Intentionen sind jedoch auf eine mysteriöse Weise miteinander verwebt. Man denke an einfache Erfahrungen, die jedermann beim Tennisspielen oder Autofahren machen kann. Wer ein guter Tennisspieler werden will, musz lernen, den Ball in der richtigen Weise wahrzunehmen; daraus erwachsen ihm die richtigen Bewegungen. Umgekehrt gilt aber, dasz die richtige Haltung zum Ball das genaue Lokalisieren des Balles ermöglicht. Dies ist eigenartig. Dennoch liegt hier keine mysteriöse "harmonie préétablie" vor. Vielmehr müssen wahrnehmende und motorische Intentionen als Momente eines einzigen existentiellen Motus aufgefaszt werden. Mich bewegend und wahrnehmend wende ich mich den Dingen zu, die Aufforderungscharakter besitzen, bezw. weiche ich Dingen aus, die mir drohend scheinen. Wenn mir ein Ball zugeworfen wird oder ein Auto auf mich zufährt, dann reagiere ich unmittelbar. Ich brauche mir dabei nicht das Auto vorzustellen, das im Begriffe ist, mich zu überfahren. Die Dinge sind nicht zuerst für mein Bewusztsein, das dann dem Mechanismus meines Leibes Befehle erteilt. Auf dem elementaren Niveau bestehen die Dinge direkt für mich als Leibsubjekt, und alle ihre anderen "höheren" Weisen, für mich zu sein — ihr Vorgestellt-, Phantasiert-, Gedachtsein — setzen ihren primären Appell an mich als Leibsubjekt voraus.

Es ist ferner völlig wahr, was Husserl und Sartre festgestellt haben: der Leib ist sein eigenes "hier". Das bedeutet aber nicht, wie Husserl meinte, dasz der Eigenleib a priori das zentrale Objekt für mich ist. Das Umgekehrte gilt: Von meinen subjektiven leiblichen Möglichkeiten hängt die Struktur des Raumes ab, den ich bewohne. Merleau-Ponty illustriert dies durch Krankheitsbilder, die er der Psychiatrie entlehnt. Ich möchte diese komplizierten Analysen durch die Erinnerung an eine einfachere Erfahrung ersetzen. Wenn ich völlig erschöpft zu Bett liege, dann ist das Glas Wasser, das auf dem Nachtkästchen steht, "weit entfernt". Für den Kräftigen, der sich erheben kann, ist es jedoch "ganz nahe". — Da aber meine leiblichen Möglichkeiten in jedem Fall beschränkt sind, ist der Raum, den ich bewohne, endlich. Ich "weisz" — ohne es thetisch zu wissen — dasz ich "dies" erreichen kann, zur Not auch noch "das", dasz sich aber Dinge im Raume befinden, die für mich unerreichbar sind. Analoges gilt für die Wahrnehmung. Von *diesem* Standpunkt sehe ich "das", von *jenem* erblicke ich "jenes". Es sind aber immer noch andere Standpunkte möglich. So kommt es, dasz die räumliche Welt, die ich konkret erfahre, nicht "abgegrenzt", nicht "mit Brettern vernagelt", wohl aber von einem Horizont umgeben ist. Sie ist *endlich*. — Diese und ähnliche Analysen Merleau-Ponty's sind nicht als Argumente gegen den unendlichen Raum der Geometrie gemeint. Unser Philosoph wird allerdings betonen, dasz der gedachte Raum des Geometers die Erfahrung des konkreten Raumes voraussetzt.

Wir haben auch von den Ausdrucksintentionen des Leibsubjektes gesprochen. Wenn wir im wachen Zustand etwas tun, ohne es zu wissen und zu wollen, dann ist dieses "etwas" zumeist eine expressive Geste. Wir lächeln, wir lachen, wir stöhnen, wir weinen, wir verziehen die Miene, wir zucken, wir erröten "unwillkürlich". Was bedeutet Letzteres? Jenes "unwillkürlich" kommt einem Hinweis auf das Leibsubjekt gleich. In dem spontanen Ausdruck spricht sich die vorpersönliche Leiblichkeit aus. Die gesamte Sprachphilosophie Merleau-Ponty's beruht auf diesem Grundgedanken. – Man könnte aber auch sagen: Mein Leib ist expressive Räumlichkeit. Keine andere "res extensa" verrät gegen meinen Willen meine Intentionen. Hierfür ein einziges Beispiel: Ich weisz noch nicht, ob ich spazierengehen werde oder nicht; ich zögere, weil schlechtes Wetter herrscht. Mein Hund aber "weisz", was ich will; er bellt freudig und springt an mir empor. Ich musz irgendeine expressive Bewegung vollzogen haben, ohne noch bewuszt einen Entschlusz gefaszt zu haben.

Dank meiner Leiblichkeit bin ich unmittelbar zur Welt, so lautet Merleau-Ponty's wichtigste These. Die Welt weist jedoch zwei verschiedene Aspekte auf: sie ist Naturwelt und sie ist Kulturwelt. Als Naturwelt stellt sie einen Bereich von Bedeutungen dar, die der absoluten Vergangenheit der Natur entstammen. Als Naturwesen existiere ich, insoferne ich mit Lungen atme, mich von Kohlenhydraten nähre, schreie, wenn ich Schmerz empfinde usw. Alle diese elementaren Verhaltensweisen und Reaktionen habe ich geerbt, ich verdanke sie der Vergangenheit der menschlichen Rasse. Als solche sind sie allgemein, namenlos, unpersönlich. Sie haben Bezug auf eine Leiblichkeit, die ich als Exemplar einer bestimmten Spezies passiv empfangen habe und durch die ich in einem physischen Kosmos verwurzelt bin.

Von diesem Niveau biologischer Allgemeinheit unterscheidet sich das spezifisch menschliche. Was dieses betrifft, führt Merleau-Ponty eine weitere Unterscheidung ein. Zunächst empfangen die biologischen Gesten und Verhaltungsweisen vom Menschen einen neuen Sinn, der gleichsam wie ein Edelreis direkt auf den wilden Stamm gepfropft wird. Man könnte in diesem Zusammenhang an das stilisierte sich-bewegen, den Tanz, an die erotische Zärtlichkeit denken, die keineswegs mit dem sexuellen "behavior" identisch ist. In zweiter Linie wären die Bedeutungen zu berücksichtigen, die nicht unmittelbar aus leiblichen Intentionen erwachsen. Solche Bedeutungen entdeckt der Mensch nur infolge der "Instrumente", die er sich geschaffen hat. Unter "Instrumenten" im weitesten Sinne versteht Merleau-Ponty einerseits konkrete Gegenstände (Werkzeuge, Waffen, Produktionsmittel, Tauschmittel, Meszapparate usw.); andererseits Denkinstrumente (Klangsymbole, Sprachen, Begriffe, Kategorien, Denkmethoden, Systeme usw.). Konkrete und abstrakte Instrumente des Menschen sind aufeinander abgestimmt und machen zusammen seine Kulturwelt aus. Beide Arten von Instrumenten haben dies gemein, dasz sie vom Menschen nur infolge seiner

habituellen Erwerbe gebraucht werden können. Die "Habitualitäten" (so
würde Husserl sagen) sind Mittel, sich einer Kulturwelt zuzuwenden, wie
die körperlichen Organe Mittel sind, uns in der Natur zu verankern. "Das
Bewusztsein entwirft sich in eine Naturwelt und hat einen Leib so, wie es
sich in eine Kulturwelt entwirft und seine Habitualitäten besitzt", stellt
Merleau-Ponty fest; "Denn es kann Bewusztsein nur sein, indem es mit den
in der absoluten Vergangenheit der Natur oder den in seiner personalen
Vergangenheit gegebenen Bedeutungen spielt".[27] Dasz hier Elemente zu
einer Philosophie der Rezeptivität und Passivität vorliegen, die im Gegensatz
zu Sartre's aktivistischem Ethos stehen, ist deutlich.

Merleau-Ponty's Philosophie des leiblichen Ausdrucks ist von groszer
Bedeutung. Auf ihr beruht nicht nur — wie bereits angedeutet — seine höchst
originelle Sprachphilosophie, sondern auch seine Philosophie des Denkens
und der idealen Gebilde. Das Denken ist nicht unabhängig vom Sprechen.
Die Idee — etwa die des pythagoräischen Lehrsatzes — besteht nicht fix und
fertig im Kopfe des Geometers und sucht in einer zweiten Phase nach einem
Kleide von Worten. Die Idee bedarf vielmehr des Ausdrucks, um überhaupt
zustande zu kommen, sich zu klären, feste Formen anzunehmen. Das Wort
vermittelt zwischen der Wahrnehmung und dem Gedanken, und ohne diese
Vermittlung kommt der Gedanke nicht zu sich. Das "cogito" ist ursprünglich
ein gesprochenes "cogito", das in einem bestimmten historischen Augenblick
vernehmbar geworden ist. Das vergiszt jedoch der stolze Denker; er glaubt
im nachhinein, eine ewige Wahrheit im Reiche der Ideen entdeckt zu haben.
Merleau-Ponty ruft ihn zur Ordnung, mahnt ihn zur Bescheidenheit. Ein
Denken, das um Klarheit ringt, kann des Instruments der Sprache nicht
entraten, betont er. Dies ist für ihn eine letzte Tatsache, "un fait dernier".[28]
Hinter diesen Sachverhalt zurückfragen zu wollen, ist sinnlos. Man könnte
die Sichtweise *dieses* Merleau-Ponty als eine Art von phänomenologischen
Positivismus charakterisieren.

In seiner Spätphilosophie gibt sich unser Denker mit dieser Beschränkung
nicht mehr zufrieden. Er will jetzt tiefer schürfen. Er strebt danach, die
Dialektik des Subjekt-seins und Objekt-seins, vom phänomenalen Leib und
vergegenständlichten Körper zu übersteigen und nach dem zu suchen, was
jener Spaltung vorhergeht. Dieses vor-subjektive und vor-objektive "etwas"
nennt Merleau-Ponty in seiner Spätphilosophie "Être" und schreibt es mit
einem Groszbuchstaben. Dieser Seinsbegriff deckt sich jedoch nicht mit dem
"*ov*" des Aristoteles, noch mit dem "esse" der Scholastiker, am allerwenigsten
mit dem — wie immer geschriebenen — "Seyn" Heideggers. Das Sein, von dem
der ältere Merleau-Ponty spricht, ist dasjenige, was ich niemals von auszen

zu betrachten vermag, dem ich niemals gegenüberstehe, womit ich immer bereits verbunden, ja verflochten bin.

Auch in dieser "zweiten Philosophie" spielt die Leiblichkeit eine wichtige Rolle. Sie wird jedoch auf eine gänzlich andere Weise beschrieben. Wenn mein Leib überhaupt imstande ist, die Dinge wahrzunehmen, dann ist dies der Tatsache zu danken, dasz er bereits vor allem Fungieren und Aktiv-sein mit ihnen solidär ist. Die Solidarität des Leibes mit den Dingen kommt in dem Umstand zum Ausdruck, dasz er, der sehende, selbst sichtbar ist; dasz er, der fühlende, selbst fühlbar ist. Merleau-Ponty führt hier u.a. den Term "entrelac" ein und bezeichnet damit eine Verflechtung, die auf einer wechselseitigen Implikation beruht. Diese Implikation ist keineswegs einem Zufall zuzuschreiben, ihr ist vielmehr die Intelligibilität des Seins zu verdanken. Dadurch, dasz sich etwas Sichtbares auf sich selbst bezieht, entsteht allererst Sichtbarkeit; und der Leib ist der "Ort", an dem Sehen und Sichtbarkeit zusammenkommen.

Aus all dem geht hervor, dasz für den älteren Merleau-Ponty das eigentliche Problem das der Sichtbarkeit ist, und nicht mehr so wie früher das des Sehens. Der Titel des Spätwerkes *Le Visible et l'invisible* steht in einem gewissen Gegensatz zu dem Programm, das die *Phénoménologie de la perception* ankündigt. Im Mittelpunkt des philosophischen Interesses steht nicht mehr die existierende Subjektivität und deren Leistungen. Die Frage nach der Sichtbarkeit und Wahrnehmbarkeit musz ontologisch geklärt werden.

Der Titel des soeben genannten Spätwerkes besagt bereits, dasz darin auch ausdrücklich das Unsichtbare, das Nicht-wahrnehmbare bedacht wird. Das Unsichtbare katexochen ist aber das lautere Denkgebilde, das reine "ens rationis", das nur ideal Seiende. Das Ideale befindet sich jedoch nicht in einem Reich der Ideen, das von dem irdischen und fleischlichen Sein streng getrennt ist. Im Gegenteil: das Unsichtbare erwächst aus der Sublimierung einer Tendenz, die dem Sichtbaren, dem, was "es gibt", bereits innewohnt. M.a.W. Merleau-Ponty zufolge ist das Unsichtbare nichts anderes als die Kehrseite des Sichtbaren; es ist keineswegs dessen "Nichtung", auch nicht dessen Ueberstieg, dessen Transzendenz. Unser Philosoph drückt dies folgendermaszen aus:

Unser Problem besteht darin zu zeigen, dasz der Gedanke im engeren Sinne (reine Bedeutung, Bewusztsein, zu sehen und zu fühlen [im Sinne Descartes'], nur als Erfüllung durch andere Mittel des Wunsches verständlich ist, der dem 'es gibt' innewohnt, durch eine Sublimierung des 'es gibt' und die Verwirklichung eines Unsichtbaren, das die genaue Kehrseite des Sichtbaren ist, die Potenz des Sichtbaren.[29]

Auf diese wenigen Hinweise musz ich mich beschränken. Nicht aber will ich ein Bedenken verschweigen, das Merleau-Ponty's Spätphilosophie bei mir erregt. Es ist wahr, dasz der ältere Merleau-Ponty gründlicher als andere Phänomenologen mit allen Dualismen abrechnet — sie mögen nun platonischer, neuplatonischer, scholastischer, rationalistischer oder idealistischer Herkunft sein. Fällt er jedoch nicht der Charybdis zum Opfer, während er bemüht ist, der Scylla zu entgehen? Ist seine Ontologie nicht *monistischer Natur?* Werden hier nicht alle Einwände laut, die seit jeher gegen den Monismus erhoben wurden? — Beschränken wir uns etwa auf Merleau-Ponty's Bild von der "Auszenseite" und der "Kehrseite" des Seins. Husserl würde sicherlich bemerken, dasz das wahrnehmende Subjekt einfach dadurch, dasz es umhergeht, die Auszenansicht auf kontinuierliche Weise in eine Innenansicht übergehen lassen kann. Ist etwas Derartiges auch dann möglich, wenn es sich um eine Wahrnehmung und ein Denkgebilde handelt? Wird hier nicht ein essentieller Gegensatz bagatellisiert?

Ich möchte diese kritische Erwägung nicht weiterführen und statt dessen als Historiker eine Zusammenfassung unserer Einsichten in der Form einiger Thesen bieten. Ich bin mir dabei des Umstandes bewuszt, dasz ich viele Autoren, die hier behandelt hätten werden müssen, nicht genannt habe: etwa Gabriel Marcel, Max Scheler, Helmuth Plessner, Frederik Buytendijk, Paul Ricoeur, Emmanuel Levinas. Doch wollte ich bewuszt eine enzyklopädische Aneinanderreihung von Namen und Titeln vermeiden. Andererseits glaube ich, dasz sich das Wirken der genannten Autoren zwanglos in das Gesamtbild einreihen liesze, das ich in den folgenden sehr allgemeinen Thesen zu umreiszen versuche.

(1) Die Vertreter der phänomenologischen Bewegung distanzieren sich im Laufe der Jahre zunehmend vom phänomenologischen Idealismus und der Transzendentalphilosophie Husserlscher Prägung.

(2) An die Stelle des *einen* transzendentalen Bewusztseins oder der *einen* transzendentalen Monadengemeinschaft tritt bei ihnen eine Vielheit von Subjekten, die miteinander, aber auch gegeneinander in einem natürlichen Kosmos, in einer sozialen und historischen Welt wirken und leben. Im Zusammenhang damit erregen Probleme der auszersprachlichen und der sprachlichen Kommunikation das Interesse phänomenologischer Philosophen.

(3) Das Bewusztsein wird nicht mehr als ein Absolutum der Welt gegenübergestellt. Es ist als verleiblichtes oder als leibliches Bewusztsein in der Welt zu Hause; es bewohnt die Welt und gliedert den mundanen Raum.

(4) Die Subjektivität gilt nicht mehr als ein konstitutiver Bewusztseinsstrom. Dadurch treten Probleme der Endlichkeit, der räumlichen und

temporalen Horizonte, der Historizität, aber auch die der Faktizität, der Rezeptivität und Passivität in den Vordergrund.

Zusammenfassend könnte man vielleicht sagen, dasz die Vertreter der phänomenologischen Bewegung bewuszt oder unbewuszt der Losung Jean Wahls Folge geleistet haben: "Vers le concret!" "Hin zum Konkreten!" Die geistigen Söhne und Enkel Edmund Husserls bringen einen Stil konkreten Philosophierens, der von dem Begründer der Phänomenologie lediglich inauguriert worden war, zu einer hohen Blüte.

University of Nijmegen

NOTEN

[1] *The Problems of Embodiment. Some Contribution to a Phenomenology of the Body*, Nijhoff, Den Haag 1964.

[2] *Ideen zu einer reinen Phänomenologie und phänomenologischen Philosophie, Erstes Buch*, Nijhoff, Den Haag 1950, Husserliana III, § 54, S. 133.

[3] Loc. cit. S. 133/4.

[4] *Ideen zu einer reinen Phänomenologie . . . , Zweites Buch*, Nijhoff, Den Haag 1952, Husserliana IV, § 10, S. 24.

[5] Ibid.

[6] Loc. cit., § 36, S. 144.

[7] Loc. cit., § 45, S. 163/4.

[8] *Cartesianische Meditationen und Pariser Vorträge*, Nijhoff, Den Haag 1973, Husserliana I, § 44, S. 126–130.

[9] Loc. cit., § 54, S. 148.

[10] Loc. cit., § 11, S. 65.

[11] Vgl. *Ideen . . . , Zweites Buch*, loc. cit., S. 143–162.

[12] *Cartesianische Meditationen*, loc. cit., § 44, S. 128.

[13] *L'Être et le néant. Essai d'ontologie phénoménologique*, Gallimard, Paris, 1950 – im Folgenden abgekürzt: EN.

[14] Uebersetzung: *Das Sein und das Nichts. Versuch einer phänomenologischen Ontologie*, Rohwolt, Hamburg 1970 – im Folgenden abgekürzt: SN.

[15] SN, S. 405; " . . . *la forme contingente que prend la nécéssité de ma contingence.*" EN, S. 371.

[16] S. 399; "Ou bien il est chose parmi les choses, ou bien il est ce par quoi les choses se découvrent à moi. Mais il ne saurait être les deux en même temps", EN, S. 366.

[17] SN, S. 425; " . . . , le corps ne peut être que le dépassé: . . . ", EN, S. 390.

[18] SN, S. 44L; " . . . un point de vue sur lequel je peux prendre un point de vue, un instrument que je peux utiliser avec d'autres instruments." EN, S. 406.

[19] SN, S. 464–521, EN, S. 431–507.

[20] SN, S. 770; "l'homme est une passion inutile" EN, S. 708.

[21] 1942, 4. Auflage 'Presses Universitaires de France' Paris 1960, im Folgenden abgekürzt: SC.

22 Gallimard, Paris 1945 – im Folgenden abgekürzt: PP.
23 Posthum von Claude Lefort herausgegeben, Gallimard, Paris 1964.
24 1961 in *Les Temps Modernes* 17 (1961), 195–227, später bei Gallimard, Paris 1964 erschienen.
25 Vorwort zu SC.
26 Vergleiche u.a. Amadeo Georgi: *Psychology as a Human Science. A Phenomenologically Based Approach*, Harper & Row, New York 1970.
27 Der Verfasser ist von Rudolf Boehms Uebersetzung ein wenig abgewichen. Vgl. *Phänomenologie der Wahrnehmung*, der Gruyter, Berlin 1966, § 19, S. 166. Merleau-Ponty's Text lautet: "La conscience se projette dans un monde physique et a un corps, comme elle se projette dans un monde culturel et a des habitus, parce qu'elle ne peut être conscience qu'en jouant sur les significations données dans le passé absolu de la nature ou dans son passé personnel" PP, S. 160.
28 PP, S. 447.
29 "Notre problème est de montrer que pensée au sens restrictif (signification pure, pensée de voir et de sentir) ne se comprend que comme accomplissement par d'autres moyens du voeu du 'il y a' par sublimation du 'il y a' et realisation d'un invisible qui est exactement l'envers du visible, la puissance du visible". *Le Visible et l'invisible*, loc. cit., S. 190, Anmerkung.

ANGELA ALES BELLO

SEELE UND LEIB IN DER KATEGORIALEN UND IN DER ORIGINÄREN PERSPEKTIVE

Es ordnen sich gemäss unseren Betrachtungen rechtmässig *zwei Arten der realen Erfahrung* nebeneinander, die *'äussere' Erfahrung*, die physische, als Erfahrung von materiellen Dingen, und die seelische Erfahrung als *Erfahrung* von seelischen Realitäten. *Jede dieser Erfahrung ist grundlegend für entsprechende Erfahrungswissenschaften, die Wissenschaften* von der materiellen Natur und die Psychologie als Wissenschaft von der Seele.[1]

Ich zitiere diese Überlegung Husserls, weil mir scheint, dass sie seinen Gedanken über das Argument, das Gegenstand unserer Untersuchung ist, zusammenfasst.

Gerade in ihrer Prägnanz kann diese Stelle in zwei Richtungen zu einem gültigen Ausgangspunkt werden: um die Bedeutung der Husserlschen Darlegung des hier zur Frage stehenden Themas zu erfassen und um eine weitere Analyse einzuleiten. Dieses letztere Erfordernis ergibt sich aus den Schwierigkeiten, denen sich Husserl gegenüber sieht und aus der nicht vollständigen Überwindung dieser Schwierigkeiten.

Husserl unterscheidet nicht nur zwischen realer 'äusserer' oder physischer Erfahrung und psychischer Erfahrung, sondern behauptet, dass sich darauf die entsprechenden Wissenschaften gründen. Es kann ja tatsächlich beobachtet werden, wie sich seine Analyse über Seele und Leib auf zwei zusammenhängenden, miteinander verschlungenen Ebenen abwickelt: auf der Ebene der wissenschaftlichen und auf der Ebene der erfahrungsmässigen Deskription. Ausgehend von der von der Wissenschaft dargebotenen 'Vermittlung' wie sie sich historisch dargetan hat in ihrer Scheidung zwischen Natur- und Geisteswissenschaft neigt er jedoch gleichzeitig dahin, beide Realitäten, die innere und die äussere, als direkte Erfahrungsobjekte zu untersuchen. Dadurch wird eine doppelte Arbeit vollzogen: die klare Herausstellung der methodologischen Strukturen der Wissenschaft und die Untersuchung des Objekts, auf das sich die Wissenschaft bezieht. In diesem Zusammenhang sieht sich Husserl dann in die realistische Voraussetzung: wie das Objekt, so die Methode, verstrickt, wenngleich er dann die Instrumente liefert, diese Voraussetzung zu überwinden. Und gerade deshalb nähert er sich dem Problem auf zweifache Art: von den bereits konstituierten Wissenschaften und von ihrem Forschungsobjekt her. Die Bedeutung des Objekts zu klären bedeutet auch das Warum der wissenschaftlichen Annäherung zu verstehen.

37

A-T. Tymieniecka (ed.), Analecta Husserliana, Vol. XVI, 37–48.
Copyright © *1983 by D. Reidel Publishing Company.*

All das ist gerechtfertigt im Hinblick auf den kulturellen Hintergrund, vor dem sich sein Werdegang abspielt, einerseits stark beeinflusst von der Psychologie, andererseits von den mathematischen und physikalischen Wissenschaften. Gerade seine Einstellung gegenüber der Psychologie, seine "Seitenwahl" zugunsten von Stumpf, Lozte und Brentano lassen bereits verstehen, dass, wenn für ihn die Psychologie eine Wissenschaft sein muss, sie dies auf eine andere Weise zu sein hat als die Naturwissenschaft. Er übernimmt von der antipositivistischen Reaktion die Unmöglichkeit, die ganze Realität mit den von der Naturwissenschaft angebotenen Instrumenten zu untersuchen — daher die Notwendigkeit, die Wissenschaften durch das Objekt zu charakterisieren —, doch bemerkt er gleichzeitig, sie nicht allein durch das Objekt rechtfertigen zu können.

Husserl ist sich klar bewusst, dass die Wissenschaften in ihrer Gesamtheit an einen Kategorisierungsprozess gebunden sind. Wenn es aber darum geht, die einzelnen Wissenschaften zu untersuchen und zu unterscheiden, so führt ihn die Notwendigkeit, einen Bereich des Geistes zu umreissen, der seine eigenen Charakteristiken habe, dazu, den Wissenschaften, die sich damit befassen, ein 'anderes' epistemologisches Statut zu geben. Andererseits ist er davon überzeugt, dass die authentische antipositivistische Aktion dann bis zum Äussersten geführt wird, wenn die Exemplarität der Naturwissenschaften in Krise gesetzt wird. So setzt sich eine Antinomie zwischen den Geisteswissenschaften und den Naturwissenschaften fest, die auf die Beziehung zwischen *Innerlichkeit* und *Äusserlichkeit* zurückführbar ist.[2]

So kann man die Exemplarität der Physik innerhalb der zweiten, wie auch die Exemplarität der Psychologie für die ersten verstehen,[3] auch wenn diese Psychologie noch zu konstruieren ist. Daher sein Beharren auf der Unterschiedlichkeit von Bereich und Zweck zwischen Geistes- und Naturwissenschaften[4] und auf der besonderen Rolle, die von den ersteren ausgeübt wird, um zum Bewusstsein zu gelangen, da sie doch verstanden sind als privilegierter Weg, der zur universalen Wissenschaft führt.[5]

Fragt man sich jedoch, ob die für die Wissenschaft im allgemeinen aufgefundene Struktur auch für die Psychologie von Wert ist, und wenn die Antwort darauf, wie Husserl es an manchen Stellen tut,[6] bejahend ist, so entfällt die Eigentümlichkeit der Geisteswissenschaften.

Es ergeben sich daher folgende Fragestellungen: (1) Besteht eine Verschiedenheit zwischen Naturwissenschaften und Geisteswissenschaften? (2) Und diese Verschiedenheit, vorausgesetzt sie bestehe, bezieht sie sich auf das Objekt ihrer Forschung oder auf die Methode?[7] Wenn sie sich auf das Objekt bezieht, so bleibt man weiter in einer realistischen Auffassung der

Wissenschaft, eine Auffassung, die Husserl bezüglich der Naturwissenschaften
und an manchen Stellen auch bezüglich der Geisteswissenschaften überwindet.
Es handelt sich also um eine methodische Verschiedenheit, auch wenn an
der Basis der Verfahren, sowohl der ersteren als auch der zweiten, immer
ein Idealisierungs-[8] und Induktionsprozess[9] steht, der in gewisser Hinsicht
erlaubt, sie zu assimilieren.

Trotz der Schwierigkeiten Husserls, den Unterschied zwischen den Geisteswissenschaften und der Phänomenologie festzulegen, ist jedoch zu bemerken,
dass das Interessanteste, das aus der Husserlschen Perspektive zu gewinnen ist,
in der Feststellung des Kategorisierungsprozesses liegt, der sowohl in den
Naturwissenschaften als auch, alles in allem, in den Geisteswissenschaften
gegenwärtig ist. Freilich trägt die in den Kategorisierungsprozess hineingelegte
Unterscheidung zwischen den beiden Disziplinen dazu bei, dass die beiden
Realitäten, die des 'Geistes' und die der 'Natur', gerade in Bezug auf die
Verschiedenheit der entsprechenden Objekte weiter differenziert werden.
Die Begriffe Seele und Leib, die das Ergebnis einer dualistischen Perspektive
sind, werden durch Abgrenzung der entsprechenden Wissenschaftsgruppen
weiter vertieft und getrennt.

WISSENSCHAFTEN UND PHÄNOMENOLOGIE

Der Dualismus, charakteristisch für die philosophische Tradition der westlichen
Welt und Resultat eines Kategorisierungsprozesses, der jeden Wissensbereich
berührt, findet eine weitere Bestätigung in den Wissenschaften. Er stellt sich
aber als sehr problematisch dar: nicht nur, dass sich gerade in der phänomenologischen Perspektive die Verschiedenheit der beiden Wissenschaftsgruppen
als zweideutig erweist, sondern dass insbesondere di Beziehung zwischen
Geisteswissenschaften und Phänomenologie diese Zweideutigkeit klar zu
erfassen gestattet.

Wenn die Wissenschaft das Resultat eines Objektivierungsprozesses ist, wie
sollen dann die Geisteswissenschaften vor einem solchen Prozess gerettet
werden? Aufgrund ihrer grösseren Affinität mit der Phänomenologie, behauptet Husserl. Handelt es sich aber um eine Affinität in der Methode oder
im Objekt? Im ersten Fall versteht man die Eigentümlichkeit der phänomenologischen Methode nicht mehr. Besteht dagegen eine Affinität im Objekt,
wie es eher den Anschein hat, die in der gemeinsamen Bezugnahme auf die
Untersuchung der Subjektivität liegt, so entfällt die von der Phänomenologie
vorgeschlagene besondere Näherung an die Subjektivität ebenfalls.

Die Frage kompliziert sich noch mehr, wenn man zwischen einer naturalistischen Psychologie und einer neuen, auf der Basis des phänomenologischen Ansatzes noch ganz aufzubauenden Psychologie unterschiedet. Muss diese denn gegenüber der Phänomenologie propädeutisch oder "konsequent" sein? Das Element, das sie mit der Phänomenologie gemein hat, ist in jedem Fall die Untersuchung der Subjektivität; aber dann erweist es sich sowohl im ersten, als auch im zweiten Falle schwierig, die beiden Gesichtspunkte, den psychologischen und den phänomenologischen, zu 'trennen'.

Ist die von der Wissenschaft durchgeführte Forschung objektivistischer Art, so charakterisiert und bezieht sie *alle* Wissenschaften ein. Der Versuch, eine Gruppe zu retten, geht über den Zweck der methodischen Differenzierung eines Verfahrens hinaus und zieht die Gültigkeit des phänomenologischen Vorschlages selbst in Mitleidenschaft.

Hier tut sich die wichtige Frage auf, was man unter 'Grundlegung' der Wissenschaften durch die Phänomenologie zu verstehen hat. Wenn diese nämlich Fundament der Wissenschaften nicht in dem Sinne sein soll, dass sich auf sie ein wissenschaftliches Verfahren gründen soll,[10] sondern vielmehr, so meine ich, in dem Sinne, dass sie erlaubt, die gnoseologischen Mechanismen aufzudecken, die an deren Basis stehen, so muss sich ihre Methode radikal von der Methode der konstituierten Wissenschaften unterscheiden, und das ist ja tatsächlich auch der Fall.

Worin liegt also die Spezifität des phänomenologischen Verfahrens und der Terminus der Untersuchung? Wenn das 'Prinzip aller Prinzipien' wirklich das führende Moment der phänomenologischen Analyse ist, so ist die Dimension die sich in diesem 'Gegebenen' auftut, die der *Erlebnisse*, und an diesem Punkt ist es möglich, von der Kategorialität zur Originarität überzugehen, nach einem von Husserl mehrmals aufgezeigten Weg.[11]

DIE ORIGINÄRE SPHÄRE

Auf originärer Ebene wird dann jedes Begriffsgebilde auf seine letzten es konstituierenden Elemente reduziert: nicht nur die wissenschaftlichen Formationen werden in ihrer Konstitution als Idealisierungs- und Objektivierungsprozesse untersucht, sondern es müssten alle fundamentalen Begriffe einer Analyse unterzogen werden. Eber hier, in der originären Dimension, zeigen nicht nur die Naturwissenschaften und die Geisteswissenschaften, sondern auch die Begriffe Seele und Leib ihre kategoriale Konstitution. Sie zu studieren heisst zu den Erlebnissen zurückzugehen, die ihnen zu Grunde liegen. Welches sind aber solche Erlebnisse, wie können sie festgestellt und in zwei getrennte Gruppen unterschieden werden, wie ist diese Scheidung erfolgt, wie ist die Koexistenz zweiter Standpunkte bezüglich der Realität Mensch möglich? Und auch die Realität Mensch selbst, wie stellt sie sich auf der Basis der Erlebnisse dar? An dieser Stelle bedarf es einer eingehenden Analyse der Erlebnisse.

Jenseits der bereits konstituierten Schemata und der kulturell festgesetzten oder überlieferten Scheidungen — die auch weiter bestätigt werden können, die sich aber grundsätzlich ausschliessen — ist eine Untersuchung der Begriffe Seele als bleibende *Substanz* und Leib als lebende *Substanz* notwendig. Sie nehmen innerhalb einer progressiven Unterscheidung zwischen verschiedenen Bereichen Gestalt an: auf der einen Seite im Bereich Ding-Körper, auf der anderen Seite im Bereich des Geistes. Innerhalb dieser Differenzierungen bilden Leib und Seele eine psychophysische Einheit, aber auch eine Dualität.

Es ist der zwischen dem leblosen und dem belebten Körper bestehende Unterschied, der herkömmlich zur Unterscheidung zwischen Körper und Seele führt. Nicht nur die Aktivität des Menschen im Wachzustand, sondern auch der Körper beim Schlaf zeigen an, dass es 'etwas mehr' gibt. Dieses Etwas ist zwar auch im Tier da, doch der Unterschied zwischen dem Menschen und dem Tier, das sich bewegt, sich fortpflanzt, dessen Handlungen absolut wiederholend, also nicht auf ein einheitliches Projekt ausgerichtet sind, dem die Überlegung, die artikulierte und komplexe Kommunikation, die kulturelle Ausarbeitung und die kreative Konstruktion des Menschen fremd sind, eben dieser Unterschied veranlasst dazu, die Anwesenheit des Geistes im Menschen zu behaupten.

Was wird nun aber 'festgestellt', das von Ding, Körper, Seele, Leib und Geist zu sprechen erlaubt? Welches sind die Erlebnisse, die diesen Begriffsgebilden zu Grunde liegen? Die Unterscheidung zwischen materiellen und geistigen Realitäten beruht doch letztlich auf einem einzigen Realitätsbegriff, der als "Einheit bleibender Eigenschaften mit Beziehung auf zugehörige Umstände"[12] definiert wird.

Hier wirkt jener Mechanismus der Varianz/Invarianz, der Husserl von seinen mathematischen Studien der Variationsrechnung eingegeben wird, und den er deskriptiv verwendet, um zu verdeutlichen, dass jede Realität, von der angenommen wird, sie besitze bleibende Eigenschaften, im wesentlichen von dieser Beziehung bestimmt wird, ja sie sogar das Ergebnis der Ausschaltung der Variationen zugunsten der Invarianz ist.[13] Das ist das kategoriale Verfahren, das der wissenschaftlichen Reduktion der Objekte zu Grunde liegt.

Das Verdienst Husserls ist es, die Existenz eines derartigen Prozesses festgestellt zu haben. Er zeigt aber nur den *Weg* an, der zu Originarität führt.

Wenn es darum geht, das, was sich gibt in den Grenzen, in denen es sich
gibt zu beschreiben, so muss man über die begrifflichen Konstruktionen
hinausgehen, und zwar nicht, weil diese nicht mehr bestehen sollen, sondern
weil man sich bewusst sein soll, dass sie das Produkt eines intellektuellen
Verfahrens sind, das sich nach und nach in immer komplexere Ausarbeitungs-
ebenen schichtet.

Was man durch die 'Reduktion' ausfindig macht, ist eben die Sphäre der
in der Immanenz wahrnehmbaren Erlebnisse, den immanenten Fluss der
Erlebnisse, in dem sich das transzendente Sein offenbart;[18] auf diese Weise
wird die authentische Bezeihung *Immanenz-Transzendenz* entdeckt, die auf
phänomenologischer Ebene den Leitfaden für das Verständnis der beiden
Realitäten, Seele und Leib, darstellt.

Wenn der Immanenzbegriff auch den Transzendenzbegriff voraussetzt
und auf ihn verweist, so entsteht er doch nicht so sehr unmittelbar aus
der Wechselbeziehung mit diesem letzteren — man fragt sich ja anfangs
nicht, was transzendent und was immanent ist —, als vielmehr als Rückschlag
(Gegenstoss) der Reduktion, die, da antinaturalistisch, auch antiexistentiell ist,
und die die Dinge, wie sie sich in ihrer 'Natürlichkeit' darstellen, in Klammer
setzt und sie dadurch in der Weise sieht, in der sie sich durch die Erlebnisse
konstituieren. Dieser Übergang, will man einen Ausdruck aus dem nunmehr
festen Sprachgebrauch benützen, dem man tatsächlich nur schwer und vielleicht gar unmöglich entgehen kann (hier tut sich das Problem der Schichtung
der Sprache auf, die die Schichtung des Dinges als Erfahrungs- und Erkenntnis-
objekt nicht nur begleitet, sondern mit ihr eins ist), ist der Übergang zur
Immanenz, jedoch: was bleibt 'draussen'?

Es zeichnet sich so die Notwendigkeit ab, eine Analyse der Begriffe Ding,
Körper, Leib, Seele und Geist anzustellen. Hier soll nur ein Hinweis darauf
gegeben werden, wie man phänomenologisch verfahren sollte, um ihren
'Inhalt' hervorzuheben.

(a) *Ding*

Bekanntlich ist für Husserl das Ding, als Sinneseinheit, aus aufeinanderfolgen-
den Abschattungen gebildet und immer mit der Subjektivität in Beziehung,
doch nie sich vollkommen in sie auflösend.[20] Schreitet man zur höheren
Konstitution des Dinges, so fordert es, da es mit einem logischen Subjekt
in wechselseitiger Beziehung steht, eine logische Bestimmung, die dem Ding
ein 'bleibendes' Sein vorschreibt, welches sich dann in ein Gefüge bleibender

mathematischer Eigenschaften umsetzt. Auf diese Weise ist eine Verwissenschaftlichung des Dinges möglich.

Hier beeilt sich Husserl nun aber, den Unterschied zwischen der Substanz des mathematisierten Dinges und der Substanz der Psyche zu behaupten. Es stimmt zwar, dass der ganze Prozess der erkennenden Organisation 'drinnen' erfolgt. Warum heisst aber dieses 'Drinnen' Psyche und warum unterscheidet es sich von der mathematischen Natur des physischen Dinges und von der schematisierten Natur des Dinges der Intuition? Wird hier im wesentlichen das Resultat eines historischen Prozesses akzeptiert, der die progressive Mechanisierung der Natur erlebt hat, um sie sodann als in re existierend und notwendig zu behaupten, und will ein Teil von ihr, der Teil des Geistes, der nicht auf dieselben Kriterien reduzierbar scheint, gerettet werden? Wäre es nicht zweckmässig, die Perspektive 'umzukehren', wie es der Husserlschen Analyse implizit scheint — die in diesem Zusammenhang explizit gemacht werden muss —, und den Dualismus als 'Landepunkt' eines Prozesses zu sehen? Andernfalls kommt der Verdacht auf, dass von der Ausarbeitung der bereits konstituierten und in Natur- und Geisteswissenschaft geteilten Wissenschaft ausgegangen wird, um einen Unterschied zu rechtfertigen, der, wenn er als authentisch und gültig entdeckt wird, er dies ist, weil er es auf originärer Ebene ist, ohne es aber sein zu *müssen*, weil er bereits vom Kategorialen Gesichtspunkt her als solcher auftritt. Es ist ja nicht die logisch-mathematische Struktur, die in der originären Sphäre auftaucht.

Dann stellt sich das Problem in anderen Termini: es geht nicht darum, eine 'vorausgesetzte' Unterscheidung zu rechtfertigen, sondern darum, den Menschen zu untersuchen. Seine Erlebnisse beziehen sich auf das Verhältnis zu den Dingen, zu den anderen und zu sich selbst, doch impliziert schon diese Unterscheidung eine auf ihre ursprüngliche Erkennung folgende Gruppierung. Es muss deshalb, wie im Falle des Dinges, die Art und Weise bestimmt werden, in der man zur Bildung der anderen angeführten Begriffe gelangt.

(b) *Körper*

Der Körper ist ein Ganzes von körperlichen Elementen, von physischen Teilen, die ein Ganzes bilden, wie eben physische *Dinglichkeiten* ein Ganzes bilden, — und hier kann und muss von einer inneren Kausalität, einer Kausalität der physischen Teile in bezug aufeinander gesprochen werden. Die körperlichen Elemente sind selbst wieder Substanzen.[21]

Der Körper gleicht dem Ding, identifiziert sich aber nicht mit ihm,[22] vielleicht

weil es keine Autonomie in sich selbst hat, sondern nur als Privation in Bezug
auf den Leib gesehen wird.

(c) *Leib*

Der Leib konstituiert sich tatsächlich als physisches Ding, d.h. er hat reale
Extension und Qualitäten (Farbe, Glattheit, Härte, Wärme) und gleichzeitig
als das, was Empfindungen hat. Aus diesem Grund ist das, was den Leib
charakterisiert, das Empfinden. Eben deshalb bildet der Leib den zentralen
und neuralgischen Punkt der ganzen Rede über den Menschen, denn nur
ausgehend vom Empfinden kann man von 'Erkenntnis' sprechen, kann man
eine überlegte Rede anstellen, kann man denken, entsteht das Bewusstsein.
Im Unterschied zwischen unbelebt und belebt stellt sich der Leib als ein
Körper dar, der als 'etwas mehr' erfahren wird, sich also unmittelbar sowohl
vom Körper als auch vom Ding unterscheidet, nicht nur in Bezug auf uns
selbst, sondern auch auf die Umwelt, in der die Affinität mit dem fremden
Leib und die Differenz zu den Dingen durch Einfühlung erfasst wird.

(d) *Seele*

Die Erlebnisse der Empfindung sind also grundlegend, um die Unterschiede
zwischen dem Unbelebten und dem Belebten zu erfassen. In dem letzteren
stellt man so einen Teil fest, der Körper ist, und einen anderen Teil, der
'empfindend' ist. Bisher ist der empfindende Teil mit dem Leib identifiziert
worden, wie aber zeichnet sich die Anwesenheit der Seele ab? Auch wenn die
herkömmliche Unterscheidung zwischen Leib und Seele die gewöhnlichste
und unmittelbarste Unterscheidung zu sein scheint, so folgt aus dem Gesagen
doch, dass eine Bestimmung der Grenze zwischen den beiden Realitäten
schwierig ist. Wenn man dem Leib das Empfinden zuschreibt, wie soll man
sich etwas vorstellen, das, selbständig lebend, 'empfindet', d.h. die Seele?
In einer rein naturalistischen Interpretation kann sie unmöglich vom Leib
unterschieden werden. Man darf aber nicht so verfahren, dass man von
einer bereits als sicher vorausgesetzten Trennung ausgeht, sondern muss
zu den Erlebnissen zurückkehren. Die Dualität läuft so letztlich auf einen
Komplex von Erlebnissen hinaus, die mit dem Empfinden in Bezug stehen.
Das Berühren ist ein immanentes Erlebnis in Bezug auf die Tastempfindung;
das berührte Objekt löst sich eben in die Tatsache auf, berührt, zu sein,
als Terminus der Empfindung. Doch gibt es ja nicht nur das Berühren, es
gibt Erlebnisse, die sozusagen keiner objektiven Beziehung im Sinne von

extrasubjektive bedürfen, und die auch nicht mit den Taktilempfindnissen, den Sehempfindnissen usw. in Beziehung stehen; eben hier beginnt eine psychische Realität Gestalt anzunehmen, die autonom erscheint, deren Autonomie freilich darin zu sehen ist, dass die Erlebnisse nicht in Bezug zu einer äusseren Realität stehen und nicht unmittelbar an die Körperlichkeit gebunden sind.

(d) *Geist*

Die Entscheidung zu einem gewissen Verhalten, die Begründung einer Handlung, die Bedeutung einer Geste oder der Sinn, der irgend etwas zuzuschreiben ist, sind, wenngleich von der Konkretheit und von der Körperlichkeit vermittelt, so doch mit Absichten belastet; die Untersuchung der entsprechenden Erlebnisse führt nun eben dazu, diese 'Absicht', die sich als das zeigt, was das einfache Zeichen charakterisiert, zu fokussieren. Der Komplex dieser bedeutenden Realität wird mit geistig definiert.

Es stellt sich jedoch wieder eindringlich die Frage nach der Rechtmässigkeit der zuvor eingeführten Unterscheidung zwischen den Erlebnissen. Haben sie als solche alle dieselbe Struktur oder ist es möglich, sie auf der Grundlage der hyletischen und noetischen Komponenten zu unterscheiden, die in einigen anwesend sind (in denen, die sich auf die Empfindung beziehen), in anderen dagegen nicht? Die Entwicklung dieses Themas würde eine globale Revision des ganzen originären Bereichs erfordern. Auf diese Revision kann nur hingedeutet werden: das Erlebnis als Moment des originär Gegebenen muss beschrieben werden können, ohne sich darum zu kümmern, a priori Unterschiede festzulegen, was nicht bedeutet, dass das noetische, intentionale Moment in allen Erlebnissen dasselbe ist, denn eine Klassifizierung auf der Grundlage der verschiedenen Modalitäten ihres Gegebenen ist ja möglich.

Eine solche Analyse rückt auch die Funktion des kategorialen Moments ins Licht, und da versteht man, wie das Verbleiben der Sustanz nichts weiter ist als die Invarianz, die man nach Ausschaltung aller kontingenten Aspekte erhält. Was nicht bedeutet, dass man in eine empiristische Haltung zurückfällt, dass man das Problem der Substanz auf lockesche Weise löst als etwas, das 'hinter' dem Phänomen steht, sondern dass der Empirismus selbst zur äussersten Konsequenz geführt wird, indem man ihn radikalisiert, wie Husserl behauptet.[23] Es geht nicht so sehr um das Verbleiben der Substanz 'Seele' oder der Substanz 'Körper', es geht vielmehr darum, jenseits der Diskussion zwischen Spiritualismus und Materialismus, die Existenz 'geistiger' Fähigkeiten oder Handlungen und 'körperlicher' Dimensionen, in die das menschliche

Leben gegliedert ist, in ihrer Verflechtung, doch auch in ihrer Unterscheidungsmöglichkeit anzuerkennen. Die Unterscheidung ist einzig an die Analyse der entsprechenden Erlebnisse gebunden, woraus man die Unauflösbarkeit, doch auch die Auflösbarkeit der beiden Momente ableitet: die mögliche Erfahrung eines Überlebens ohne Körperlichkeit, jedoch die notwendige Wiedereroberung einer Einheit, wenn das Leben triumphieren und sich in seiner ganzen Fülle wieder zeigen soll. Das Verbleiben identifiziert sich also nicht mit der Substantialität − oder zumindest ist diese Identifizierung das Resultat einer in einem präzisen kulturellen Kontext entstandenen Interpretation −, sondern ist zu verstehen als Möglichkeit der 'ewigen' Wiederholung, das was Husserl, zwar noch in einer wesensmässigen Sprache, die Allzeitigkeit und die Zurückrufbarkeit des *Eidos* nennt.

Centro Italiano di Fenomenologia

ANMERKUNGEN

[1] E. Husserl, *Ideen* II, Husserliana Bd. IV, II, par. 30, S. 125.

[2] Die Antinomie zwischen dem Standpunkt der Naturwissenschaften und dem der Geisteswissenschaften liegt darin, dass die ersten das Subjekt als ein Objekt innerhalb der Natur, die zweiten die Natur in Beziehung auf das Subjekt betrachten. So kommt es zur Antinomie zwischen Subjektivität und Objektivität, zwischen Wissenschaft der Lebenswelt und Wissenschaft von Natur: diese zwei Anschauungsweisen verlaufen parallel und treffen nie aufeinander (*Ms. trans. A VII 13*, S. 253−257).

[3] *Phänomenologische Psychologie*, Husserliana Bd. IX, S. 221.

[4] Man fragt sich ja tatsächlich, ob die den Naturwissenschaften eigene Idee der objektiven Realität und der objektiven Wahrheit analoger Weise für den Bereich des Seelischen gilt, oder ob es sich nicht um eine ganz andersartige Objektivität handelt (*Ms. trans. F I 32*, S. 133−135). Was vermieden werden will ist allerdings, in einen Dualismus zu verfallen, der durch einen naturalistischen Methodenmonismus einen psychologischen Naturalismus gelten lässt, denselben, in den auch Rickert und Windelband verfallen sind.

[5] *Erste Philosophie* II, Husserliana Bd. VIII, *Idee der Wissenschaft* (1926): was für eine Art Theorie und Wissenschaft kann die Psychologie sein? Können die geometrischen und physikalistischen Theorien als Prototypen aller Theorien betrachtet werden? Während Husserl an anderen Stellen seiner Forschung dazu neigt, der Psychologie ein eigenes Statut zu geben, sagt er hier, dass, da jedes Ding der Welt eine Natur hat, es ihm deswegen zusteht, "zu einer definiten Gesetzlichkeit und Mannigfaltigkeit" zu gehören (S. 17). Andererseits besteht das Einheitsmoment aller Wissenschaften darin, einem Idealisierungsprozess unterworfen zu sein (*Ms. trans. K III 22*).

[6] *Ms. trans. A IV 14 und K III 22*.

[7] In der Diskussion um das epistemologische Statut der Geisteswissenschaften wird die Frage bezüglich des Objekts, auf das sie sich beziehen, noch heute lebhaft erörtert: "Le

domaine d'application de la notion de science s'est tellment élargi depuis quelque temps, qu'il est devenu impossible d'attribuer cette qualification à une discipline quelconque sur la base des ses 'contenus'" (E. Agazzi, *Problems Epistemologiques des Sciences Humaines*, in *Epistemologia* 2 (1979), 39–66). Wenn bei der Umgrenzung der laut Agazzi für die Wissenschaften charakteristischen Objektivität man sowohl in den Natur- als auch in den Geisteswissenschaften das Vorhandensein eines methodischen Verfahrens vorfindet, so liegt die Verschiedenheit der beiden Bereiche, wenn diese aufrechterhalten bleiben soll, entweder im Innern der Methode selbst (er betont ja, dass im Falle der Geisteswissenschaften die Objektivität und Strenge nicht mit der den Naturwissenschaften eigenen Mathematisierung übereinstimmt) oder sie muss sich auf den Inhalt beziehen.

[8] In der *Ersten Philosophie* z.B. werden die Geisteswissenschaften für 'positiv' im Sinne von 'abstrakt' gehalten: "Die Noetiken (die normativen Vernunftlehren) als positive Wissenschaften teilen den Grundmangel der Abstraktheit mit allen positiven Wissenschaften, auch den Geisteswissenschaften". (S. 288).

[9] *Ms. trans. K III 22.*

[10] *Ms. trans. A VI 21.*

[11] Husserl ist sich ja des Unterschiedes zwischen den zwei Weisen, die Subjektivität zu verstehen, wohl bewusst: "Die Geisteswissenschaften als personale Wissenschaften haben es mit den Personen als handelnden und ihren Handlungen (im weiteren Sinn) zu tun – nicht mit dem eigentlich konstituierenden Bewusstsein" (*Erste Philosophie*, S. 287, Anmerkung 2).

[12] Die von J. D. Robert (*La fonction de la phénoménologie a l'égard des sciences de l'homme et des anthropologies philosophiques*, 'Laval Théologique Philosophique', Oct. 1977, S. 273–308) kann nur akzeptiert werden, wenn der trans-disziplinäre Dialog, den die Phänomenologie seiner Ansicht nach anregen sollte, als qualitative Überwindung der Wissenschaft selbst verstanden wird, aber nicht im Sinne der 'Essenz-Analyse' oder der hermeneutischen Untersuchung. Es geht ja tatsächlich nicht so sehr darum, das der realen Erfahrung Implizite, die den Menschen und das Menschliche zum Objekt hat, explizit darzulegen – die erste Aufgabe der Phänomenologie laut S. Strasser (*Phénoménologie et Sciences de l'homme – Vers un nouvel esprit scientifique*, traduit par A. L. Kelkel, Editions Béatrice-Nauwelaerts, Paris 1967), auf den Robert zurückgreift, noch die vom empiristischen, objektivistischen und szientistischen Vorurteil geläuterten Wissenschaften in eine "Gesamtanschauung" einzuordnen – die zweite Aufgabe –, noch die Resultate der Wissenschaften für eine Interpretation der Existenz zu verwenden, und das weil (i) eine Analyse der Methode der Wissenschaften vom Menschen zu einer Interpretation führt, die, wie bereits gesagt, nicht realistisch sein kann; (ii) wenn die Aufhebung des Empirismus, des Szientismus und des Objektivismus Gültigkeit besitzt, so wird sie nicht erreicht, indem man die wissenschaftliche Forschung in eine "Gesamtanschauung" einfügt, sondern indem man den globalen Standpunkt gegenüber dem Standpunkt der Wissenschaft als 'alternativ' betrachtet und folglich; (iii) sie nicht zweckbestimmbar sind auf eine 'wahrere' Kenntnis des Menschen, sondern nur auf eine Kenntnis, die in einer gewissen Weise 'organisiert' und 'ausgerichtet' ist.

[13] Man vergegenwärtige sich das bereits angeführte *MS A VII 13*, in dem die Analyse der Natur und des Geistes vor der Wissenschaft, in der blossen Erfahrung, umrissen wird.

[14] *Ideen*, II, par. 33.

[15] Siehe *MS. A VII 13.*

16 "Diese höhere Dingkonstitution schreibt dem Ding ein verharrendes Sein zu, einen Bestand verharrender mathematischer Eigenschaften, aber so, dass die allgemeine Struktur des Dinges, die Form der Realität-Kausalität erhalten bleibt" (*Ideen* II, Husserliana Bd. V, S. 132). Bedeutet das also, dass jedesmal, wenn man von Substanz spricht, eine mathematische Struktur dazwischentritt? Husserl verneint es; "wenn es für das Seelenreale eine physiopsychische Seite (. . .) gibt, die ihre Umstände in dem Leib und seinen Leibkausalitäten findet" (ibidem, S. 135), so ergibt sich auch ein "Bewusstsein (. . .) als abhängig sozusagen von sich selbst" (ibidem, S. 135), aber von welchem Typ von Untersuchung ist es Objekt?

17 *Ideen* I, Par. 24.

18 *Ideen* II, Par. 32.

19 Dazu: D. Conci, *L'universo artificiale – Per una epistemologia fenomenologica*, Ed. Spada, Roma 1978.

20 *Ideen* II, par. 32.

21 *Zur Phänomenologie der Intersubjektivität* II, Husserliana Bd. XIV, S. 68.

22 *Ideen* II, Par. 35.

23 *Phänomenologische Psychologie*, Husserliana Bd. IX, S. 301.

PIERRE TROTIGNON

L'OEIL DE LA CHAIR

> ... einen sinnlichen Leib für einen im Verstehen
> erfassten geistigen Sinn. "Leib" und "Geist" ist
> erscheinungsmässig in eigentümlicher Weise einig.
>
> Husserl [1]

La phénoménologie est à la fois théorie des essences et théorie des vécus qui donnent l'expérience des essences. Le jeu de l'essence et de l'exemple est le noyau de sa méthode. Or ce jeu met en oeuvre des variations imaginaires qui se déploient dans un temps originaire et constitutif par lequel la conscience s'ouvre de l'intérieur à la transcendance de ses objets. Le champ des variations est infini. Cette infinité est de fait limitée pour mon être empirique, de sorte que la temporalité dans le moi pur exige une infinité de consciences et je vois ainsi se dessiner une multitude de corps animés qui communiqueront entre eux pour réaliser cette infinité.[2] Par l'irruption naturelle de l'*alter ego* dans la *chair* le *temps pur* fait accéder le sujet transcendantal au monde de la *vérité réelle*. L'expérience de la chair (*Leib*) est la condition nécessaire pour que la phénoménologie transcendantale puisse présenter le règne de la raison[3] comme la finalité réelle de la nature. Or nous verrons qu'en ce point les difficultés de la théorie se redoublent et font nettement apparaître une aporie très significative.

Tout instant contient en soi la virtualité de la réflexion, chaque moment vaut comme possibilité de l'extension de l'esprit. Par cette extension le présent reprend le flux du temps, et, lorsque nous procédons à la réduction éidétique, le rabattement sur l'instant ouvre les variations éidétiques de la réflexion. Dans l'expérience de la localisation des sensations[4] la chair se révèle comme l'organe de la volonté.[5] Par cette expérience le moi distingue en lui la facticité et le centre de la subjectivité pure, ce qui lui permet tout à la fois d'acquérir la capacité de se voir lui-même et de constituer pour la connaissance des objets du genre "chose". Cette double mise en relief de la conscience de soi et de la connaissance des choses efface la perception de la chair par elle-même. La chair est l'oeil de l'esprit. Si l'oeil veut se voir, il doit se mirer en un autre oeil. La chair contient en ses déterminations essentielles l'apparition d'un autre sujet que mon propre moi. La conscience de soi de la monade concrète suppose l'apprésentation de l'*alter ego*, qui est ainsi une

49

A-T. Tymieniecka (ed.), Analecta Husserliana, Vol. XVI, 49–61.
Copyright © 1983 by D. Reidel Publishing Company.

nécessité a priori pour le sujet pur si ses objets de connaissance doivent avoir une valeur objective certaine.[6] Nous sortons du solipsisme et l'universalité effective de la vérité est fondée.[7]

La constitution temporelle d'un pur donné de la conscience et, plus profondément, l'autoconstitution du temps phénoménologique résident dans le lien entre le temps et l'intentionalité de la conscience. Or l'intention est un pouvoir du moi vigile, capable de renverser son attitude naturelle et de modifier les rapports de fond et de forme qu'y mettaient en évidence les modes d'actualité du perçu.[8] Le passage du simple flux vécu à la temporalité fondatrice de la philosophie requiert donc que les propriétés de l'espace vécu soient modifiées.[9] Cette modification est possible parce que le remplissement concret du temps est toujours donné en corrélation avec une extension spatiale et corporelle du vécu.[10] La capacité de se mouvoir en déplaçant avec lui le point de vue du moi confère à mon corps le pouvoir d'étendre la sensation en corrélation avec le sentiment du temps et avec l'aperception des autres sujets.[11]

Dans la chair le temps et l'espace se répondent, dessinant les linéaments d'un monde intersubjectif où le moi rencontre les autres sujets. La liaison du divers suppose une relation entre des sujets telle que ma conscience, intentionnellement liée à la conscience étrangère, revienne sur soi-même lorsqu'elle rencontre l'autre sujet. Encore faut-il que ce rebroussement réflexif ne soit pas conçu comme un choc inopiné. Il doit être l'effet de l'aperception que je prends d'une vie transcendantale, où la pluralité des monades était déjà principiellement inscrite.[12] La conscience intuitive que j'ai des autres sujets présente ainsi deux aspects: c'est un caractère transcendantal du moi pur, mais je dois y voir aussi une spatialisation du flux temporel vécu par le biais de ma propre chair.[13] Tout de même que le surgissement de l'instant ouvre, au coeur du moi pur, une source ontologique où l'intentionalité puise la certitude que son corrélat est bien une réalité transcendante, tout de même la chair dessine, dans la sphère d'appartenance de la monade concrète, un chemin vers une altérité.[14] Le temps et la chair ont partie liée. Le temps du moi constitue originairement la noèse: après réduction, j'accède au temps originaire lui-même, le moi apparaît comme pôle actif et vigile de la monade concrète qui vise des choses dans l'espace objectif du monde. Comment le peut-elle si ce n'est par la chair, qui est une apprésentation à déchiffrer par le jeu qui, sur une autre chair que la mienne, livre à mon regard des phénomènes marqués par d'autres regards?.[15]

Si je ne considère que les relations de la conscience pure et des noèmes, le temps sera un pur présent, même si j'en ressens l'écoulement intime. Par le biais de la chair, l'apprésentation crée une mémoire dont les traits se dessinent comme des chemins dans le monde. Nous déchiffrons sur les objets les potentialités déjà tracées qui sont pourtant l'expression de la vie de la conscience.[16] Le jeu des potentialités est la vie intentionnelle. Tout le possible n'est pas donné, un tremblé et une esquisse virevoltent. Certes un objet intentionnel sert de guide noético-noématique pour ordonner la diversité,[17] mais il faut également que le moi persiste et subsiste à travers l'aventure des phénomènes. Le moi doit donc exister pour soi. Cette condition doit être identique à elle-même aussi bien dans l'attitude naturelle que dans l'attitude phénoménologique. Nous devons donc supposer une évidence continue du moi. Par cette évidence continue le moi se constitue comme existant dans le courant de sa pensée. Le cogito *instantané* traverse le *temps*, ce qui entraîne deux conséquences: d'une part le moi possède un pôle d'unité identique, d'autre part il doit pouvoir reconnaître cette identité à travers la diversité des expériences. La conscience intentionnelle se rapporte à *un* objet qui synthétise la multiplicité phénoménale, mais cette synthèse de l'objet unique, parce qu'elle renvoie à l'unicité du moi qui pense, suppose des synthèses subjectives par lesquelles le moi existe pour soi, dans des "habitus" que je peux ressaisir dans la conscience que j'ai de mes déterminations actuelles:

Je me décide, l'acte vécu s'écoule, mais la décision demeure − que je m'affaisse, en devenant passif, dans le sommeil, ou que je vive d'autres actes − la décision demeure continuellement en vigueur et, corrélativement, je suis désormais déterminé d'une certaine façon. (...) *je me transforme moi-même*, moi qui persévère dans ma volonté permanente, lorsque je "biffe", lorsque je renie mes décisons et mes actes.[18]

Or l'unité formée par le moi polaire, qui soutient l'identité référentielle des habitus subjectifs, *et* "ce sans quoi il ne serait pas concrètement",[19] forme la *monade*, l'ego concret au sens plein. La conscience de soi de l'ego transcendantal intentionnel renvoie non seulement à des objets, mais aussi, *en tant qu'il est transcendantal*, à la monade concrète. La chair sera le nom de ce complément du moi transcendantal qui en fait une monade. Remarquons bien que la chair appartient à la sphère du moi, elle ne deviendra corps que par la rencontre avec la nécessité d'un autre sujet, là-bas dans le monde. La chair est le champ transcendantal pour des objets qui n'ont pas le statut de "chose".

Considérons la monade concrète. Elle ne peut apparaître, comme objet de pensée, que pour la conscience qui a procédé à la réduction et qui, de ce fait, se donne comme identique à l'ego transcendantal, lié à ses états de conscience purs et aux synthèses objectives qui se dessinent à travers les actualités et les potentialités des états de conscience. Parmi les esquisses directrices, qui apparaissent dans l'objectivité transcendantale, il en est une qui va jouer un rôle singulier: *l'autre sujet*, qui est un objet du monde doté de la particularité d'*être un corps physiologique animé*, doté de comportements intentionnels. L'autre est *un objet qui indique une subjectivité dans le monde*, dans le même monde où je constitue mes objets. Toute théorie du monde objectif passe ainsi par l'existence d'autres sujets pour lesquels ce que je vois comme objet est aussi un objet. Par la médiation de l'existence de l'autre, l'objet est renforcé dans son objectivité. Ce renforcement de l'objet se marque par le fait que la réduction du moi à sa sphère d'appartenance transcendantale, quand elle le laisse "seul", ne supprime pas le renvoi au monde, car le sens du monde est inhérent au moi, à l'être transcendantal et absolu de l'ego autour duquel gravite la sphère d'appartenance.[20] Ce qui revient à dire que l'*altérité* et la *pluralité* appartiennent a priori au moi pur, d'où il suit que la réduction ne peut mettre ces déterminations entre parenthèses. Cette altérité n'a donc jamais été une altérité et l'on peut douter que la phénoménologie puisse sortir du solipsisme transcendantal.

Car les objets-hommes sont dans ma sphère d'appartenance. Sans doute ils transgressent le mode d'être de *mon* ego, en ce que leur réflexion, au lieu de revenir sur ma racine monadique, indique un autre point transcendantal de rebroussement, qui est pourtant perçu dans ma propre sphère d'appartenance. A l'intérieur de mon être propre se constitue une pluralité de points nodaux du transcendantal, dont *un seul* est toujours en concordance avec mon moi transcendantal. Ce qui revient à dire que n'est pas moi ce que je n'identifie pas maintenant comme moi, sans que cela donne une solution de principe à la différence de moi et de l'autre.

Si donc je mets de côté ce qui m'est étranger pour ne conserver que ce qui m'est rapporté en propre (*das Mir-Eigene*), je dois éliminer les choses physiques, puis l'animalité, les valeurs culturelles, sociales et historiques, car tout m' est *étranger*. Reste seulement le *phénomène de monde* comme caractère transcendantal et a priori des relations entre le moi pur et la monade que je suis. Mais dans l'attitude naturelle nul n'a jamais confondu sa jambe, la tête d'autrui, un cheval qui passe dans la campagne, la Vénus de Milo et l'économie capitaliste. On peut se demander par conséquent si ce n'est pas l'intention rectrice de la phénoménologie qui rassemble en une masse étrangère globale un réel qui était très diversifié et très complexe avant même que n'intervienne la réflexion du philosophe. Si la réduction ne crée pas réellement cette transformation, la prétention de redécouvrir le réel est absurde, puisqu'on ne l'a en fait jamais perdu de vue, et si la réduction a vraiment placé la pensée dans une situation de solitude telle qu'elle doive tirer d'elle la justification de tout ce qui lui appartient en propre, on peut douter que le moi y parvienne jamais. On comprend alors la fonction de la chair. Elle sera l'envers de la réduction phénoménologique.

La conséquence la plus extrême de la réduction nous conduit à ce rapport du moi et de la monade dans le phénomène du monde. Telle est la racine ultime, la matrice de tout rapport entre eidos et phénomène. Comment définir ce phénomène du monde? Quelle relation entretient-il avec le thème de l'*alter ego*, avec la chair? Nous pouvons peut-être entrevoir une réponse en partant d'une remarque marginale de Husserl au § 44, *V° Méditation*,[21] à propos justement d'un texte où l'autre sujet et la chair apparaissent dans la sphère propre de la monade. Husserl a écrit:

Die Totalerscheinung der Welt — im Strömen immer gemeinte Welt.
Die Totalerscheinung der Natur.
Die Totalmeinung der Welt, die Einzelmeinung — Einzelerscheinung des Einzelwelt-lichen, aber die Meinung schichtet sich, ich kann abstrahieren. Dingerscheinung, Schichte der Kultur oder Schichte des menschlichen Daseins als (*eine Stelle freigelassen*) in der strömenden Gegenwart. Der Strom der Welt-"Erscheinungen", der "Wahrnehmungser-scheinungen" Ontologisch-Gemeintes. Cogito-Schichten, gemäss denen jede Schichte eine (entsprechende) Schichte des cogitatums hat. Das Ich gerichtet auf das Gemeinte.

On voit le sens de ces remarques. Le monde est le corrélatif transcendantal premier et constituant, mais la totalité du monde n'est saisissable que par ce qui se donne à penser dans le flux de conscience. La subjectivité individuée est l'immanence primordiale que le flux de ses sensations tourne vers le monde. Mais le phénomène adéquat du monde ne peut être dégagé que par la réduction. Or si la réduction suppose la capacité de distinguer le moi pur du moi factuel, elle n'a de sens que si la différence entre mon ego propre et les autres personnes est conservée dans la sphère transcendantale comme détermination a priori. Sinon, nous ne pourrions jamais sortir du solipsisme transcendantal. Reste à savoir si la conservation et la répétition de la différence entre moi et les autres, dans la sphère pure de la réflexion, est une imperfec-tion indépassable de la méthode phénoménologique ou la découverte d'une fondation absolue de la pluralité des personnes concrètes dans un même monde d'expériences dont elles peuvent se communiquer la vérité. Il est certain que Husserl choisit la seconde solution. Mais alors, comme il serait insensé de vouloir déduire la réalité des monades indépendantes à partir de la conscience transcendantale de soi, la seule question qui se pose porte sur la manière dont le monde manifeste la distinction transcendantale du moi propre et de l'autre moi.[22] Il s'ensuit que le monde manifeste une différence

qu'il ne fonde pas, ou, ce qui revient au même, que le moi transcendantal
doit être ouvert de l'intérieur par le surgissement d'un *même* ou d'un *soi*
qui est l'unité radicale de la *nature*. La nature, comme donation ultime et
universelle des synthèses, comme horizon pour tout remplissement hylétique
du monde pour n'importe quel moi.[23] Ce que le moi découvre en lui comme
ouverture du temps, la communauté des monades concrètes le rencontre
comme nature. Si donc une apparition totale du monde comme phénoménalité
est impossible, en revanche nous rencontrons une présupposition du monde
sur fond de nature. Grâce à cette présupposition je peux déterminer le sens
du monde pour la monade que je suis. Par l'enracinement des significations
dans le réseau temporel de la nature — ce qui se déploie dans le jeu des
impressions psychiques à travers ma chair — je peux abstraire ce qui relève,
dans l'expérience mondaine, du rapport avec les objets physiques, les animaux,
les formes symboliques de la culture. Les deux concepts de "nature" et de
"monde" ne peuvent donc être développés en parallèle, leur réalisation
simultanée est impossible, de sorte que, par un jeu de va-et-vient, chacun des
deux concepts peut servir à une réduction ou à une purification de l'autre.

La réduction, au sein du monde qui m'est propre, des significations qui
renvoient à d'autres choses que moi, renvoie au moi en tant que ce moi est
dirigé sur les significations dans le flux du présent, ou, *ce qui revient au
même*, renvoie à la *nature pure et simple de ce qui m'appartient*:

Betrachten wir das Ergebnis unserer Abstraktion näher, also das, was sie uns übrig lässt.
Es scheidet sich am Phänomen der Welt, der mit objektiven Sinn erscheinenden, eine
Unterschichte ab als *eigenheitliche Natur* . . . .[24]

La nature n'est pas ici un être objectif pour une science possible, elle n'est
pas "une couche abstraite du monde", mais la nature dans son essence pure:

So gehört zu meiner Eigenheit als von allem Sinn fremder Subjektivität gereinigte, ein
Sinn *blosse Natur*, der eben auch dieses *Für-Jedermann* verloren hat, also keineswegs
für eine abstraktive Schicht der Welt selbst, beziehungsweise ihres Sinnes genommen
werden darf.[25]

Le noyau de cette nature absolue, autour duquel ses mouvements sont
organisés, est la *chair de mon corps* qui tranche sur tous les autres objets par
son animation interne spontanée:

Unter den eigenheitlich gefassten Körpern dieser *Natur* finde ich dann in einziger Auszeichnung *meinen Leib*, nämlich den einzigen, der nicht blosser Körper ist, sondern
ein *Leib*, das einzige Objekt innerhalb meiner abstraktiven Weltschichte, dem ich erfah-
rungsgemäss Empfindungsfelder zurechne . . . .[26]

De ce que nous avons dit, il suit que la chair de mon corps n'est pas un objet dans le monde: elle reproduit dans la texture du monde phénoménal la scission qui refend le sujet dans la réflexion. De même que le moi est une perpétuelle activité de scission, qui se rapporte à elle-même dans le flux du temps, la chair se rend à soi-même étrangère dans le rapport de mon corps avec un autre corps dont le centre réflexif m'échappe. Il suffit pour s'en convaincre de comparer les textes suivants:

Ich sehe, dass *Ichleben in Aktivität* durchaus nichts anderes ist als ein *Sich-immerfort-in-tätigen-Verhalten-spalten* und dass immer wieder ein allüberschauendes Ich sich etablieren kann, das (ein) alle (jene Akte und Aktsubjekte) identifizierendes ist, oder vielmehr, und in ursprünglicherer Fassung: Ich sehe, dass ich selbst mich als in höherer Reflexion ein überschauendes Ich etablieren kann, dass Ich in evidenter synthetischer Identifizierung der Selbigkeit aller dieser Aktpole und der Verschiedenheit ihrer nodalen Seinsweise bewusstwerden kann. Und so sage ich: Ich bin hier überall derselbe, ich als Reflektierender derselbe, der sich als Unreflektierten im Nachgriff erfasst, der als Selbstwahrnehmender mir als Hauswahrnehmenden (z.B.) zusieht usw.[27]

Pour un tel moi, ce qui est absolument nécessaire, c'est uniquement d'être posé comme le sujet identique qui se désigne comme étant le même à travers toutes ses expériences. Une telle position d'identité et de continuité ouvre le temps dans lequel le moi se scinde et se retrouve: elle détermine l'ensemble de ma vie comme moi humain.[28] Mais — et voici le second texte que j'invoquais, comment décrire ce qu'est un moi humain, ce qui veut dire un moi qui ordonne autour de l'ego transcendantal une sphère d'appartenance, qui assure à la fois son individualité et ses rapports à d'autres sujets reconnus? Il est clair que c'est par la reconnaissance de la chair comme *nature venant au monde dans la conscience*:

Wie aber finde ich *"mich" als Menschen*? Nur in der Weise, dass ich meinen Leib irgendwo draussen im Raum denke, wie wenn es ein fremder Leib wäre, und in der Weise einer Einfühlung. Nur wenn ich meinen Leib sozusagen entfremde und dann doch wieder als denselben ansehe, der mir wirklich gegeben ist in der einzigartigen Weise des zentralen Dingphänomens, um das sich alle Welt als aüssere gruppiert, nur dann fasse ich mich als Menschen und lege in dieser mittelbaren psychophysischen Apperzeption mein Ich als eingefühltes dem Leib ein. *Ich finde mich selbst als Menschen auf dem Wege über eine Selbstentfremdung meines Leibes.*[29]

Il s'ensuit que l'image spectrale de mon corps, devenu étranger par une variation imaginaire, introduit dans la chair une scission entre l'aperception de la chair comme telle (*Leibapperzeption*) et l'aperception de l'objet charnel qu'est mon corps, pris comme objet pour une perception en général, la mienne ou celle d'un autre (*Leibkörperapperzeption*).[30] Sujet devenant objet pour soi et pour d'autres, sujet pour qui certains autres objets se donnent comme signes de subjectivités certaines, le moi découvre dans la chair que l'entité perçue est *éclosion de la nature*:

Wahrnehmend tätig erfahre ich (oder kann ich erfahren) alle Natur, darunter die eigne Leiblichkeit, die darin also auf sich selbst zurückbezogen ist.[31]

Examinons ce que signifie cette correspondance entre l'analyse du temps originaire, dans la scission réfléchissante du *moi* transcendantal, et l'analyse de la chair comme scission de l'*être* en monde et nature.

Considérons la réduction à l'appartenance (*die eigenheitliche Reduktion*). Je fais abstraction, dans un objet déjà soumis à la réduction phénoménologique, de tout ce qui n'est pas une propriété exclusive du moi. Cela revient à suspendre momentanément le caractère projectif de l'ouverture au monde, son intentionalité foncière. Si je réduis à l'appartenance les objets animés que sont les "autres hommes", j'obtiens des *choses* inertes. Si je réduis mon propre corps à l'appartenance, "so gewinne ich meinen *Leib* und meine *Seele*, oder mich als psychophysische Einheit, in ihr mein personales Ich, das in diesem Leib und *mittelst* seiner in der *Aussenwelt* wirkt ... ".[32] Ce moi réduit n'est plus un moi au sens naturel (*im natürlichen Sinn*), je n'appartiens plus au "monde", au sens naturel de ce terme (*alle meine Weltlichkeit im natürlichen Sinn*), et pourtant il me reste une "espèce de monde qui est une nature réduite" (*eine Art Welt, eine eigenheitlich reduzierte Natur*), qui forme une unité cohérente et concrète (*konkret einig*). Donc, dans l'attitude naturelle, le monde apparaît comme l'ensemble des objets parmi lesquels je suis, et cette attitude "naturelle" nous fait perdre en réalité la vraie Nature, qui est la liaison temporelle cohérente des essences dans le jeu en apparence hasardeux des perceptions. Par le renversement phénoménologique, le vrai monde apparaît, corrélat noématique de l'intention du moi pur, et c'est pour ce moi pur, dans la scission et la reprise de soi par la temporalité radicale, que la Nature réapparaît, de sorte que l'expérience de la chair est l'expérience originaire du champ transcendantal.

Mon corps se découpe dans la chair comme le seul vivant originaire de la sphère d'appartenance et c'est par assimilation (*verähnlichende Apperzeption*) avec lui que j'opère une transposition aperceptive (*apperzeptive Uebertragung*), qui me fait concevoir l'autre corps comme un autre organe volontaire pour un autre moi pensant. Cette transposition aperceptive n'est pas une analogie, parce qu'elle ne procède pas par induction. Remarquons que ce genre de transposition vaut pour tous les objets familiers de la perception et de la vie quotidienne:

Jede Apperzeption, in der wir vorgegebene Gegenstände, etwa die vorgegebene Alltags-welt mit einem Blick auffassen und gewahrend erfassen, ohne weiteres ihren Sinn mit seinen Horizonten verstehen, weist intentional auf eine *Urstiftung* zurück, in der sich einen Gegenstand ähnlichen Sinnes erstmalig konstituiert hatte.[33]

Cette création première, à laquelle renvoie toute familiarité avec le monde, indique comment la structure du monde existe *pour soi*, indépendamment de notre attitude ou de notre conviction: le monde subsiste en soi en tant que la Nature sédimente en lui des actes spirituels,[34] sans lesquels l'unité cohérente de mon intentionalité ne constituerait jamais la clotûre d'un monde qui fasse sens:

Denken wir uns nun, es wären in meiner Umwelt nie Leiber aufgetreten, so dass ich keinerlei Ahnung von fremder Subjektivität hätte. Dann wäre für mich in der Tat jede objektive Realität, die ganze Welt; die jetzt ein Lebloses wäre, nichts anderes als eine Zusammenhängende Vielheit von intentionalen Polen, als Korrelateinheiten für Systeme meiner möglichen und wirklichen Erfahrungen . . . .[35]

L'*être vrai* se réduit alors à la pure *construction* de signes formels, il n'accède jamais à l'évidence de la *présence vivante*, "en chair et en os", ce qui se dit chez Husserl: "leibhaft".[36] La différence entre la science et la philosophie a son fondement dans cette différence qui les sépare, dès l'origine, lorsqu'il s'agit de définir la *vérité de l'être*. Pour la science, la vérité de l'être est atteinte dans la conformité aux règles de construction d'un objet universelle-ment intelligible, alors que pour la philosophie il n'est pas de différence entre le dévoilement de son objet et l'activité par laquelle une monade décide de changer sa vie en lui donnant une fin absolue:

Ganz anders steht die Sache bei dem Philosophen. Er bedarf *notwendig* eines eigenen, ihn als Philosophen überhaupt erst und ursprünglich schaffenden *Entschlusses*, sozusagen einer Urstiftung, die ursprünglichere Selsbstschöpfung ist.[37]

La fondation la plus originaire, la création de soi toujours recommencée dans un temps qui est ouvert par une libre décision, où pouvons-nous mieux les trouver que chez le philosophe, auquel nous pouvons appliquer tout ce que nous avons dit de la nature? En somme les difficultés que nous pressen-tions s'évanouiraient si nous considérions le philosophe: en lui la temporalité qui fait exister l'ego comme for intérieur, où la réflexion reprend sans fin la scission de l'ego, s'accorderait avec la nature, puissance d'organisation et de clotûre de l'être.

En fait les difficultés vont devenir inextricables. Considérons la dualité originaire de la chair et de l'ouverture du for intérieur, non comme si elle était le propre du philosophe, mais une caractéristique de toute conscience. Dans le flux du for intérieur la clotûre de la conscience s'ouvre continûment sur le monde sans jamais recontrer de bornes, parce que toute détermination est vécue dans la chair comme une connexion appariée (*Paarung*) à d'autres consciences. Les liaisons temporelles de l'âme sont infiniment offertes par la chair à l'épreuve du monde. Mon corps est le lieu où le temps intime de la conscience s'étale et s'expose en un maillage intersubjectif. Mais si l'analyse phénoménologique croit échapper au solispsisme en découvrant dans la chair l'interconnexion des monades,[38] du monde objectif et de l'ego, elle est victime d'une illusion, car la réduction de mon être à la sphère du propre répète le procès général des réductions, de sorte que *la chair est à sa façon un regard*. Husserl s'est explicitement posé la question: s'il est vrai que "chaque élément de notre expérience quotidienne recèle une transposition par analogie du sens objectif, originellement créé, sur le cas nouveau",[39] faut-il en conclure que l'apprésentation des autres sujets par l'accouplement originaire de mon corps "serait une simple aperception par transfert, comme n'importe quelle autre"?.[40] La réponse de Husserl est peu satisfaisante. Décrire l'autre moi comme une suite de phases psychophysiques concordantes qui ne relèvent pas de l'activité de mon propre moi dans mon propre corps,[41] revient à se donner ce que l'on recherche, en faisant du corps l'"objet premier en soi".[42] Donc le corps de l'autre serait le premier objet, l'objet premier en soi (*das an sich erste Objekt*). Mais il s'ensuit qu'un objet inerte du monde physique est un objet primaire qui aurait perdu sa couche apprésentative. Déjà l'animal est "une variante anormale de mon humanité".[43] Descendons encore la série. Le végétal, puis le minéral apparaîtront comme des dérivés, par variation anormalisante, de mon humanité. L'originaire absolu, valant à la fois pour la nature et pour la réflexion du philosophe, serait cette apprésentation de l'être à soi-même comme autre indéfiniment multiplié dans la chair. Il suffit qu'un autre corps soit en rapport avec le mien pour que l'indéfinité des sujets apparaisse, et par eux, le monde des objets. A vrai dire, il suffirait d'énoncer que nos deux corps, le mien et celui de l'autre sont présents l'un à l'autre par des *signes*, qui ne se confondent jamais avec des phénomènes naturels, pour que la difficulté s'évanouisse. Mais Husserl ne dit rien de tel. Entre les monades, la perception d'un autre que moi constitue une pénétration intentionnelle (*ein intentionales Hinreichen*) dont résulte une *communauté* (*Gemeinschaft*), non une simple communication (*Kommunikation*):

Ist jede Monade reell eine absolut abgeschlossene Einheit, so ist das irreale intentionale Hinreichen der Anderen in meiner Primordinalität nicht irreal im Sinne eines

Hineingeträumtseins, eines Vorstellig-seins nach Art einer blossen Phantasie. Seiendes ist mit Seiendem in intentionaler Gemeinschaft.[44]

C'est sur le fondement de cette communauté inconditionnée et naturelle des hommes que pourront se constituer les communautés relatives, où l'accessibilité passe par une communication. L'essentiel, de notre point de vue, c'est l'affirmation que l'intersubjectivité forme une communion naturelle qui a une fonction ontologique radicale:

... von der absolut unbedingten Zugänglichkeit für Jedermann, die wesensmässig zum konstitutiven Sinn der Natur, der Leiblichkeit und damit des psychophysischen Menschen, letzterer in einer gewissen Allgemeinheit verstanden, gehört. Allerdings reicht in die Sphäre der unbedingten Allgemeinheit noch dies hinein (als Korrelat der Wesensform der Weltkonstitution), dass Jedermann und apriori in derselben Natur lebt, und einer Natur, die er in notwendiger Vergemeinschaftung seines Lebens mit dem Anderer in individuellem und vergemeinschaftetem Handeln und Leben zu einer Kulturwelt, einer Welt mit menschlichen Bedeutsamkeiten gestaltet hat – mag sie auch noch so primitiver Stufe sein.[45]

Tout comme la pensée, dans la saisie d'un noème, vise l'unité transcendante et idéale d'une essence, la nature vise un telos ultime à travers la multiplicité des corps humains. Mais la nature, système unique, ne peut viser autre chose que soi. La chair est l'expérience où la nature se donne à voir à elle-même dans le *regard* qu'un individu porte sur un autre individu, également voyant et vu. La chair est l'objet originaire de la nature. Rencontre de deux regards appariés, elle est le visible par excellence. La suffisance de la nature et l'autonomie de la conscience du philosophe sont la même chose: la possibilité de tout tirer de soi.

Que conclure? La phénoménologie veut se replacer au commencement absolu d'où la conscience pure du philosophe pourrait justifier et fonder le champ entier de l'expérience. Dans la solitude de son intimité, le moi pur se saisit, et cette réflexivité suppose une coupure, une scission indéfiniment reprise de soi-même. Cette scission est redécouverte dans la monade concrète comme expérience de la chair. La chair joue pour la monade le même rôle que le regard pour la réduction et la méditation philosophiques: saisir c'est extraire, "das Fassen ist ein Herausfassen",[46] puis introduire des variations imaginaires à travers lesquelles l'essence se maintient dans le temps. La chair est le regard de la nature sur elle-même, et l'infinitude des corps animés des vivants est le jeu des variations où se profile la fin ultime, la raison. L'univers des monades forme ainsi un tout éternel où, dès la vie organique et psychique la plus simple,[47] l'univers vise la perfection d'une humanité et d'une sur-humanité,[48] où la nature totale deviendra consciente de soi:

Und diese Weltkonstitution ist Konstitution eines immer höheren Menschen- und Uebermenschentums, in dem das All seines eigenen wahren Seins bewusst wird und die Gestalt eines frei sich selbst zur Vernunft oder Vollkommenheitsgestalt konstituierenden annimmt.[49]

On conçoit assez bien comment Leibniz pouvait penser une semblable harmonie de l'univers, puisque pour lui Dieu ordonnait avec sa puissance et son infinie sagesse le concert de toutes choses. L'origine radicale est enracinée en un Etre parfait, existant par la nécessité de son essence.[50] Peut-on vouloir rechercher l'origine radicale dans le moi pur? N'est-ce pas par un ironique retournement que la chair devient alors, en dépit de son épaisseur et de ses hasards, le lieu de ce regard absolu? Quand le philosophe veut occuper la place de Dieu, il ne peut retrouver la réalité mise en suspens qu'en la redécouvrant d'un bloc, comme puissance dont il suppose que tous les effets concourent obscurément au but qu'il avait assigné à la philosophie.

La chair est le signe unique d'une pensée qui préfère la réflexion à l'interprétation des signes.

Université de Lille III

NOTES

[1] Edmund Husserl, *Ideen II*, Beilage VIII, in *Husserliana*, Bd. IV, S. 320.
[2] *Die räumlich-zeitliche Unendlichkeit der Welt fordert Endlosigkeit der in Kommunikation stehenden absoluten Bewusstseine* (1909), in *Husserliana*, Bd. XIII, S. 14–17.
[3] *Ideen II* Beilage XIV, in *Husserliana*, Bd. IV, S. 377–80.
[4] *Ideen II*, §§ 36ff., in *Husserliana*, Bd. IV, S. 144ff.
[5] *Ideen III*, Beilage I, § 4, in *Husserliana*, Bd. V, S. 118f.
[6] *Zur Einfühlung* (27/29, Januar 1932), in *Husserliana*, Bd. XV, S. 444ff.
[7] *Ideen III*, Beilage I, § 6, in *Husserliana*, Bd. V, S. 128f.
[8] *Ideen I*, § 35, in *Husserliana*, Bd III, S. 61–4.
[9] *Aus der Vorlesungen: Grundprobleme der Phänomenologie, Wintersemester 1910–1911*, 2. Kap., § 12, in *Husserliana*, Bd. XV, S. 141ff.
[10] *Wahrnehmung und ihre Selbstgebung*, 2, in *Husserliana*, Bd. XI, S. 295f.
[11] *Ding und Raum, Vorlesungen 1907*, in *Husserliana*, Bd. XV, S. 278ff.
[12] *Zum Problem der Intersubjektivität in den "Cartesianischen Meditationen"* (1907), d, in *Husserliana*, Bd. XV, S. 76–7.
[13] *Einfühlungsproblem: Die Apperzeption meines Leibes.* (1934), in *Husserliana*, Bd. XV, S. 648–57.
[14] *Leiblichkeit als Vermittlung der Geister* (um 1912), in *Husserliana*, Bd. XIII, S. 279ff.
[15] *Monadische Zeitigung und Weltzeitigung* (Januar 1934), in *Husserliana*, Bd. XV, S. 634–41.

[16] *Cartesianische Meditationen*, II, § 19, in *Husserliana*, Bd. I, S. 82.

[17] Ibid., § 20, in *Husserliana*, Bd. I, S. 83ff.

[18] Ibidem, IV, § 32, in *Husserliana*, Bd. I, S. 101.

[19] Ibidem, IV, § 33, in *Husserliana*, S. 102.

[20] Ibidem, V, §§ 43–44, in *Husserliana*, Bd. I, S. 122–130.

[21] Stenographierter Text, in *Husserliana*, Bd. I, Textkritischer Anhang, S. 241.

[22] Ibidem.

[23] *Die Probleme der definiten Bestimmbarkeit der Welt* (1923), in *Husserliana*, XI, S. 433–7.

[24] *Cartesianische Meditationen*, V, § 44, in *Husserliana* Bd. I, S. 127.

[25] Ibidem, S. 128.

[26] Ibidem.

[27] *Erste Philosophie II*, 40. Vorlesung, in *Husserliana*, Bd. VIII, S. 91.

[28] *Erste Philosophie II*, Beilage XVII, in *Husserliana*, Bd. VIII, S. 410ff.

[29] *Die naturalisierte und die reine Subjektivität und die korrelativen Erfahrungsarten* (Juni 1920), § 3, in *Husserliana*, Bd. XIII, S. 442–3.

[30] *Leib, Ding, Einfühlung – Anknüpfung Seele-Leib*, § 2 (1921), in *Husserliana*, Bd. XIV, S. 57.

[31] *Cartesianische Meditationen*, V, §44, in *Husserliana*, Bd. I, S. 128.

[32] Ibidem.

[33] Ibidem, § 50, in *Husserliana*, Bd. I, S. 141.

[34] *Erste Philosophie II*, 49. Vorlesung, in *Husserliana*, Bd. VIII, S. 151–2.

[35] *Erste Philosophie II*, 54. Vorlesung, in *Husserliana*, Bd. VIII, S. 186.

[36] Ibidem.

[37] *Erste Philosophie II*, 30. Vorlesung, in *Husserliana*, Bd. VIII, S. 19.

[38] *Konstitution meines Leibes in der Phantomstufe durch Kompräsentation* (1922), in *Husserliana*, Bd. XIV, S. 281–5.

[39] *Cartesianische Meditationen*, V, § 50, in *Husserliana*, Bd. I, S. 141.

[40] Ibidem, § 51, in *Husserliana*, Bd. I, S. 145.

[41] Ibidem, § 54, in *Husserliana*, Bd. I, S. 147–9.

[42] Ibidem, § 55, in *Husserliana* Bd. I, S. 153.

[43] Ibidem, S. 154.

[44] Ibidem, § 56, in *Husserliana*, Bd. I, S. 157.

[45] Ibidem, § 58, in *Husserliana*, Bd. I, S. 160.

[46] *Ideen I*, § 35, in *Husserliana*, Bd. III, S. 62.

[47] *Das Kind. Die erste Einfühlung* (Juli 1935), in *Husserliana*, Bd. XV, S. 604ff.

[48] *Monadologie* (aus dreissiger Jahre), in *Husserliana*, Bd. XV, S. 610.

[49] Ibidem.

[50] Leibniz, *De rerum originatione radicali* (23.11.1697), in C. I. Gerhardt, *Die philosophischen Schriften von Gottfried Wilhelm Leibniz*, Bd. VII, S. 303.

PART II

THE RECURRENT QUESTION OF DUALISM

FRANÇOISE DASTUR

HUSSERL AND THE PROBLEM OF DUALISM

> ... il fallait pousser jusqu'au bout le portrait d'un
> monde sage que la philosophie classique nous a laissé,
> — pour révéler tout le reste .... Bon gré mal gré,
> contre ses plans et selon son audace essentielle, Husserl
> réveille un monde sauvage et un esprit sauvage.
> Merleau-Ponty, 'Le philosophe et son ombre'.[2]

I have chosen as a theme for this paper Husserl's struggle against dualism, discernible in all of his works. This struggle was, finally, in vain, but it nevertheless remained exemplary for those thinkers who elected to follow Husserl, for example, Heidegger and Merleau-Ponty. I have, therefore, adopted as my title "Husserl and the *problem* of dualism"[1] because, even if transcendental phenomenology remains dualistic in spite of Husserl's efforts toward monism, its purpose is not to assert dualism dogmatically, but rather to demonstrate, in line with the phenomenological way of thinking, that unity can only be given pretheoretically (*vortheoretisch*): the awakening of thought splits this unity irrevocably into pieces. That is why, for Husserl, dualism never ceases to be a problem — a problem which pointed to itself as the most thought deserving. But Husserl's greatness as a thinker resides also in his acceptance of the limited validity of dualism, and even of its "truth" not only for reflexive thought but also for life itself, as Merleau-Ponty did later in his own way.[2]

We can find this "truth" of dualism primarily in the foundations of the transcendental attitude (*Einstellung*). The most "idealistic" passage of Husserl, § 49 of book 1 of *Ideas* (*Ideen I*) concerning the hypothesis of world annihilation, seems to be also the most "dualistic," since Husserl intends here to destroy the natural consensus existing between consciousness and the world and separates their respective meanings by "a veritable abyss."[3] Consciousness and reality thus belong to two radically opposed species, whose difference is like that of the absolute to the relative being. The essential ambiguity of this well-known paragraph lies in the fact that, on the one hand, Husserl refutes the ontological equivalence postulated by the Cartesian tradition between *res cogitans* and *res extensa*, while, on the other hand, in an effort to equate reality with the subject (for this rupture of the natural harmony between consciousness and the world happens by thinking and

65

A-T. Tymieniecka (ed.), Analecta Husserliana, Vol. XVI, 65–75.
Copyright © 1983 *by D. Reidel Publishing Company.*

in fiction only and does not lead to the loss of nature, which is restituted as noema), he continues to define the subject in the language of tradition, i.e., as substance, since "immanent being is therefore without doubt absolute in this sense that in principle nulla 're' indiget and existendum."[4] The view that the aim of transcendental reduction is not so much the opposition of two modes of being as the revelation of the intentionality which binds them together is presented in book 2 of *Ideas* (*Ideen II*). There, Husserl declares, among other things, that the "thought-experiment" (*das Gedankenexperi-ment*), which is nothing else but the solipsism mentioned in § 49, results only in a "constructed subject" (*ein konstruierte Subjekt*)[5] which cannot know anything about the relationships a living subject has with things and other subjects through the medium of its body, precisely because it is only a construction of the mind. However, the analyses of *Ideen II* — it is no coincidence that Merleau-Ponty, in trying to determine Husserl's unthought-of element,[6] refers principally to this book — are focused on the reciprocal and simultaneous constitution of nature, body, and soul.[7] In fact, there is a possibility of mediations[8] between the two "regions," the world of nature and that of mind. It is phenomenology's task to understand the transition from a "personalistic" attitude, which constitutes the course of our daily life, to a naturalistic attitude, which opens up the world of nature, the universe of *blossen Sachen*. The last passages of Husserl's *Crisis* once again insist on this return to the prescientific world. It is in these passages that dualism has been explicitly refuted. Husserl shows that dualism is the result of the Galilean "act," understood as a dividing factor in a previously united world. This is what distinguishes the universe of bodies from the universe of persons.[9] Dualism can therefore be considered as the milieu in which modern rationality developed and which found its philosophical expression in the Cartesian duality of substances. What can be criticized in this conception of the world, or better, duality of worlds, is not just a particular idea of the relation between body and soul, but also what Husserl calls "naturalism," a conception which considers nature and mind as realities of the same ontological status. It is in naturalism that Husserl sees the origins of the crisis of European humanity, the specialization of the sciences and the dissemination of knowledge.[10] Husserl condemns dualism in the most violent terms: it is the origin of "the incomprehensibility of the problem of reason" (title of § 11); it is an aberration;[11] the old rationalism which came to be through dualism is "an absurd naturalism"[12] and "the absurdity of giving equal status in principle to souls and bodies as realities" must be denounced.[13] What Husserl wants to refute is the naturalization of the soul, an obstacle that even the most ardent defenders of the *Geisteswissenschaften*, like Dilthey, Windelband, and Rickert,[14] could not overcome in their efforts to found a new nonobjectivistic psychology. At this stage, it should be noted that Husserl's praise of Dilthey is in agreement with Heidegger's acknowledgment in 1927 of Dilthey as being the first thinker of historicality (*Geschichtlichkeit*).[15] Without question, Dilthey was for both of them the philosopher who questionned the foundation of naturalistic epistemology and demonstrated that it would be absurd to integrate "the subjectivity which accomplishes science" to the natural world and give it an objective status.

Husserl's intention was to found a "pure" psychology, distinct from the positive science of the same name, and which would be identical to transcendental phenomenology, i.e., to the science of transcendental subjectivity.[16] Those who contribute to the elaboration of this new psychology are no less than true phenomenologists who understand that it is not possible to apply the ideal of *mathesis universalis* to the whole world, because the world is not only nature, but also something spiritual. We must therefore renounce Spinoza's ideal of a universal ontology,[17] because for the world "being-in-advance" is an absurdity and "a Laplacean spirit is unthinkable."[18] *Ideen II* defined the world by its "openness" (*Offenheit*) and suggested that its infinity is not transfinite because it is related to a subjectivity which keeps discovering new objects for its constituting activity.[19] We can here see the emergence of a new concept of world which seems to herald the birth of Heideggerian "worldhood" (*Weltlichkeit*). However, for Heidegger the openness of the world is not identical to the potentiality of ever-new acts of consciousness, but on the contrary is the basic condition for all encounters with singular beings. As far as Husserl is concerned, he contents himself by saying that the world is not an objective being and that as such, it cannot be the object of an ontology. The world is actually a spiritual reality, and there is no ontology of spiritual realities, for ontology, in modern times, requires a mathematical a priori and implies the naturalization of the spiritual element. Only phenomenology can "free us from the idea of an ontology of the soul which could be analogous to physics."[20] The division of the world in two universes having identical ontological values and subject to the same causal laws has led to the equalization (*Gleichstellung*) of bodies and souls, to such an extent that they are thought to be "integrally and really connected similarly to, and in the same sense as, two pieces of a body."[21] This is, as Husserl says, due to "conceptual superstructures" added to the "pure experience" to which we need to return in order to bring to light the *Lebenswelt*'s level.[22] As pure experience teaches us, all spiritual objects, especially souls,

are only indirectly and improperly (*uneigentlicher Weise*) "embodied." They
are not themselves part of the spatial realm, but are "coextended with their
bodies" (*mitausgedehnt mit ihrer Körpern*).[23] As far as the soul, in particular,
is concerned, the realization of its embodiment (*Verkörperung*) is strictly
personal and implies experiencing one's own body, i.e., experiencing the
power of the soul over the body — its instrument. However, this experiencing
of one's own body (the *Leiblichkeit*'s level) happens in the first person and
cannot be an objective experience, because it is not part of the spatiotemporal
world. It is certainly possible to find visible traces of this experience in the
world, as the soul is coextended within it: I *see* my arm moving in space, but
I can only feel the kinesthetic reality of this movement deep inside myself.
Finally, the experience of one's own body provides the only true refutation
of psychophysical parallelism. We must now evoke the analyses of *Ideen
II* concerning the constitutive role of one's own body, which had been
completely overshadowed by the dualistic style of *Ideen I*, in which Husserl
affirms that "certainly an incorporeal and, paradoxical as it may sound, even
an inanimate and non-personal consciousness is conceivable . . .".[24]

The novelty of *Ideen II* is undoubtedly the recognition that body and
consciousness are intimately intertwined. This clarifies the phenomenon of
perception, which was still enigmatic in *Ideen I*. The relativity of worldly
reality and the essential necessity that a thing offers itself only through
successive *Abschattungen* — through a series of perspective variations —
can only be explained through the incarnation of *all* consciousness and
must be put in relationship with one's own body, which is the point of origin
— the *Nullpunkt* of all experience. However, *Leiblichkeit* is not only the
fundament of our habitual rapport with the environment; for Husserl's merit
consists of giving a more "concrete" — or less intellectualistic — image of
perception. Here it is not only a question of broadening the traditional
concept of sensibility by integrating into it the role of kinestheses considered
the most important element of perceptive consciousness. In addition, Husserl's
objective is to show that the physical thing, which is the object of natural
science, is firmly rooted in the perceived thing and that the constitution of all
objective being, including ideal entities, implies an embodied consciousness
as its base, even though, as soon as these entities appear, the activity of consciousness which was their origin obliterates itself. Husserl alludes repeatedly
in *Ideen II* to this kind of "forgetting of self" (*Selbstvergessenheit*)[25] that
the subject must accomplish so that the constitution of material nature and
ideal entities can be achieved. Nevertheless the basic condition of physical
and ideal entities is that "he who is knowledgeable of such objects must

experience things which remain *the same things*, and, if he is to know this identity, he must be in a *Einfühlungs* relationship with other knowledgeable people and must also belong to the same world and have *Leiblichkeit* . . . ."[26] It is clear from this that there is no objectivity without an intersubjectivity which itself is based upon *Leiblichkeit*, for "the *Einfühlung* goes from body to mind," as noticed by Merleau-Ponty.[27] Others appear at first as part of the same *Leiblichkeit* which I feel unto myself when I am experiencing my own body as a "perceiving thing." Not only spiritual dialogue is impossible for Husserl without the sensible being as a mediator, but it is also impossible outside an incarnate consciousness. This leads him to affirm that even the Absolute Spirit, God himself, "should have a body" and "would be dependent on sense-organs"[28] if there should be communication with Him. Husserl pointed out, in *Ideen I*,[29] that to posit an Absolute Being as correlate of an Absolute Spirit was to dismiss evidence, because this would be an admission that, to speak in Kantian terms, there is, besides the *intuitus derivativus* which characterizes man, an *intuitus orinarius* which would be the distinctive mark of an infinite consciousness. According to Husserl, it is nonsensical to attribute a general meaning to intuition and then to differenciate two kinds of intuition. Only derivative intuition exists, and it is impossible to view reality differently. Eidetical laws are valid even for God who can perceive things only in their perspective variations – and here Husserl considers God as "a necessary limiting concept in epistemological reflection" and does not want to carry this debate "into the domain of theology."[30] Husserl reminds us with even greater insistence than Kant of our finitude, which is not to be considered as an imperfection in relation to an infinity, which itself can be postulated only through transgression of the limits of our intuition. The ultimate base of *all* being is *Leiblichkeit*, which is the source of all reality and as such cannot be a reality itself: one's own body cannot be localized anywhere in the objective world, for it is the absolute "here" which defines all "over there's." It is the immovable center, the spatiotemporal point of origin relative to which all change in the objective world is measured.[31]

Now, if this centrality necessarily defines one's own body as a nonobject, the actual signification of the questions addressed by *Ideen II*, as also the language it uses, indicate that in reality Husserl is preoccupied with the transcendental conditions of the constitution of objectivity, rather than with the revelation of *Leiblichkeit* as a preobjective stratum. For Husserl, it is important to retrace the perceived being back to the perceiving act, the thing to its somatical conditions, nature to consciousness. As Paul Ricoeur very justly says in his analysis of *Ideen II*: "*Ideen II* does not endeavor to dispel the prestige of the ideas of nature, reality, or of objective sciences. On the contrary, phenomenology justifies them by firmly anchoring them to an activity (*Leistung*) of consciousness."[32] In this matter, Husserl's methods are diametrically opposed to Heidegger's. The concepts of nature and reality are not for Heidegger starting points, but results, "derivatives," which require a comprehension of more fundamental structures. " 'Nature' ... can never make worldhood intelligible"[33] but, on the contrary, it is derived from worldhood and sets as a precondition a definite modality of Being-in-the-world, i.e., the theoretical attitude. It is through a specific phenomenon called *Entweltlichung* — being deprived of worldhood — that one passes from "the world as a totality of equipment ready-to-hand" to nature as "a context of extended things which are just present at hand and no more."[34] In the same way, the concept of reality "has no priority among the modes of Being of entities within-the-world" because traditionally it means "Being in the sense of the pure presence-at-hand of things" and as such has its ontological foundation in worldhood.[35] Heidegger does not begin with these concepts which point to, in Husserl's terms, the "naturalistic attitude" (which is described in the *first* two sections of *Ideen II*),[36] but is entrenched right from the start in what Husserl calls the "personalistic attitude," which he names *Besorgen* ("concern"). In the context of this "concern" *all* beings are encountered as tools ("equipment") and not as mere things, including the natural phenomena themselves besides the artifacts. The traditional modes of access to beings, experience of the senses and rational knowledge, are therefore not the *original* modes for Heidegger. Undoubtedly that is why the moment of the senses is not a thematic in *Being and Time* and equally, why, as A. de Waelhens observed,[37] *Leiblichkeit* has not been analyzed. We cannot draw the conclusion, however, that experience of the senses is absent in our daily dealings with things; the hammer, for example, is certainly perceived, seen, touched, at the same time as it is being used. Neither sight nor touch *as such* give access to the tool. Only circumspection (*Umsicht*) and daily concern can do this. As such, for Heidegger, the finitude of existence does not result from the *fact* of incarnation, from the necessity for consciousness to reside in spatiotemporality, but, on the contrary, the necessity of existence to be incarnate derives from a finitude even more original — finitude of *existence itself* as "being dependent upon What-is" and being "handed over it."[38] In other terms, finite existence can globally be defined as care (*Sorge*). Even though the analyses of Husserl and Heidegger converge in many points, especially in the description of the "theoretical" or "naturalistic" attitude,[39] one can easily distinguish where they diverge. While remaining inside the bounds of the philosophy of consciousness, Husserl tries to go beyond the limits of tradition and can only find the opening, at the end of his travail, leading to this insecure and wild[40] terrain where there is an intermingling of constituted and constituting, objective and subjective beings. Even though the questions he asked can lead to the problematic of existence, its concept remains totally foreign to him. This is best illustrated by his analyses of *Leiblichkeit*: his goal is not to avoid the concept of an object-body (*Leibkörper*), but rather to understand the process by which *Körper* (physical body) becomes *Leib* (living body), that is, how a thing which is a thing among other things had acquired the privilege of sensibility and life. For Husserl, one's own body is, on the one hand, a physical thing, part of matter and space, and, on the other hand, a sense organ. The body is therefore doubly constituted right from the origin[41] and escapes only partially objective determination. Even though Husserl is attentive to the specificity of *Leiblichkeit*, he defines body from an objectivistic viewpoint as "a thing of a very unique species" and finally as "a surprisingly incompletely constituted thing."[42] One's own body is considered to be the locus of a fundamental ambiguity, for it is simultaneously caught up in the objective texture of things and is the center from which the world emerges and is shaped. In this manner it manifests that it belongs at the same time to transcendental subjectivity and empirical reality. *Leiblichkeit* is the limiting experience of this "project to gain intellectual possession of world"[43] which is constitution.

A certain analogy can be found between the status reserved to one's own body in *Ideen II* and earth's role as seen by Husserl in a 1934 manuscript entitled "Downfall of Copernicus's theory" (*Umstürz der kopernikanischen Lehre*),[44] quoted by Merleau-Ponty and discussed by Jacques Derrida in a long footnote in his introduction to *L'origine de la Géométrie*.[45] Just as one's own body is the point of origin of solipsistic experience, so is earth the archorigin (*Urarche*) of mankind as an historical community. Like one's own body, the earth is an absolute "here," immovable, and all movement can make sense only in relation to it. Exactly like one's own body, the earth as *Urarche*, as experienced reality, cannot be understood objectively and thus shuld not be confused with the planet earth, the object of the natural sciences. Like one's own body, the earth benefits from an ambiguous status: it is not merely one physical body among other astronomical bodies, but "ground-body" (*Bodenkörper*), the physical "home" of mankind. Just as *Leiblichkeit* is the transcendental condition of all objectivity, so is being part of the same earth, of the same original ground, the basic condition of all history. Husserl's suggestion to think together of one's own body and earth certainly belongs to this "unthought-of element" of phenomenology which opens out on something else, unveiling new horizons outside the realm of the philosophy of subjectivity and allowing phenomenology to rejoin secretly the apparently totally different views of Nietzsche and Heidegger.[46]

To return to what Husserl really says, it appears that those things which are unintelligible — such as experiencing one's own body and being part of the earth — even though they have been thought through as limiting experiences, are still too easily resolved in a transcendentalistic sense. The experience of one's own body is treated as a subject only to clarify the constitutive activity of consciousness, and the earth merely provides a formal home for mankind, whose unity itself remains teleological. The riddle of "living" in the form of *this* body on *this* earth remains unresolved. For this reason we have finally to ask ourselves if Husserl has really uncovered the real *Lebenswelt*, the world of "living." Husserl's objective in *The Crisis* is to unearth the roots of the *mathesis universalis*, to uncover the foundations of modern reason, to unveil the first layer of consciousness preceding all theorizing. The regressive inquiry (*die Rückfrage*), however, is actually made to permit transcendental subjectivity to overcome the *Selbstvergessenheit* inherent in all constitutive activity. Husserl's point of view has not undergone any fundamental modification between *Ideen II* and *The Crisis*, as can be seen from the fact that in both cases science is considered as a definite though indirect continuation of perception. Between *ordo vivendi* and *ordo cognoscendi* there is neither a clean break nor total opposition. On the contrary, a certain structural simularity becomes apparent between the life-world and the objective world, as in, for example, the transition from perceived shapes to ideal geometrical shapes which appears as a permanent possibility, as if this idealisation was always foreseeable in life-world.[47] Finally, to live, for Husserl, means "straightforwardly living toward whatever objects are given."[48] The original means by which an object is given, and which "presents and apprehends a self in its bodily presence,"[49] is perception. All the objects of the life-world ought to be given "in the mode of self-presence (*selbstgegenwärtig*)."[50] That is why the perceptible world constitutes the fundamental layer of the life-world, "for everything that exhibits itself in the life-world as a concrete thing obviously has a bodily character, even if it is not a mere body, as, for example, an animal or a cultural object, i.e., even if it also has psychic or otherwise spiritual properties."[51] Cultural objects in the broader sense — from tools to works of art — are not just material bodies; they also include aesthetical, spiritual, practical, and axiological, "significations." But these require a material vector. The same is true for *animalia*, including man. *Res extensa* is the material foundation of the life-world on which spiritual significations and value predicates are superimposed, making things part of the life-world.

Dualism, highly criticized as the source of the deviations of modern rationalism, surfaces again insidiously in Husserl's own way of thinking, certainly because Husserl remains faithful to all foregoing metaphysics, even though he attempted to shed light on their foundations by putting them in question. Heidegger once said of these metaphysics that they have not as yet attained even the beginnings of understanding what *Leiblichkeit* is in man, as "the body phenomenon is the most difficult problem."[52]

University of Paris I (Sorbonne)

NOTES

[1] I understand here under dualism a doctrine which acknowledges *two* irreducible principles of reality.

[2] "Notre corps n'a pas toujours de sens, et d'ailleurs nos pensées, dans la timidité par exemple, ne trouvent pas toujours en lui la plénitude de leur expression vitale. Dans ces cas de désintégration, l'âme et le corps sont apparemment distincts et c'est la vérité du dualisme" (M. Merleau-Ponty, *La Structure du Comportement* [Paris: P.U.F., 1960], p. 226).

[3] *Ideas*, trans. W. R. Boyce Gibson (London: Allen and Unwin, 1931), p. 153.

[4] Ibid., p. 152.

[5] *Ideen zu einer reinen Phänomenologie und phänomenologischen Philosophie*, vol. 2, *Husserliana*, vol. 4 (1952), p. 81.

[6] "At the end of Husserl's life there is an unthought-of element in his works which is wholly his and yet opens out on something else" (Merleau-Ponty, *Signs*, trans. R. C. McCleary [Evanston: Northwestern University Press, 1964], p. 160).

[7] "Die *'Natur'* und der *'Leib,'* in ihrer Verflechtung mit diesem wieder die *Seele, sich in Wechselbezogenheit aufeinander*, ineins miteinander *konstituieren*" (passage written in 1912 and cited by Marly Biemel in the introduction to *Ideen*, 2: xvii).

[8] "Wir kennen ja sonst kardinale Unterschiede von 'Welten,' die doch durch Sinnes- und Wesensbeziehungen vermittelt sind" (*Ideen*, 2: 210).

[9] Cf. *The Crisis of European Sciences and Transcendental Phenomenology*, trans. D. Carr (Evanston: Northwestern University Press, 1970), §10.

[10] Ibid., §11.

[11] "Eine Verkehrtheit," not only a "mistake," as the English translation says (p. 297).

[12] Ibid., p. 298.

[13] Ibid., title of §62.

[14] Cf. *Crisis*, p. 297 and *Ideen*, 2: 172.

[15] Cf. *Being and Time* (New York: Harper and Row, 1962), §77.

[16] "There is only a transcendental psychology, which is identical with transcendental philosophy" (*Crisis*, p. 257).

[17] Ibid., p. 65.

[18] Ibid., p. 265.

[19] "Besagt die "Unendlichkeit" der Welt statt einer transfiniten Unendlichkeit (als ob die Welt ein in sich fertig seiendes, ein allumfessendes Ding oder abgeschlossenes Kollektivum von Dingen wäre, das aber eine Unendlichkeit von Dingen in sich enthalte), besagt sie nicht vielmehr eine 'Offenheit'?" (*Ideen*, 2: 299).

[20] *Crisis*, p. 265.

[21] Ibid., p. 215.

[22] Ibid., p. 216.

[23] Ibid.

[24] *Ideas*, § 54.

[25] For example *Ideen*, 2: 82.

[26] "Jeder Erkennende solcher Objekte . . . muss die Dinge und *dieselben Dinge* erfahren, muss, wenn er diese Identität auch erkennen soll, mit dem anderen Erkennenden im Einfühlungsverhältnis stehen, er muss dazu Leiblichkeit haben und zur selben gehören etc" (*Ideen*, 2: 82).

[27] *Signs*, p. 169.

[28] "Natürlich müsste der absolute Geist zu Zwecken der Wechselverständigung auch einen Leib haben, also wäre ja auch die Abhängigkeit von Sinnesorganen da" (*Ideen*, 2: 85).

[29] *Ideas*, § 43.

[30] Ibid., § 79, n. 1.

[31] Cf. *Ideen*, 2: 158.

[32] "*Ideen II* ne travaille pas à dissiper le prestige des idées de réalité, de nature, ni le prestige des sciences objectives, bien au contraire: en les enracinant dans un travail (*Leistung*) de la conscience, la phénoménologie les justifie" (P. Ricoeur, "Analyses et Problèmes dans Ideen II de Husserl," *Revue de Métaphysique et de Morale* [1951]: 360).

[33] *Being and Time*, trans. J. Macquarrie and E. Robinson (New York: Harper and Row, 1962), p. 94.

[34] Ibid., § 24, p. 147.

[35] Ibid., § 43 (c).

[36] "In diesen Abschnitten waren die Untersuchungen bezogen auf die naturalistische Einstellung" (*Ideen*, 2: 174).

[37] "On ne trouve pas dans *Sein and Zeit* trente lignes sur le problème de la perception; on n'en trouve pas dix sur celui du corps" (M. Merleau-Ponty, *La Structure du comportement* [Paris: Gallimard, 1960], p. vi).

[38] "Existenz bedeutet Angewiesenheit auf Seiendes als ein solches in der Ueberantwortung an das so angewiesene Seiende als ein solches" (M. Heidegger, *Kant und das Problem der Metaphysik* [Klostermann, 1973], p. 221–22).

[39] The convergence is such that Merleau-Ponty wrote: "The whole of *Sein und Zeit* springs from an indication given by Husserl and amounts to no more than an explicit account to the 'natürlicher Weltbegriff' or the 'Lebenswelt' which Husserl, towards the end of his life, identified as the central theme of phenomenology" (*Phenomenology of Perception*, trans. C. Smith [London: Routledge and Kegan Paul, 1962], p. vii).

[40] According to Merleau-Ponty's expression, cf. "exergue."

[41] Cf. *Ideen*, 2: 145.

[42] Cf. ibid., 2: 159.

[43] Merleau-Ponty, *Signs*, p. 180.

[44] Published by M. Faber in *Philosophical Essays in memory of E. Husserl*, pp. 307–25.

[45] Cf. Merleau-Ponty, *Signs*, p. 177 and *Phenomenology of Perception*, p. 429. See also J. Derrida, *L'Origine de la Géométrie* (Paris: P.U.F., 1962), p. 79.

[46] Cf. for example: "Ihrem Leibe und dieser Erde nun entrückt wähnten sie sich, diese Undankbaren. Doch wem dankten sie ihrer Entrückung Krampf und Wonne? Ihrem Leibe und dieser Erde. ... Redlicher redet und reiner der gesunde Leib, der vollkommene und rechtwinklige: und er redet vom Sinn der Erde" (F. Nietzsche, *Also sprach Zarathustra*, I, Von den Hinterweltlern); "In der Mundart spricht je verschieden die Landschaft und d.h. Erde. Aber der Mund ist nicht nur eine Art von Organ an dem als Organismus vorgestellten Leib, sondern Leib und Mund gehören in das Strömen und Wachstum der Erde, in dem wir, die Sterblichen, gedeihen, aus der wir das Gediegene einer Bodenständigkeit empfangen" (M. Heidegger, *Unterwegs zur Sprache* [Neske, 1959], p. 205).

[47] "The things of the intuitively given surrounding world fluctuate, in general and in all their properties, in the sphere of the merely typical" (*Crisis*, p. 25).

[48] Ibid., p. 144.

[49] *Ideas*, p. 137.

[50] *Crisis*, p. 105.

[51] Ibid., p. 106.

[52] Heidegger and Fink, *Heraclitus Seminar 1966–67*, trans. C. H. Seibert (University of Alabama Press, 1979), p. 146.

TADASHI OGAWA

"SEEING" AND "TOUCHING"*

or

Overcoming the Soul–Body Dualism

> The intelligence is created in order to operate the
> material, and therefore . . . it has no definite purpose
> to touch the depth of the material.
>
> — Bergson[1]
>
> A subject only with eyes would not possess the
> appearing body.
>
> — Husserl[2]

It is no wonder that today we acutely feel the distance that exists between
the body as an object of physiology and the body in which each of us is
living at each moment. This distance may be the modern version of the
traditional difference between the lifeless and empty (*kara*) *Karada* and the
Lively *Mi.*[3] The former rather should be called the "body-thing" (body as a
thing) and the latter the "living flesh." In German the former is called *Körper*
and the latter *Leib. Körper* deives from the Latin word *corpus* and it means
a thing or a lifeless corpse. On the other hand, *Leib* has the same etymology
as that of "life" in English and *Leben* in German. Then it seems that *Leib*
adequately convers the meaning of each body of ours living at each moment.[4]
In fact, it is not that the body takes two different modes of existence, but
that there are two different modes of approach to the body. *Körper* can
be approached by the physical apparatuses such as scalpel, pincette, and
oscillograph, as an object of the natural-scientific observations. *Leib* exists on
its own account without any medium of apparatuses, and it means exactly
what I am in the direct way of my living before the approach by me. Actually
even the body-thing which is physiological, chemical, and physical should
implicitly presume my body as a subject of perception. Without my body
that sees, calculates, and gives meanings to the body-thing, even the physiological body could by no means appear in the world.

Ontologically speaking, the distance described so far derives from the
Cartesian thought that the mechanical nature should be substantially separated
from the "I." Descartes, too, regarded the body in which I am living as
the unification of body and mind. However, it was merely grasped as the
occurrence of the practical "daily talkings." In other words, it was regarded

77

A-T. Tymieniecka (ed.), Analecta Husserliana, Vol. XVI, 77–94.

as an event on the level of sensibility, but never on the level of understanding which makes possible theoretical thinking. Descartes thought it right that in the theoretical attitude body and mind are separated; he thought that the distinction between the extensional substance including my body and the thinking substance is right. For in Cartesian thought the unification of body and mind is not a "clear and distinct" idea, but a primitive idea or the fact of life.

Apropos, the substantial distinction between body and mind includes the following thoughts: (1) the clear distinction between the private "internal world" of mind-spirit and the "external world" of extension; (2) the ontological, epistemological framework of subject-object-difference; (3) to construct body as a medium of the causational chain of the external stimuli; (4) that the objectification of the world prevents my body from being understood as an existence embedded in the world, and that the body is one-sidedly reduced to a mere tool for conquering Nature, and that the nature in which the world is in gear with the body ("the basis of subjectivity — the basis of nature [my nature]")[5] is forgotten. Therefore it is only after we enclose mood, feelings and thoughts in this private, internal world, and then the internal world is introjected into the external world that climate, weather, mood of life, and atmosphere can be grasped.[6]

Today it seems hard to accept the conclusions of Cartesian thought. It is well known that even in those days, many philosophers after Descartes — among them Princess Elizabeth — made one effort after another in order to solve this weak point of Cartesian philosophy, that is, the question of how the nonextensional spirit which is conscious of the body as an objective extension works upon this extensional body. However, so far as they move only within the area of the question explored by Descartes, they can never get out of the whirlpool of the Cartesian substantial thought. What is important for us is to "discover a new way of putting a question" (Bergson) which is different from Descartes', to destroy the framework of the dualism of body and mind, and to deepen and expand our experiences of the body. The outlines of the horizon of deepening and expanding the experience of the body are as follows: (1) to think "intentionality," which concept Husserl inherited from Brentano and which Husserl regarded as an essence of consciousness, arises in corporeality — and thus to release corporeal experiences from the framework of phenomenology of consciousness; (2) to clarify the structure of "figure-ground-difference"; to see what figure this intentionality as the corporeal arising constitutes on the ground of the world; (3) to understand that the corporeal experience is the arising of "existence-in-the-world," which way of thinking makes fluid the entire boundary of "subject-object-relation"; then to clarify that the corporeal experience is the very place where subject holds intercourse with object, and that my corporeality exists within the world as a fundamental place for my encounter with the world, and at the same time my corporeality vertically intersects the world.

I

We cannot describe the body without the description of "my body." For it is no other than "my body" that exists within the cosmos and the world. Moreover, without "my body" the world does not appear, nor does the time-space openness spatially open. To understand this, Bergson's famous experiment of thought such as the amputation of an afferent nerve seems unnecessary. Then, when I close my eyes, the visual world does not appear to me. Both the appearance of the things that are in the world and that of the world entirely depend on my body. Therefore, when, for example, Bergson dealt with the question of body and mind,[7] or when Valéry simply reflected on the body,[8] they could not help but begin by securing the onto-logical, privileged status of "my body" in the world. Obviously "my body" is a privileged object which enables me to exist within the world at each moment, and for each of us "my body" is neither interchangeable with nor commensurable to the other existence (the other person or the other thing). However, it is Valéry and Sartre that recognized the privilegedness of "my body," but after all, they succumbed to the Cartesian dualistic conception. It seems to me absolutely necessary to examine how they kept dark the fundamental dimension of my corporeality, in order to purify us from Cartesianism.[9]

In Valéry, my body is the most important for each of us, and it is at once subordinate and *opposed to the world*. When my right hand grabs my left hand, my left hand is "non-I" and it is the same act as to grab an outer thing (the refusal of the double sensation). "My body" is "my own" but it is at the same time "my enemy." And my body is momentary. In other words, my body has no past but it is the present itself. Thus Valéry thought.

Sartre, too, denies the double sensation. In Sartre there is no essential difference between my *sense of sight* about my body and that about the other's body, nor is there any difference between the other's *perception* of my leg and my perception of my own legs: "When I touch my leg with my fingers, I feel my leg touched by my fingers. But this double sensation is not the essential phenomenon .... 'Touching' and 'being touched' are two different dimensions which possess no common basis."[10] Sartre defined perception exclusively as the sense of sight, and since sight is inclined to objectify, he cannot understand that the body constitutes one entire system (*être-pour-soi=cogito*, "seeing"). Sartre's ignorance of this matter is plainly shown in that he discarded the *sensuelle hylē* of 'consciousness' which Husserl thought provides perception with materials. This is because Sartre regarded it as the "hybrid existence" which is neither *être-en-soi* nor *être-pour-soi*. Sartre discarded the *sensuelle hylē* mainly because the *être-en-soi* ("thing") and the *être-pour-soi* ("consciousness") constitute the dualistic ontology, and they are in the relation of hostile conflict with one another. Moreover, in *Ideen I* Husserl had already suggested the kinaesthetic function by "the functional concept of *hylē*" (*funktionierende Stoffe*). So it is appropriate here to point out Sartre's misunderstanding about Husserl, but what is more fatal is his blindness to the phenomenon.

That the world, the body, and consciousness are in hostile conflict with each other in Valéry and Sartre proves that the Cartesian motives are strongly working upon them. In order to elucidate this point we must thoroughly study the "double sensation" which they ignored.

II

When I am absorbed in meditation, I put my hand on my forehead or I support my cheeks with my hands. At this moment I feel that my hand touches my forehead but at the same time I feel that my forehead touches my hand too. This is a phenomenon which has been traditionally called "double sensation." However common this phenomenon may seem, we shall be at a loss if we try to clarify the scope of its meaning.

Generally speaking, when I perceive a thing, I grasp a thing as something. In such a situation of perception, the body works as an organ of perception and a thing faces my body in a certain phase. In *Ideen I*, however, Husserl does not explicitly deal with the role of the body in perception; instead, he educes intentional *morphē* and material *hylē* as a noetic moment of consciousness, and he conceives that by the scheme of form-content he can apperceive the meaning of an object and that the perceived will appear; namely, the sensational *hylē* as sense-data or sensational experience provides materials of perception, and the material being activated by the higher, sense-giving stratum is the perception which is the unity of noematic meaning. However, if we understand the perceptional grasp of a thing by the mosaic schema of form and material, we fail to adequately consider corporeality which works in perceiving a thing, and because of the deepening of passivity, *hylē* later must appear as affection.[11] The deepening of passivity means the deep stratum of intentionality that restricts the achievement of the active intentionality of operation. Namely, to perceive is already an act of the "I," but each object given prior to this act of objectification as well as the field of perception pressingly advance toward me as a unified meaning of object structured before being objectified. For example, when a red spot on a sheet of paper strongly stimulates me, the whiteness of the paper as well as the red spot are pregiven and both of them undergo the structurization of "figure-ground." It is the preobjectifiedly given prior to me as the unity of meaning that works upon and affects the ego. The pregiven passively structured and the "I" are grasped under the relation of "to evoke" and "to be evoked," and the affection in the "I" results from the first prereflexive stimulus or the evocation of object to the "I." It is this affection that becomes a condition for the arising of, or turning toward (*Zuwendung*), an object which is the first activeness in the "I." Namely, the pregiven pressingly advances toward the "I," and evokes the active, awakening "I" from the prereflexive doze by the affectional stimulus. Therefore, the passivity of ego is regarded as the innermost stratum of activeness of ego. Then what does "ego's being-affected" (*Affiziert-Sein*) mean? First, it acts as medium between the affectional stimulus which presses the pregiven toward me and which is the most primordial ability of the "I." Second, *Affiziert-sein* is no other than the corporeal arising, because the first acceptor that accepts the pregiven which stimulatively affects me is, among others, the sense organ such as eyes, ears, skin, and my body which unifies them as one whole system. To stare at the red spot on the paper or to listen to a whistle of a train penetrating silence is based on the fact that "I" is awakened by this *Affiziert-sein* which lies on its deepest stratum. This dozing, nonpositing, and prereflexive dimension indeed belongs to "the dimension of pre-being,"[12] partly because every objectification and positing reflection presupposes the dimension, and partly because their possibility is conditioned by it. To see this *Affiziert-sein* in connection with Husserl's theory of body, it goes as follows: the body originally fills the condition for making the subject (in Husserl, mind or spirit) appear in the world, and the subject gets localized and becomes a being-in-the-world by somehow combining with the body. In this sense we can say that I am anchored in the world by the body. Husserl grasps this body as a performer of localizing sensation and of *Sich-Bewegen* ("self-movement"). The localization of sensation means that in the body which is present and given in sensing something, sensation is corporealized and

therefore gets some seeming expanse. Husserl calls this localized sensation
Empfindnis. Of course, this expanse is qualitative unlike the extensional
substance of a thing, for *Empfindnis* have no profiles or gradations (*Abschattungen*). Therefore, *Empfindnis* is "an event of the body" affected by an
external thing, that is, "the quality of being influenced (worked upon)."[13]
Naturally eyes and ears cannot function as localizing the visual phantom and
the aural phantom (sound) in the body. Eyes do not turn red on seeing blood,
nor is music conceived in ears. Therefore we can say that the localization of
sensation is plainly seen in the sense of touch. When, for example, I touch
my desk with my fingers, the sensation and the appearance of a desk are
exactly the same phenomenon (perceiving the tactile quality of the smooth
surface). On the other hand, eyes and ears are localized only by being touched
with hands or fingers. Moreover, it is in the sense of touch that sensation and
the function of kinaesthesis (that I move myself and I conquer my body as
I wish) are most closely united, and this can be understand from the fact
that the tactile quality of smoothness in letting my fingers slide on the desk
is given simultaneously at the same place by a series of sensational orders
which consist of the motion of fingers and the contact with the desk. In this
context, Husserl writes as follows: "The body is ... primordially constituted
only by the sense of touch and by what is localized with it such as warmth,
coldness, ache and so forth [*Empfindnis*]."[14] From what I have described
so far, it is evident that the phenomenon of double sensation belongs to the
dimension of the localization of sensation in the tactile sense, that is, to the
dimension of the simultaneous coexistence of a person to touch and a thing
to be touched.

                                        III

When I see my left hand with my eyes, or when I touch my left hand with
my right hand, we can say that in both cases my left hand is constituted by
another limb of my body, in spite of the important difference in its medium,
visual or tactile. Next, let my left hand touched my forehead. This time, my
left hand becomes a constituting element and my forehead is constituted as
a part of my body. In other words, whenever my body and a part of my body
are constituted, my body coexists anonymously as the already constituting
body; namely, it is impossible that the whole of my body is seen through by
me as if applied the X-ray. As is often said in Buddhism, the eye cannot see
the eye itself. My body cannot perfectly appear visually, partly because I
cannot see my body from every aspect. But the deeper reason rather lies

in that my co-functioning corporeality cannot be immune from opacity and anonymity; as Husserl points out, "My back cannot be seen but it is co-experienced."[15]

Husserl writes as follows to the effect that in constituting the limbs of my body, my body which functions always and already is presupposed as a silent witness to my identification: "To experience the limbs of the body which are presupposed as being functioning, presupposes, however, *the corporeality which is co-present* with each experience."[16] But it is not true that the corporeality presupposes the functioning body regressively without limitation: on the one hand, no other than the phenomenon of double sensation exposes the always present, co-functioning body in its anonymity, because the perceived part of my body is recurrently related to the functioning body; on the other hand, by making this anonymous, co-functioning body "the original corporeality," the double sensation applies a brake to the unlimited regression of the functioning corporeality.

Apropos, as I described at the end of the previous chapter, in the tactile system the tactile field and the tactile function of the body coexist as correlative; in other words, the tactile appearance of a thing and the corporeal function to touch, directly touch each other and cover each other. The double sensation is the phenomenon peculiar to this tactile system. Namely, it is the phenomenon that the discontinuing parts of the tactile field touch and cover each other. Husserl states in his posthumous manuscripts as follows: "The tactile field (which is different from the field of sight) has the following fundamental character: every data is undoubtedly 'outside one another' [*im Aussereinander*], but the two data being outside one another and their place can nevertheless cover each other. Undoubtedly, they are *fused* in some way or other, but they are *never mixed (Mischen)* actually. Locality remains mutual externality. . . . The given which are locally different *touch* each other and *cover* each other, but still they never hide oneself from each other. This is a kind of event phenomenologically unique, and it should not be replaced by the act of hiding myself which is characteristic of the visual phantom."[17] In the double sensation a circuit is constituted in which the position-data (*Stellungs-Datum*) of touching and the tactile aspect-data (*Aspekt-Datum*) given by being touched are reciprocally circulating. But to what dimension of being does this circuit belong? We must recall now the aforementioned "original-corporeality," that is, the preobjectifing, co-functioning anonymous corporeality which accompanies the objectifying function of the body. For, this circuit of "touching – being touched" arises in the same body. Evidently, in this circuit a hand which touches a forehead constitutes the forehead and at the same time it is constituted by the forehead. Judging from their character of simultaneous coexistence in the tactile system, the circuit of "touching – being touched" arises simultaneously. It is not until then that we can say the body constitutes (functions) and is constituted in the most primordial sense.

IV

Although Husserl took a glance at the original corporeality mentioned earlier, why could he not grasp it fully? It may be partly because he stuck to dealing with the double sensation by the schema of "constitution-objectification." That the constitution of the limb of the body is grasped as objectification (*Objektivation*), is due to the character of "phenomenology of ego-ray" peculiar to Husserl, in the sense that the ego-ray gives meaning to an object.[18] Therefore, when a right hand touches a left hand, he understands that *touching* is one mode of *seeing*. And so we might as well say that my right hand "sees and seizes" my left hand.

Another reason concers the ambiguity (*Zwei-Deutigkeit*) of the body, which constitutes and is constituted as described earlier. The ambiguity of the body is a phenomenon that, on the one hand, the body is constituted visually, namely, "in perspective and by kinaesthesis," and, on the other hand, it is constituted "by the tactile kinaesthesis."[19] It is Claesges who educed the duality of corporeal constitution from Husserl's posthumous manuscripts.[20] According to him, the body is egoic (*ichlich*) and ego-strange (*ichfremd*) at once. In the egoical body, the field of tactile sensation is localized on the surface of the corporeal phantom, and the whole of the body is *organized* in the motion of organ as an egoical achievement of the kinaesthetic function. Thus, on the one hand, the constituting body is grasped as a freely-working, egoical organ of perception. On the other hand, the ego-strange body is constituted by the visual *perspectification* as an existence that is spatial, bodily, and ego-strange (*res extensa*). However, unlike Claesges, we had better not unify this body of double meaning by interpreting it as "self-mediating directness." We should rather grasp the inevitability of the occurrence of this ambiguity by exploring each performer of organization and perspectification in the corporeal experience. We might as well consider that the perspectification and the organization of the body are almost equivalent to Hoche's discrimination of "noematic body-thing" from "functioning body," and Schmitz' discrimination of "organ (each isolated island of the body)" from "function (corporeal titubation)."[21] But what should not be confused is that perspectification is mainly visual, while organization lies in that I always move my body as I wish and I and the organ come to be unified in its motion. Evidently the actual movement of my thought does not occur in my head. Nor do we say, "My body walks," but we say, "I walk."[22] Or even when my car runs against an electric pole, we still say, "I ran against an electric pole." Thus I and my body are unified in this kinetic, functioning body. When, for example, I move my hand, it looks not at all different from the ego-strange movement of a thing if we see it purely visually. However, by the tactile, kinaesthetic function opposed to this movement, I let the movement of this hand belong to my body which is always and already constituted tactilely. Namely, the constitution of my body as *res extensa* is completed by the co-working of the eyeball movement on which perspectification is based and of the tactile, kinaesthetic function. On the other hand, the organization of my body, too, is constituted by the kinaesthetic function not perspectified when I move my body directly as I wish.[23] In order to throw light upon the inevitability of ambiguity of the body, it is necessary to make sure of the true nature of the tactile sensation and the visual sensation.

What is important primarily is the tactile function of the body that conditions the sense of sight as its auxiliary function. For example, when I see a thing, I hold my body erect against the earth, carry my body near to the thing to be seen and keep a certain distance from it. Then I turn my head and eyes to a certain direction, open my eyes, and start the kinaesthesis of eyeball movement. Such a simple description as this will easily make us understand that what conditions the starting of the function of sight is the tactile, kinaesthetic function including the kinaesthesis of walking. It goes without saying that the eye itself is constituted by the double sensation, such as by being touched with fingers, but the movement of eyeballs is also a tactile, corporeal event. What matters next may be the question of keeping a certain distance, which is a part of the visual function (of perspectification). Namely, when I see a thing, what gives rise to this seeing is the distance between I and a thing, if we leave the tactile function of the body out of consideration. When the distance between a person to see and a thing to be seen is bridged, the seeing will no longer arise. For in the space of sight, what appears visually either disappears at the farthest extremity, or appears at the optimum distance. Anyway, whatever intends to appear fluctuates between these two limits, and the actual appearance requires the discrimination of perspective by the optimum distance. In short, what gives rise to the seeing is the distance between a person to see and a thing to be seen. In the space of touch, however, as I mentioned earlier, the difference of perspective is unnecessary; a person to touch and a thing to be touched simultaneously share the same place, and there is no alienation by distance between the two. Although, purely visually speaking, my corporeal phantom does not recede from me, it is nothing but a visual phantom alienated from me. Claesges calls this state "an abyss between the body and the external world which cannot be bridged within the scope of the visual constitution."[24]

My argument so far makes me conclude that it is the appearance of the world by the tactile system that unifies the disintegration between a person to see and a thing to be seen and that solidifies the circuit of "to see – to be seen."[25] At least, in spite of Merleau-Ponty's criticism, it is not because we take the Cartesian viewpoint that we try to seek the source of the primordial meaning of the world not so much in the sense of sight as in the sense of touch.[26] According to Merleau-Ponty, Descartes considered that light has a function of touch and that the model for sight is touching. However, it should be rather put as follows: it is the sense of sight that is a prototype for the scientific knowledge and the common sense. For the oblivion of the tactile corporeality and the exfoliation of ego from the world by the disunion between subject and object (which is based on the disintegration of "to see – to be seen") are exactly the Cartesian, dualistic world-intuition. As a seer the eye foresees danger and what is to arise. It is often said that science sees in order to foresee (*voir pour prévoir*).[27] And it seems that the disunion between subject and object in the sense of sight becomes a model of its fundamental method for the science that wants to extract benefits from Nature by foreknowing and foreseeing danger and by operating things usefully. The eye is a skillful scout for discerning things of a mechanical nature.

We can thus characterize the sense of sight, while touching and hearing *do not* originally require the disunion between subject and object. But what truly opens the temporality and the spatiality of the world is the auditory sense which, for example, listens to a melody, and the tactile sense which touches the expanse. This is also understood from the fact that before we "seize by seeing," we can find the most primordial perception of the body. This perception arises where the atmosphere which fills the world (*Atmosphäre*) and my body touch one another directly.[28] For example, as soon as we enter a neighbor's room, we may immediately feel the unusual atmosphere that the people in the room produce. This is because my skin directly touches the unusualness of the atmosphere. Or, as I shall refer to later, we perceive the change of weather, first by tactilely grasping humidity that is generated from the contact of my skin with moisture. At the end of our rainy season when a hot summer is about to set in, the unusual humidity destroys even the spontaneity of my body and changes every action of mine into dull motion. This dull motion and humidity touch one another and are generated simultaneously.

In spite of my assertion so far, it may be very difficult to uproot the prejudice that it is the tactile body itself that uses a thing, makes it into a tool, and then operates things. However, the sense of touch of which we have an idea in our mind is not the sense of touch which should be possessed by the active, technical body that transforms a thing and reconstructs the world. Instead, it is the corporeality which must be forgotten at that moment, that is, the tactile corporeality before the world and I are grasped as a different substance respectively. In other words, since in transforming a thing and reconstructing the world my body is grasped as existence heterogeneous to and alienated from the thing or the world, the sense of touch which we call in question is the tactuality possessed by my corporeality in the sense that as a condition for this transformation and reconstruction I am an existence that is settled to inhabit the world.

Thus we can discover the deeper dimension of the primordial body that exists prior to seeing and being seen. Namely, it is the deeper dimension of corporeality which exists before the distance is generated between a person to see and a thing to be seen, that is, before seeing arises, and which assists the generation of sight.

If we take Schmitz' thought into consideration, this corporeal experience prior to the generation of sight will be more easily explained. Schmitz often quotes Hegel's remark: "Being subject and object, that is, the opposition between a person to see and a thing to be seen ... is left out in seeing itself."[29] But we should rather say that this "seeing itself" is grasped neither by a person to see nor by a thing to be seen, but it constitutes the scope on which their opposition is based. This is equivalent to what Schmitz calls "plain perception."[30] For him plain perception means that "the distance between subject and object, therefore between a person to see and a thing to be seen disappears."[31] It seems to us that the foundation of this plain perception is no other than tactile corporeality. When the tactile perception performs its function, I touch the world and meet with the world fundamentally. And at this moment I am neither identified with the world nor totally different from it. "I touch the world" means "the world touches me," and the arising of this "reciprocal touching" co-generates the world and the "I" simultaneously. Naturally my body may become "figure" and the world "ground" at this moment.[32] However, this "figure-ground-difference" is not the difference in the sense of the antonym of identity, but it is the difference which is generated from the plainly indifferent contact which is neither identity nor difference by way of disjunction.[33] As I have mentioned already, a person to touch and a thing to be touched are fused but not mixed, and they *can be separated*. That the body and the world become disjoined from the indifferent arising of "the reciprocal touching" signifies, to put it in Schmitz' terms, that the body and the world are in the chaotic relation. What he means is that the body and the world are neither identical nor different, but they are in an undecidable, amalgamative relation. Namely, it is the fundamental relation of the body and the world in which "which is the body" and "which is the world" are not clear. Moreover, from within the circulation and inter-change of meaning between the world and the body which are neither clear nor definite, the discrimination rises between a person to touch and a thing to be touched. Junichiro Tanizaki writes as follows: "The color of the sweet jelly beans turns ... all the more meditative when soaked in the darkness. When a man keeps that icy, smooth thing in his mouth, he feels as if the darkness of the room becomes one sweet lump, which melts at the tip of his tongue. ... It seems that the unusual depth is added to the taste."[34]

V

The simultaneous arising of my body and the world founded on the reciprocal touching is an experience neither unusual nor special. Merleau-Ponty, for example, described such corporeal experience, though inadequately. The sense of sight which he thought of is different from the "eyebeam" of Sartre and Husserl, and from the beginning it was strongly characterized by the inhabitation into an object. Prior to his *L'Oeil et l'esprit*, Merleau-Ponty had already mentioned in his doctoral dissertation, *La phenomenologie de la perception*, that "seeing" means plunging into an object and inhabiting it.[35] This thought was deepened in *L'Oeil et l'esprit*, but the term, "in-habiting an object" is an operational concept in Fink's sense. In the first place, Merleau-Ponty never defined this term clearly. According to Fink, an operational concept is a concept opposed to the thematic concept that an original philosopher continuously makes for intentionally; it is opposed to the concept that "a thought fixes what is thought and keeps it." Namely, the operational concept is the concept which is "fully used usually" and is "penetrated with the philosophcal thinkings," but which is "not taken into consideration." Fink names this operational concept "a shadow of philosophy." [36] Actually, although Merleau-Ponty used this concept of *habiter* to indicate such a decisive passage as the opening of *L'Oeil et l'esprit* [37] he merely used it multivocally and did not pay much attention to it. This may suggest that we must once more consider "a shadow of philosopher," Merleau-Ponty, and the matter which is thoroughly considered by him, from a new point of view.

In *L'Oeil et l'esprit* Merleau-Ponty puts both touching and seeing at least on the same level. [38] To repeat, this is because he did not thoroughly consider the concept of *habiter*.

What I can see actually is at least what I can touch by right. On the other hand, there is a thing which cannot be seen but nevertheless can be touched. For example, it is the humid weather, an unusual atmosphere, the lukewarm soiled air, and such. "When we regard silence as feeling, we find how rich and varied silence is. The solemn silence in the Nature or in the high-ceilinged hall. Or the serene silence is equipped with expanse, weight, and thickness, and provides an atmosphere that fascinates our heart. There is something weirdly menacing and overwhelming in the silence which is depressing, heavy and 'stuffily enveloping.' " [39] The silence thus described by Schmitz is not merely the auditory and physical silence which is nothing but a lack of sounds; it is also the silence which implies nuances and which cannot be grasped except by contact with our entire body. In summary, the sense of sight arises by being assisted by the sense of touch, and the former is based on the latter. When Merleau-Ponty stated that one's look touches a thing and *inhabits it*, it seems he implicitly meant the sense of sight is a subspecies of the sense of touch. In *Le visible et l'invisible* (which can be called his posthumous manuscripts), Merleau-Ponty himself writes that the tactile perception by the eye is the "distinguished variant" of touch. [40] Then is it not true that "to see" or "to inhabit an object" means to touch an object, not figuratively but primordially speaking? Otherwise how can we say that the concept of "inhabit" means to inhabit time and space, the world, the earth, and the Being (*l'Etre*)? It seems to us that "to inhabit" is a way of existence in which the body ("being-in-the-world") encounters the fundamental world.

VI

I have earlier described the co-generation of "I touch the world" and "the world touches me," and by discussing this point from a different angle, I intend to expose the kinetic character of the corporeal experience. Judging from the nature of the tactile sensation, it is clear that my touching a desk is simultaneously my being touched by a desk. But what should never be overlooked is that a desk touched by me is homogeneous to the limb of my body. For my body belongs to the inside of Nature, and instead of being made of the material totally different from the material world and Nature, the material world penetrates and fills my body. That my body is not made of composition and materials particularly different from the so-called external, material nature, is understood from the fact that after death my body "returns unto dust" and from the phenomenon that my body expands functionally. For instance, although a thing belonging to the internal world (the glasses, for example) can be removed from my body, it can still be unified and organized within the system of my functioning body by the functioning body. The generation of the corporeal expansion may indicate that the glasses and my body in the narrower sense are not heterogeneous to one another. Then in what way does the expansion of the body arise? As I have already described, my body is the world-experience itself in the contact with the world or the environment. In perceiving atmosphere or weather, it is hardly possible to definitely discern weather from the sentiment of states by my body, because they touch one another directly. Schmitz writes: "Weather is treated as *the spatial* among the impressions which are direct and which have not yet separated from me by reflection. . . . Such atmosphere of weather or climate is, in many cases, feeling, or at least is akin to feeling."[41] Here the sentiment of states by my body and the atmosphere of weather or climate are far from being private; they are rather public and constitute the spatial expanse. Simply, the stimulation which weather gives the body is so close to us that this feeling fails to be caught by our eyes, and it is perceived only by way of the reciprocal contact in my body. In this sense we say that the expansion of the body is based on the contact of the world with me in our context, and that it is the amalgamation of subject (my corporeal sentiments of states) and object (weather, climate, atmosphere, etc.). Namely, the primordial experience which is called "plain perception" by Schmitz might as well be defined as a contact of the world with the "I."[42] "When I relax myself in a bath, my body expands to the entire bathtub, or while I am lying on the grass in the suburbs basking in the warm, vernal sun, I am absorbed in the environment and experience the expansion of the body." This description by Schmitz will confirm the truth that the corporeal experience which is the contact of the world with ego within the world is achieved as the expansion of my body. In Schmitz' works there are many documents cited which describe the generation of the intersection between the "I" and the world on the stage of corporeality. Let me quote one of them: "It seemed as if what had looked like being outside or around me before, had suddenly existed within me. I felt as if the entire world had existed within me. Within me the trees were shaking their green leaves. Within me the skylarks were singing. Within me the hot sun was shining. Within me a cool shadow rested."[43] Namely, while the body (I) and the world give rise to "figure-ground-difference" simultaneously, the body tries to fuse into the world (ground) and the world into the body (figure), since the material identity is included in this difference. What enables this reciprocal encroachment is the experience of contact between the primordial world and the body, which could be called their indifference.

Second, the body could expand functionally. As I stated above, when I use my hand and write a manuscript with a pen, the tip of my body expands to that of the pen, and the hand is fused with and co-works with the pen in the motion of writing. Therefore it is not until this motion, which organ and function depend on and belong to, has its systematic unification destroyed and gets objectified as *res extensa* "by the eyebeam" that the hand and the instrument are separated noematically. The functional expansion of the body was already implied by Husserl when he grasped corporeality from the aspect of the achievement of kinaesthetic function. Or, when I drive a car, my body and the car are unified in the "tactile," kinetic oneness, and both of them are integrated in the origin of directional orientation and perspective. After all, Husserl himself noticed the phenomenon that in the tactile oneness that my body in the narrower sense and the car move together, the zero-point of perspective (my body in the narrower sense) expands.[44] In other words, in the origin of perspective from which perspectification starts, the body and the car are unified (organized) tactilely.

I believe that in this essay I grasped my world-experience as my corporeal experience and that I could elucidate the corporeality of the world-experience in its kinetic totality, at least theoretically. Thus I can conclude that I and the world, or subject and object, are like performers on the stage of corporeality, where they play the role of the complicated corporeal-arising, which expands at one time and shrinks at another time.

Hiroshima University

NOTES

* This paper first appeared in *Phenomenology Information Bulletin* 4 (1980).
1 Bergson, *La pensée et le mouvant* (Paris: P.U.F., 1962), p. 216.

[2] Husserl, *Husserliana* (*Hu.a.* for short), IV, 150.

[3] See *Iwanami Dictionary of Ancient Japanese*, p. 338.

[4] The verb form of *Leib* is preserved in the expression *leiben und leben* to which Husserl paid attention: "Ego . . . as the affected and as the one that experiences is localized within this body, . . . ego itself is constituted as the corporeal existence, as the *leibendlebend*." See *Hu.a.*, XV, 294.

[5] See *Hu.a.*, IV, 279–80.

[6] H. Schmitz, *System der Philosophie*, vol. III, pt. 2, p. xiii, 6–7; H. Schmitz, 'Leib und Seele in der abendländischen Philosophie' *Phil. Jahrbuch*, 2d Halbband (1978), 222.

[7] Bergson, *Matière et mémoire* (Paris: P.U.F., 1968), pp. 11–12.

[8] Valéry, 'Réflexions simples sur le corps,' *Variétés* 5 (Gallimard), pp. 67–70.

[9] For the details of the confrontation between Sartre and Valéry, see Tadashi Ogawa, 'Perception and Body,' which is to be published in *the Journal of the Area Studies* (Hiroshima University, the Faculty of Integrated Sciences and Arts).

[10] Sartre, *L'Être et le Néant* (Gallimard, 1943), p. 351.

[11] Husserl, *Erfahrung und Urteil* (Claassen, 1963), §§ 15–17.

[12] See Tadashi Ogawa, 'Time and the Primordial Nature,' *Shiso* (October 1978), 48.

[13] *Hu.a.*, IV, 136.

[14] *Hu.a.*, IV, 150.

[15] *Hu.a.*, XV, 271.

[16] *Hu.a.*, XV, 326–27.

[17] *Hu.a.*, XV, 297.

[18] For example, according to Husserl, even a melody which was thought as a clue to the analysis of time, "sees" sounds: "The eyebeam penetrates every aspect which overlaps with each other as an intentionality of a certain sound in the continuous flow of progress . . . " (*Hu.a.*, X, 80).

Moreover, Brand, too, points out the incompleteness of Husserl's phenomenology, as "philosophy of seeing." See 'Horizon, World, History,' in *The Fundamental Problems of Phenomenology*, ed. Nitta and Ogawa (Kyto: Koyoshobo), p. 231.

[19] See *Hu.a.*, XV, 295.

[20] Claesges, *Husserls Theorie der Raumkonstitution* (The Hague: Nijhoff, 1964), p. 110.

[21] Hoche, *Handlung, Bewusstsein und Leib* (Freiburg: Alber, 1973), § 28; Schmitz, *System der Philosophie*, vol. II, pt. I, pp. 396–97.

[22] It should rather be said that my body moves as a visual *res extensa*.

[23] See Claesges, p. 113.

[24] Ibid., p. 104.

[25] Gadamer tries to restrict the denotation of the seeing, for seeing means seeing always only one definite direction and it has the quality of positing an object and so of ignoring others. See Gadamer, *Wahrheit und Methode* (Tübingen: Mohr, 1960), p. 438.

[26] Merleau-Ponty, *L'Oeil et l'esprit* (Paris: Gallimard, 1964), p. 37.

[27] Bergson, *Matière et Mémoire*, p. 22. In Bergson we find a disputable point that modern natural science operates, takes advantage of, and dissects Nature which exists outside. For farther study, see Shuntaro Ito, 'Toward the Science-history of the World," *Shiso* (June 1975), p. 22. See also Shuntaro Ito 'The Theoretical Foundation of the Scientific Revolution,' in *The Method of the History of Thought and its Problems*, ed. Yujiro Nakamura (Tokyo: Todaishuppankai), pp. 102–103, 108.

I owe the concept of "controlling Nature" and of "controlling the world" to Mr. Shuntaro Ito. Bergson confirms the utility of the arithmetical and scientific intelligence by the operational quality of signs, which is almost on Berkeley's line. For the intelligence which, parallel to action, disjoints and reconstitutes Nature as it wishes, see Tadashi Ogawa, 'La réflexion bergsonienne et la réflexion phénoménologique' *Journal of Philosophical Studies* (Kyoto University), No. 532 (1977).

[28] There is a description of such primordial experience as this in Schmitz' dissertation, 'The Sentiment of States by the Body and Feelings.'

[29] Hegel, *Der Geist des Christentums* (1907), p. 316. Today, the Ullstein version edited by Hamacher is more easily available.

[30] H. Schmitz, *Subjektivität* (Bonn: Bouvier, 1968), p. 35.

[31] Ibid., p. VIII.

[32] As to the concept of "figure-ground-structure" in the Gestalt theory, see the dissertation by Waldenfels, in which it is regrasped by being related with the open dialectics, and in which he proves that it is useful for the clarification of the relation between the human existence and the world: Waldenfels, 'Moglichkeiten einer offenen Dialektik,' in *Phänomenologie und Marxismus* (Frankfurt/M.: Suhrkamp, 1977), I, 151–53.

[33] "Disjunction" means that from the primordial situation of indifference beyond opposition, a circular form is to be constituted in which one reciprocally expects the other without opposition. "Disjunction" originally derives from Schelling's *Über das Wesen der menschlichen Freiheit*. It seems that Schelling considered what is written above with the concept of disjunction, though in another context.

[34] Junichiro Tanizaki, *The Psalm of Nuance* (Tokyo: Chuokoronsha), p. 24.

[35] "To see is to enter the universe of the existence which expands itself . . . To regard an object is to come and inhabit it . . . " (*Phénoménologie de la Perception* [Paris: Gallimard, 1945] p. 82); "To see an object is to plunge oneself into it . . . " (p. 82); "A human being inhabits the world" (p. 462); "I inhabit the earth which is beyond motion and pause" (p. 491); "The Being which the eye inhabits just as a human being inhabits a house . . . " (*L'Oeil et l'esprit*, p. 27). Although "inhabitation" has only one special meaning, I intend to mean by the verb form "inhabit" "to go into an object and stay living there."

[36] Fink, 'The Operational Concept in Husserl's Phenomenology,' in *The Fundamental Problems of Phenomenology*, p. 27.

[37] "Although science operates a thing skillfully, it gives up inhabiting a thing" (Merleau-Ponty, *L'Oeil et l'esprit*, p. 9).

[38] For example, see *L'Oeil et l'esprit*, p. 21.

[39] Cf. 'The Sentiment of States by the Body and Feelings,' in *The Fundamental Problems of Phenomenology*, pp. 366–67.

[40] Merleau-Ponty, *Le visible et l'invisible* (Paris: Gallimard, 1964), p. 175.

[41] *The Fundamental Problems of Phenomenology*, p. 366.

[42] The concept of *Verquickung* is Schmitz' while that of *Verschränkung* is Merleau-Ponty's and Waldenfels'. Both of the concepts express the fundamental relation between the world and the "I." It seems that the double sensation itself exactly indicates *Verquickung* and *Verschränkung*, but it is the relation between the two items which are fused but at the same time separable. In other words, the double sensation has the undecidedness (*Unentschiedenheit*) between 'identity' and difference.

Cf. Schmitz, *Subjektivität*, p. 45; Waldenfels, 'Die Verschränkung von Innen und Aussen im Verhalten,' in *Phänomenologische Forschungen*, II, 116, 124.

[43] Schmitz, *System der Philosophie*, vol. II, pt. 1, p. 76.

[44] *Hu.a.*, XV, 276.

HANS KÖCHLER

THE RELATIVITY OF THE SOUL AND THE ABSOLUTE
STATE OF THE PURE EGO

The body-soul problem, as known from the metaphysical tradition, receives a new dimension resulting from the specific definition of the subject in transcendental phenomenology. If Husserl attempts to describe the personal I as a 'psycho-physical unity'[1] this is not to be seen as ontological realism, which could imply the equal validity of the observations of the natural sciences and of the human sciences. According to his position of transcendental idealism it would be virtually pointless "to require a psychophysical 'causal explanation' of the qualities of mind of an individual as that which is the requirement and the one and only scientific requirement to fulfil the function fulfilled by natural explanation in the physical sphere".[2] The purpose of a phenomenological description of the relationship between body and soul must lie on a different level. According to Husserl's definition "the soul" is the "unity of qualities of mind founded on the basic perceptual qualities"[3] and receives its reality due to the fact that as the unity of the life of the soul it stands in combination with the body as the unity of bodily being, which is itself part of nature.[4] This "interrelationship between the events of mind and body"[5] implies that matters relating to the soul can never be experienced as separated from the 'nature' of bodily experience.[6] To describe the multiplicity of this interrelationship is the main task of phenomenological description. In his *Ideas* Husserl tries to show that 'Nature' and the body and related to it the soul, are "interrelated with each other and together form a whole".[7] In this context he sees the body as the "meeting point of the causality of mind and the causality of nature",[8] within which causal relationships are transformed into conditional relationships between the "outer world" and the "subject consisting of the body and the soul".[9] The reality of the soul is founded on bodily existence,[10] which is, as it were, the "starting point" of the world view of the subject.[11] Thus the whole conscious mind of the human being is due to its material foundation related to the body in a certain way.[12] "My body as a point oi departure in the absolute here"[13] is the precondition of all nature constituted in an intentional relationship to the perceiving body.[14]

With the description of the concrete relationship between body and soul — in the context of transcendental phenomenology — an only seemingly

95

A-T. Tymieniecka (ed.), Analecta Husserliana, Vol. XVI, 95–107.
Copyright © 1983 *by D. Reidel Publishing Company.*

phenomenological problem has been taken into consideration. For Husserl distinguishes between a *real mental subject* — the "identical mental being, bound in reality to the respective human or animal body and thus resulting in the two-fold being, man or animal"[15] — and the pure or transcendental I, which underlies the distinction of "world" and "experience", the bodily and the mental as an ontologically prior element.[16] Whilst he attributes to the material world (within the "objective world" constituted as nature) a certain independence, "due to which it does not require the aid of other realities",[17] he also suggests that fundamentally *consciousness* itself can be imagined "*without nature*".[18] Husserl believes that consciousness, as soon as it is "purified in the process of phenomenological reduction",[19] could rid itself of its apperceptive conception as a state of mind and overcome its empirical attribution to natural causality.

The "worldly apperception" with its empirical conception of the inter-relationship between the body and the soul is to a certain extent only the first step in the transcendental phenomenological reduction to the "me as the universal, absolute and transcendental ego".[20] The natural view, which allows us to describe the concrete psychophysical unity of man, is in Husserl's phenomenological structure analysis replaced by the transcendental view, which leads to the recognition of the "absolute consciousness constituting nature",[21] on which all nature — on the basis of the relationship between the constituting and the constituted — is founded. Natural reality, which includes the bodily and the spiritual unity of the psychophysical unity, is thus deprived of its ontological significance: pure transcendental subjectivity is that which is ontologically absolute.[22] Thus the world experience of the soul is understood "as an objectivation of itself"[23] in the monad of the transcendental ego. If the basis of bodily experience — and thus also of physical presence as part of real nature — lies in causality and the resulting identifiability and distinguishability as physical individuality, then the (trans-cendental) I has an individuality "of its own",[24] and its being for itself is — in Husserl's opinion — in no way based on the psychophysical relationship to space and time: for the transcendental I time and space are not principles of individuation. According to Husserl this means that the "mind" (the transcendental subject) can always only be "naturalized" to a certain degree [25] and that thus his own analysis of the structure of the I as a "psychophysical unity"[26] are merely of relative importance insofar as they refer to the personal I as "directly governing my one and only body and thus directly influencing the primordial environment",[27] although this type of manifestation does not imply its ontological meaning. For according to Husserl "subjects cannot be encompassed by the fact that they are part of nature, as then that would be missing which gives nature its purpose".[28] Nature for him is a field of "relative spheres" leading towards the mind as the absolute principle "governing" all relative spheres.[29] The absolute qualities of the mind thus exist *in spite of* the determination by the surroundings,[30] as Husserl attempted to show in his detailed phenomenological analyses. The subject *"governs"* the body,[31] is ontologically prior. Thus for Husserl the actual, concrete, psychophysical unity has its objective purpose exclusively as "the unity of the governing subject" in the body, which thus becomes a "subjectively functioning organ".[32] According to the fundamentally idealistic conception qualities of mind (transcendental subjectivity) exist *prior* to the world in space and time. From this we must distinguish the dimension of the "soul" referring to the way "in which the mind is located in the world of space, as it were, and thus gains special representation and together with its bodily basis *reality*".[33]

II

The idealistic approach pervading Husserl's phenomenological description of the body-soul problem becomes apparent in the description of the process of death: in the "natural world view" concrete bodily presence is the precondition of possible "spiritual being", and this becomes "nothingness" as soon as the apperceptive conditions of spiritual experience are negated by death.[34] The interpretation of death as "a real event in the world"[35] refers only to the superficial level of the naive "world view", which has not succeeded in making transcendental subjectivity conscious. On the level of the pure subject the concept of "immortality" gains a new meaning, not as the liberation of the soul into its own sphere of being in the world,[36] but as the epitome of the fact that transcendental subjectivity cannot be reduced to the context of "worldly" causal relationships. The "worldly" view of death as the destruction of the soul *in* the world[37] does therefore not have to stand in contradiction to a transcendental idealist view: "The doctrine of immortality if it is not to contradict the purpose of the world as determined by objective experience would have to have a quite different meaning and can in fact have it if it is true that the natural world view of all natural and worldly life does not have to have the last word and perhaps should not do so."[38] In his phenomenology as transcendental idealism Husserl wants to show that the sum of possible objects of objective experience (what we describe as "world" in the natural sense of the word) must not be considered as being in the absolute sense of the word and that the absolute "presupposing the existence of the world" is the "mind" (i.e. the transcendental subject although not in its physical bodily concretion – as "worldly mind" – i.e. not as "the soul".[39])

Thus it again becomes obvious in what strict sense the transcendental phenomenology of Husserl distinguishes between matters of the mind and of the soul: the soul as *concrete* subjectivity, which manifests itself in the real unity with "bodily being",[40] presupposes *pure* (transcendental) subjectivity as its ontological precondition, which includes the "worldly" manifestations of "body" and "soul" as its forms of constitution. Thus Husserl transcends his originally realistic approach as manifested in his analysis of the relationship between body and soul in the *Logical Investigations*. Whilst in the *Logical Investigations* he added to the relativity of the existence of the world the relativity of the I,[41] thus assuming – in a true, phenomenological fashion – the difference between the psychological and physical to lie primarily in the different modes of *givenness*, in the *Ideas* this becomes an ontological difference which cannot be overcome in the concrete psychophysical unity of natural world experience but only be covered up. Whilst in the Logical Investigations this psychophysical unity can still be taken as the basis of consciousness, after the idealistic turn of the *Ideas* consciousness is something independent, and conscious experiences can only form a unity *among themselves*.[42] This leads to the concrete bodily being as an "intentional correlative of consciousness" and thus also one's own body as perceived by oneself (as an object of perception) stands in a constitutive relationship with consciousness. The concrete unity (and interrelationship) between body and soul proves to be only the semblance of the "natural" attitude. Also in his theory of the constitution of the body[43] Husserl failed to reject this final, idealistic ontology; this theory is rather only a complement to his constitutional analyses on the basis of these same ontological assumptions, not a revision of these assumptions. What Husserl requires of the phenomenological analysis of "objects" also applies to the analysis of the forms of the constitution of bodily existence.

By returning to absolute consciousness and the relationships of being, to be observed there, first and foremost the purposive relativity of the respective objects of this or that attitude and their respective interrelationships of being is to be understood.[44]

Interesting parallels to this fundamentally idealistic attitude can be found in the body-soul theory of the neurophysiologist Sir John Eccles.[45] He also postulated the "prime reality of conscious experience",[46] and thus the conscious component of our existence ("world 2") is not determined by the causal laws of "world 1".[47] The material world has rather only the status of a secondary reality as compared with the reality of consciousness.[48] Like Husserl assuming the necessary subjectivity of all experience[49] he formulates the theory that the external or "objective world" is a result of or a representation of various types of this personal or direct experience.[50] This leads him to the ontological hypothesis, on which any idealistic doctrine is based, that for every human being his consciousness constitutes prime reality and that all other regions possess a dependent reality, or a reality of second order.[51] This agrees with Husserl's hypothesis concerning the absolute nature of "pure consciousness"; where – as in Husserl's transcendental phenomenology – a sudden switch from the methodological to the ontological level, lacking an argumentative basis, is to be observed: for Eccles assumes that in every experience of the reality of things and the world "we" as conscious beings have to stand in the centre of all explanations as all other kinds of experience are of peripheral or secondary nature.[52] Without giving further reasons he deduces from the methodological necessity the ontological priority of "world 2" (the world of the conscious). This corresponds to Husserl's procedure, who by making the "pure I" ontologically absolute neglects the real opposition between "body" and "soul" and overcomes the opposition between both – phenomenologically clearly different – regions by means of the underlying transcendental subjectivity.

III

This ontological and not merely methodological idealism decisively determined Husserl's conception of the I, i.e. also of human subjectivity in its worldly form and directed his analysis of the body-soul problem in a monistic direction as the real independence of the bodily and material region was questioned. Therefore we shall inquire more deeply into the foundations of Husserl's idealism and his conception of human subjectivity.

The sole basis of phenomenology is *experience*. Every experience of something is experience for me. The experienced is experienced by my conscious being, and *is* thus only experienced in relationship to the subject. Thus – according to Husserl – the experiencing subject depends on that which appears *in it*, i.e. intentionality does refer to "the other" but not to the necessary *transcendence* of this other: "We do not understand how perception can take hold of the transcendent in the form of reflective and purely immanent experience",[53] as formulated by Husserl. This failure to understand is of course the result of self-reflection; in his naive state the

human being *is at one with* things, he can "claim" them in a naive sense.
If the observing I separates itself from the experiencing I then that which
Husserl calls "reduction" occurs:[54] the "given" becomes something appearing
to me and in the reflective break subject and object are seen in their epistemological relevance. The presupposed validity of the "object for itself" is
questioned. The object becomes an intentional correlative of perception
— under a critical exclusion of its purpose of being. An "epistemological
reduction" of this kind leads to an exclusion of the "transcendent presuppositions".[55] Every description of the phenomena is conducted in terms restricted
to consciousness as for Husserl this alone is "given" and cannot sensibly be
questioned: *"Immanent being is without doubt absolute being in so far as
it is basically true that* 'nulla re indiget ad existendum'. On the other hand
the world of the transcendent 'res' is dependent on consciousness, and not
as a logically constructed entity but as a present condition".[56] The basic
question — related to the definition of transcendental idealism — is: is only
the *appearance* of the world "dependent upon" consciousness (as this is
impossible without the *"immanent idea of the thing-object* in the experiencing consciousness")[57] or fundamentally in its "existence"? On the one
hand the "real world" is conceived of in the strict terms of transcendental
consciousness, this even being conceived of as the "constituting".[58] According to this conception "the real world is reduced to a universe of intentional
correlatives of possible and real experiences of the transcendental I and
cannot be distinguished from these correlatives".[59] On the other hand (particularly in his late works *Experience and Judgement* and *The Crisis of the
European Sciences and Transcendental Phenomenology*) Husserl speaks of
the "world" as being given as the "universal foundation of faith for the
experience of individual objects",[60] where a strong tendency towards realism
can be observed. Nevertheless, this unquestionable world, immediately given
to the "passive faith in the existent",[61] is interpreted as an achievement of
"functioning intentionality", as an expression of the alienation of a "living
presence" of an absolute I,[62] which experiences itself as having this world
— always "given" in naive validity. Thus "the world" is always given in a
subjective immanence solely through the medium of consciousness, whose
"form of alienation" it represents. This necessary "self-development takes
place in the form of experiencing life and its implications";[63] yet the not-I
experienced in this way is no real objectum, as I do not judge the world, but
my personal being and the "world as such"[64] implied therein. The final aim
of transcendental phenomenology is thus to found world experience in the
absolute transcendental experience of the self, and to understand that which

appears as leaving the sphere of the "surrounding" of "functional intentionality" and to prove the fact that "human existence as being for itself" is the only true being,[65] by means of intentional analysis. By this means — according to Husserl's intentions — the opposition between the I and the world, which necessarily becomes obvious with the beginning of transcendental reflection, is to be overcome in the creation of a higher unity — which can of course only be achieved on the basis of idealism. The uniting element between subject and object thus remains something subjective; a "mental monism" is to overcome the objectifying opposition between subject and object. Reality "for itself" thus becomes *immediately* given subjective experience, as

not the mundane and the world leads to the finished state of being given, and not the I as a human being in the world, but transcendental subjectivity as a world deriving its validity from presumed self-evidence and being maintained in its relativity.[66]

Here a new meaning of transcendence is to be discovered, not explicitly mentioned by Husserl, which is, however, typical of every idealistic, transcendental inward transformation: as a final possible condition an *absolute* transcendental I is assumed which precedes all conscious experience as a necessary precondition and thus — as a centralized "focus of the I"[67] — represents the most important synthetic principle of unity of all experience. This I is — according to Husserl — the "identical unity",[68] which manifests itself in all single "cogitationes". It is in principle meta-empirical and cannot be grasped in iterating reflection. It is the "identical subject of the function in all acts of the same stream of consciousness; it is the radiating centre, or rather the focus of all conscious life, of all affections and actions".[69] As a condition of all change of the empirical I it is the unchanging, the absolute (i.e. dissociated from change), beyond concrete relationships; for in Husserl's words — "*every cogito with all its components is born and decays in the flow of experiences. But the pure subject does not come into being and does not decay*", although it does in its own way make its "appearance" and "disappear".[70] It is — in opposition to the single cogitos — always transcendent and not "a real moment of these"[71] It is incompatible with the empirical I and only enters specific forms of manifestation, through which an empirical I is constituted.[72] In such a constitution of the individual cogitationes it experiences itself. Unreflected this presents itself in confrontation with the object, in reflective form as the knowledge concerning the fact that that which is confronted is constituted by functioning intentionality, not only as to how it is confronted but also in its factual nature or reality resulting purely from the way the pure I is established.

The relation of the absolute and the relative in the relationship between the pure and the empirical I [73] as presented in the problem of constitution is now concrete as that of principium and principiatum, as is to be expected from an idealist position. The single empirical I is not stationary and cut off from "primordial subjectivity" as this has entered the concrete cogitationes as the precondition of possible thought: "*I* as pure I see myself in *empirical* terms; I see myself as a being in the world and as one that is dependent and caused ... ".[74] Thus the relationship to the so-called objective world is given, the relationship to the plurality and variety, in opposition to which the pure I represents an identical unity (an idea, presented in similar form by Fichte). If the fundamental opposition between the immanent (subjective) and the transcendental in Husserl's conception of intentionality has been overcome in the intermediate medium of transcendental subjectivity (which was achieved through this ontological idealist conception), then it becomes obvious that in the basic structure of his thinking this opposition becomes evident again: the pure I as such is incompatible with the empirical I and its acts; it is a priori always already beyond this and even repeated reflection cannot reduce it to the state of immediate givenness. What happens here and now in the medium of my own and every other consciousness is only a superficial event in relationship to the "fundamental presence" of the pure I, which as an "absolute Zeitigung" in itself (in pure reflection) constitutes "immanente Zeitigung"[75] and secondly the special and temporal world of experience in its modifications: "The absolute is nothing but *absolute Zeitigung*, and already its interpretation, the absolute which I find as the original flow of being, is Zeitigung".[76] If in reflecting I wish to grasp this absolute in its purity, then the I always sees itself as an opposite, as Gezeitigtes. It has already left its state of "originality", i.e. in concrete reflection we never see the presence of the I in its pretemporal state, that which, as it were, lies before the transition into "temporal presence".[77] In comparison to the empirical, the pure, absolute I (as could also be shown in the structure of reflection) has fundamental transcendence; as the final condition of possible experience it has merely been *deduced*, postulated on the basis of mediating experience; its characteristics can only be expressed negatively, in terms of "otherness". Whilst the transcendence of the object-world in relation to the empirical I has been reduced to the immanence of "transcendent validity" in consciousness, and thus fully negated, the reflecting individual discovers in the process of the individual stages of reflection, leading him beyond that which is empirically "given", a new transcendence as against empirical consciousness (transcendence understood as the irreducibility of a region "opposed" to the empirical I to this). Here — by means of abstracting reflection — the attempt is made to discuss something which can never be fully brought into view, something which from the point of view of objective existence "as a fixed point in temporal presence" does not exist:[78] "as fundamental transcendental life, truly creative life and whose final I cannot be created out of nothing and be transformed into nothing", it is 'immortal' as dying has no meaning for it."[79] (Husserl)

IV

The "pure" (transcendental) subject is thus removed from the empirical context of the mundane body-soul relationship, it is its ontological foundation. The idea of natural determinants of this transcendental element understood as "mind" — with the aim of a strict determination of mind in the context of natural science — is thus inconceivable for Husserl.[80] Therefore the detailed description of psychophysical relationships as undertaken by Husserl is rendered relative in its philosophical importance: his analysis of the psychophysical body as the "primordial" factor in the constitution of the objective world (of space and time)[81] is negated by the principle of the "irrelativity" of the subjective in its realistic nature. The body as the point in which subjectivity is related to the reality of a thing for itself in transcending it is negated as an ontologically independent category. The difference between the spiritual and the material — which Husserl regards as "the reality beyond the purely physical thing",[82] where both dimensions of reality are psychophysically united in the body and the soul, is thus seen as a mere "layer of real events on bodies",[83] this difference now seems to be superficial and no longer fundamental to an ontological evaluation of the body-soul problem. The body, which — in its interrelatedness with the individual soul — he describes as the "subjective object"[84] and the soul which together make "conscious reality" of the personal I, which in both directions functions as the basis of causal properties,[85] are merely different manifestations of transcendental subjectivity, in which it constitutes itself as "mundanity". Ontological idealism, as we have attempted to explain by means of Husserl's transcendental phenomenology, does not allow for the opposition between the body and the soul as a real distinction of different levels of being. (This is related to a certain artificiality in the analysis of the experience of *Fremdpsychisches,*[86] as Husserl's ontological idealism finally represents a solipsist fiction, as the confrontation with the personality of the other is only possible by means of an analogical sympathy within the personal closed subjective world.)

The originally realistic approach of the analysis of the body-soul relationship to be found in the *Logical Investigations* is not only questioned by faith in the "possibly absolutely encapsuled personal being" of the soul,[87] but has been negated. Even if in the previous wording a degree of uncertainty is evident, which is related to the ambiguity and opacity of the concepts of "soul" and "body", which Husserl frequently uses with overlapping meaning,[88] the fact remains that an inductively effective typology of the soul, i.e. its causality as opposed to the material physical and the "inductive causality" of the physical[89] do not correspond to ontologically autonomous spheres of reality. The only element constituting reality is pure consciousness (pure subject). In the transcendental *epoché* the body as a special object constituted by this subjectivity and standing in relationship with empirical consciousness is thus distinguished from other objects — is thus reduced to its nature of being the object of pure intentionality. Its distinction is only one among constitutive objects and not deduced from its "inner" (ontological) structure. The realistic cause-effect relationship is thus directly inverted: not the body (constituting the relation to the material world) determines consciousness but consciousness constitutes the body as a particular object within the world. Although the body as a natural object is caught up in the chain of real causal relationships and the empirical consciousness of the body is again determined by these causal relation processes — which is also presented in Husserl's description of the *Ideas* — yet the reality of these relationships has in the light of his transcendental phenomenology to be interpreted as "nature" in the *transcendental* sense, i.e. as an intentional unity motivated by immanent relationships[90] "in pure consciousness". In his transcendental reduction, which does not only represent a methodological step towards a better description of the phenomena, but is an ontological postulation,[91] Husserl sees the actual step towards a universal and final formulation of the problem of reality, when he emphasises in the face of criticism arguing from a realistic basis: "We have in fact lost nothing but gained the whole absolute being, which understood correctly contains and constitutes in itself all the transcendent elements, as an intentional correlative of the act of habitual validity, to be realized and continued unanimously".[92] An idealistic position understood in this way does not allow the levelling process applied to the body-soul problem on the level of mundane "forms of constitution" of pure subjectivity to appear as a loss. On the other hand, a phenomenology based on the transcendental reduction (as understood by

Husserl) cannot rid itself of the constantly pressing question concerning the real ontological autonomy of consciousness related to material being — with all the existential implications which might result in answering it.

NOTES

[1] *Cartesianische Meditationen* (ed. Ströker), Hamburg 1977, p. 100.
[2] *Husserliana* (abbreviated: Ha), vol. VI, p. 479.
[3] Ha. IV, p. 120.
[4] Op. cit., p. 139.
[5] Ha. IX, p. 107.
[6] Op. cit., p. 108.
[7] Ha. V, p. 124.
[8] Ha. IV, p. 286.
[9] Ibid.
[10] Cf. Ha. IV, pp. 143ff.
[11] Op. cit., p. 158.
[12] Op. cit., p. 153.
[13] *Cartesianische Meditationen*, op. cit., p. 126.
[14] Op. cit., p. 119.
[15] Ha. IV, p. 120.
[16] Admittedly Husserl appears — in a different context — to contradict himself, when (in a criticism of Kant) he emphasizes that as soon as we distinguish between this transcendental subjectivity and the soul we get caught up in the incomprehensible and mythical. (Ha. VI, p. 120).
[17] Ha. V, p. 117.
[18] Ha. IV, p. 178.
[19] Ibid.
[20] *Cartesianische Meditationen*, op. cit., p. 102.
[21] Ha. IV, p. 179.
[22] Cf. the author's publication: *Die Subjekt-Objekt-Dialektik in der transzendentalen Phänomenologie*, Meisenheim a.G. 1974.
[23] *Cartesianische Meditationen*, op. cit., p. 134.
[24] Ha. VI, p. 222.
[25] Ha. IV, p. 297.
[26] *Cartesianische Meditationen*, op. cit., p. 113.
[27] Ibid.
[28] Ha. IV, p. 297.
[29] Ibid.
[30] Ibid.
[31] Ha. IX, p. 395.
[32] Ibid.
[33] Op. cit., p. 132.
[34] Op. cit., p. 109.

35 Ibid.

36 Ibid.

37 Ibid.

38 Ibid. – cf. also Ms. K III 6 (Husserl Archives), p. 251a.

39 Ibid. – On the problem of the terminological distinction between the concept of 'body', 'soul' and 'mind' and the corresponding different interpretation of the body-soul dualism cf. Josef Seiffert, *Das Leib-Seele-Problem in der gegenwärtigen philosophischen Diskussion. Eine kritische Analyse*, Darmstadt 1979, pp. 126ff.

40 Ha. IX, p. 393.

41 Cf. *Logische Untersuchungen* I (Logical Investigations), vol. 1, Halle a.S. 1901, p. 337.

42 On this interpretation cf. also Theodore De Boer, *The Development of Husserl's Thought*, The Hague/Boston/London 1978, pp. 227ff.

43 Cf. Ha. IV, pp. 157ff; Ha. III, pp. 117ff.

44 Ha. IV, p. 180.

45 Cf. *Gehirn und Seele*, in *Wahrheit und Wirklichkeit. Mensch und Wissenschaft*, Berlin/Heidelberg/New York 1975, pp. 211ff.

46 Op. cit., p. 238.

47 Op. cit., p. 242.

48 Op. cit., p. 61.

49 Op. cit., p. 69.

50 Ibid.

51 Op. cit., p. 77.

52 Op. cit., p. 80.

53 *Die Idee der Phänomenologie*, The Hague 1950 (Ha. II), p. 49.

54 Cf. Ha III, § 31 (pp. 63ff).

55 Ha. II, p. 49.

56 Ha. III, p. 92. – cf. also Ha. VIII, p. 488: "Immanent being in reality belonging to the transcendental sphere, the transcendental being as the unity of original presentation, always only realizing itself through appearances in subjective representations, is no absolute self."

57 *Erste Philosophie* (1923/24), II: *Theorie der phänomenologischen Konstitution*, Ha. VIII, The Hague 1959, p. 491.

58 Cf. ibid., p. 479.

59 Op. cit., p. 180.

60 *Erfahrung und Urteil*, Hamburg ²1954, p. 23.

61 Op. cit., p. 52.

62 Cf. Ms. C 3 I, p. 3 (1930) (Husserl Archives).

63 Ms. B I 5 IX, p. 8 (Husserl Archives).

64 Ms. C 3 III, p. 9 (Husserl Archives).

65 Ha. VI, p. 429.

66 Ms. K III 6, pp. 386f.

67 Cf. Ha. IV, p. 105.

68 Ha. V, p. 113.

69 Ha. IV, p. 105.

70 Op. cit., p. 103.

71 Op. cit., p. 102 (in the singular in the original).

[72] Ha. V., p. 114.

[73] Cf. also: Husserl, Ms. K III 1, pp. 23–25 (The problem of maintaining personal identity and supraindividuality).

[74] Ha. V., p. 114 (cf. Ha. VI, p. 205): The fundamental problem of the concept of the transcendental I consists in "how one and the same thing can simultaneously be constituting pure I and constituted real I" (R. Ingarden, *Kritische Bemerkungen*, in *Cartesianische Meditationen*. Ha. I, Den Haag 1950, p. 213), a problem, which is analogous to the paradox of reflection.

[75] Cf. Ms. C 5, p. 12 (1930) (Husserl Archives).

[76] Ms. C 1, p. 6 (1934).

[77] Cf. K. Held: *Lebendige Gegenwart*, The Hague 1966, p. 89.

[78] Op. cit., p. 131.

[79] Ms. K III 6 (Husserl Archives), p. 251a.

[80] Ha. IV, p. 297.

[81] *Cartesianische Meditationen*, op. cit., p. 137.

[82] Ha. IV, p. 176.

[83] Op. cit., p. 175.

[84] Ha. V., p. 124.

[85] Ha. IX, p. 133. – On the interdependence and psychophysical unity determining the personal I cf. also: *Cartesianische Meditationen*, op. cit., p. 100.

[86] Cf. *Cartesianische Meditationen*, op. cit., pp. 109ff.

[87] Ha. IX, p. 135.

[88] Op. cit., p. 138.

[89] Op. cit., p. 136.

[90] Op. cit., p. 120.

[91] Cf. the author's publication: *Die Subjekt-Objekt-Dialektik in der transzendentalen Phänomenologie*, Meisenheim a.G. 1974, pp. 79ff.

[92] Ha. III, p. 119.

LUCIANA O'DWYER

THE SIGNIFICANCE OF THE TRANSCENDENTAL EGO
FOR THE PROBLEM OF BODY AND SOUL IN
HUSSERLIAN PHENOMENOLOGY

Any description of soul and body or of their relationship in Husserl's phenomenology should begin with a description of transcendental subjectivity.
But "Who or what is this transcendental subjectivity?" This question runs
through the text of Ludwig Landgrebe's 'The problem of Passive Constitution';[1] it is posed from many different angles, and testifies both to the
centrality of this theme in Husserl's phenomenology and to the fact that its
ability to puzzle and sometimes bewilder the reader is not yet at an end. The
main difficulty is that transcendental subjectivity holds the key for the
description and solution of another large number of vital problems: the
positing of realism or of idealism or the overcoming of both, the concreteness
and the individuality of the ego as a human subject, our relationship with
others, that is, the problem of intersubjectivity, historicity, the definition of
transcendental and of transcendental life, and so on.

I start by focusing upon the question of Husserl's idealism, which is
complicated by the fact that Husserl himself called his phenomenology
"transcendental idealism" as late as the *Cartesian Meditations*.[2] Of course,
Husserl was using these two words in an entirely renewed sense, and it is
readily accepted that just as the term "transcendental" does not indicate a
return to Kant, so the term "idealism" is not to be identified either with the
idealism of Berkeley or with other forms of traditional idealism. However,
even when a general consensus has been reached upon such an understanding,
the task still remains, first, to define what exactly Husserl's transcendental
idealism is and, second, to explain why Husserl would not opt, in the alternative choice, for a renewed sense of the term "realism."

I agree with Theodore De Boer[3] that Husserl's later works from *Cartesian
Meditations* to the *Crisis*[4] do not deny or disprove the basic views about
consciousness which Husserl had reached and stated in "The Fundamental
Consideration" of *Ideas I*.[5] The absoluteness and independence of consciousness with respect to nature are strongly reaffirmed not only in *Cartesian
Meditations* and *Formal and Transcendental Logic*,[6] but also and undoubtedly
in the *Crisis*. It would not have been otherwise possible for Husserl to suggest
as a fundamental and necessary step, via the total epoché, the suspension
of the world as it is intended in the natural attitude. Transcendental

A-T. Tymieniecka (ed.), Analecta Husserliana, Vol. XVI, 109–117.
Copyright © 1983 *by D. Reidel Publishing Company.*

consciousness could not possibly view the world as a transcendental phenomenon if it had not retained the two basic attributes which it had in *Ideas I*: its absoluteness and independence with respect to the thing and its nonidentity with the psychological consciousness. Both in *Cartesian Meditations* and in the *Crisis* Husserl persists in describing psychological consciousness as "part of nature" and part of the world; hence it is quite obvious that such a consciousness could not suspend the attitude by which the world is seen as an objective reality, and thus it could not be the correlate of the transcendental phenomenon world. However, De Boer does not overlook the insistence with which Husserl protested against the accusation of remoteness from the problems of existence[7] or the importance that the phenomena of intersubjectivity and of the life-world acquire in the later Husserl. The question, then, is to find a proper definition of Husserl's idealism which would not be at odds with those so-called more "existential" themes of his phenomenology. In this respect, I am not sure that I fully understand De Boer's account of what Husserl's idealism really is and what is the way out of the at least apparent conflict among some themes of his philosophy. De Boer seems to suggest in his conclusion that since the absoluteness and transcendentality of consciousness form the necessary condition for a universal philosophy as rigorous science, as *Mathesis Universalis*, one is compelled to put aside the need to attain solutions of individual problems when and if the satisfaction of that need can only eventuate through a solution which would imply on our part the giving up of "the will to establish a rigorous philosophy."[8] While I agree that universality should remain the goal of "any true philosopher,"[9] I cannot see how one would object to such a goal if one looked for a clarification, or even for a more satisfactory solution of some vital problems which have remained unclarified in Husserl's philosophy, at least in his published works. It seems to me that the fundamental problem of how the consciousness of man stands to transcendental consciousness, which is also the problem of how the human subject stands to transcendental subjectivity, is a way of keeping alert "the will to establish a rigorous philosophy," since no human subject could direct his will toward the correct goal, as Husserl requires, without the knowledge of its role and its potential as a subject.

The statement of "The Fundamental Consideration," which De Boer maintains caused Heidegger to rebel against Husserl's position, is certainly unequivocal: Husserl *does* say, quite clearly, that "an incorporeal and even inanimate, non-personal consciousness is possible."[10] Husserl is here claiming that "it is possible for man as an empirical being" not to exist, but I do not agree that the implications of such a declaration on Husserl's part are quite clear. De Boer's comment on this point, that "the existence of man as a psycho-physical unity appears to be sacrificed to the consequence of an idealist philosophy,"[11] casts little light on the matter, since one cannot grasp clearly what "idealist" exactly means in this context. Even within "The Fundamental Consideration," there arises the problem of understanding how it is that once some human subject, like Husserl, had been able to think the thought of absolute consciousness by overcoming, to a high degree, if not entirely, its natural, psycho-physical dimension, this subject would nonetheless still be bound to language, actually to one language. I am not saying that Husserl could not abandon "the natural attitude" and think the thought of absolute consciousness because he was compelled to use language; which is one of the well-known objections to Husserl on the part of Merleau-Ponty and his followers. Rather, I am asking: how could the absolute and universal consciousness still be bound to the condition of a person, in this case Husserl, who had to think and communicate in German? I refuse to accept that Husserl, who on the evidence of his own words, was aware of this paradox, had accepted it as a point of arrival. The development of his thought after *Ideas I* suggests rather that such a paradox, that is, the possibility of conceiving one's own person as nonexisting and at the same time the impossibility of conceiving that without existing, provides the clue for a thoroughly new way of approaching the problem of what it is for a human subject to be a human subject. The interesting thing is that when the text of the fundamental consideration is compared with that of the *First Cartesian Meditation*, a very important change can be noted.[12] While Husserl insists here in saying that the ego after the epoché "is not a piece of the world"[13] and hence is not any longer this man as a psycho-physical unity, he also calls that same ego concrete, and even individual, as when he specifies in the same *First Meditation* that "This ego, with his ego-life, necessarily remains for me by virtue of the epoché."[14] Though this ego is not this man as a psycho-physical unity, it is still *this* ego. The "this" points to an individuation rather than to a one and only absolute universal consciousness.[15]

What I am suggesting here is that Husserl, in his later period, came to view the absolute universal consciousness, which is thoroughly one in its full unity, not only as the ultimate ground for consciousness as a mode of being, but also as a clue for an understanding of the human subject which would not be based on the scientific description of man as a psycho-physical unity. Such a suggestion, of course, implies that we would accept the distinction between transcendental subjectivity and transcendental ego as a fundamental one. The question of whether transcendental subjectivity is other than the transcendental ego is certainly one of the most perplexing questions in the whole of Husserl's phenomenology. According to the fourth *Cartesian Meditation*,[16] the ego in its reflective conscious life comes to apprehend itself, or constitutes itself, as transcendental consciousness, or concrete monad, as a result of performing the transcendental reduction. However, through this reduction, the ego can only know itself as absolute apodictic evidence. The character of the ego as pure transcendental universality will only be grasped through an eidetic reduction, in which I descend from what Husserl calls my de facto transcendental ego, with its accomplished typology, to eidos-ego. Here, Husserl clearly distinguished between a transcendental and an eidetic reduction, and there is no chance of identifying the two without doing violence to the text.[17] Thus, it appears that in *Cartesian Meditations* transcendental ego is the concrete ego which we attain after the transcendental reduction, while transcendental subjectivity is the eidos ego, that is, the universal pure possibility of any possible de facto ego. As such, it could be interpreted as being just a pure principle. But this interpretation is not supported by what Husserl says in the *Crisis*.[18] Here transcendental subjectivity is arrived at through a totally different path. The ego "is not reached in one leap" as in *Ideas* and *Cartesian Meditations*. Through the partial epoché of objective sciences we attain the phenomenon of the *Lebenswelt*. Through a transcendental inquiry into the life-world we come to understand the necessity of performing the total epoché which in turn will render possible the transcendental reduction. It is through this transcendental reduction that we arrive at the ultimate ground, the source of all transcendental accomplishments. It is therefore hard to deny that at this point transcendental subjectivity ceases to appear a mere absolute principle of pure possibility and acquires instead ontological connotations like the absolute source of all possible beings.

This close reference to Husserl's text reveals in its fullness the complexity of the problem. There is a clear distinction between the transcendental ego, which is very interestingly described as "my de-facto (*wie faktisch*) Transcendental Ego"[19] and the ego as absolute universality when the latter appears to be only a pure principle of universality. The distinction between transcendental ego and transcendental subjectivity is instead not clarified when transcendental subjectivity is hinted as being the universal ontological source of all possible beings, a description which seems to bring it closer to the absolute consciousness of "The Fundamental Consideration." However, it is worth while to consider that in the *Crisis*, the phenomenon of transcendental subjectivity, though it is viewed as an absolutely evident realm of experience, is not described as an absolute and apodictally establishable beginning, as Landgrebe stresses.[20] In his published works Husserl did not provide a clear and satisfactory distinction between transcendental subjectivity as universal and absolute consciousness and the transcendental ego as my de facto transcendental ego, according to the description given in *Cartesian Meditations*. Since I am in no position to show that the distinction between transcendental subjectivity and transcendental ego, with the consequent characterization of the latter as my de facto ego, prevails on the identification of the two concepts, I shall instead attempt to show that such a distinction allows for a development of the analysis on the body and soul in an entirely new direction, and that it may also help us to describe Husserl's own position as neither realist nor idealist.

When the reduction leads one to absolute consciousness, to the ultimate subjectivity, it becomes puzzling to know how one can still qualify oneself as self while having attained the absolute subject. The response of most of Husserl's readers, and even followers, to the suspension of the psychophysical unity man, is well known. While such a response is understandable and justifiable, it does not always fully succeed in unveiling the authentic nature of the problem. I take, as an example, the case of Merleau-Ponty, which is particularly interesting because Merleau-Ponty, like Husserl, was unwilling to accept a naturalistic interpretation of man, and yet he indicates the centrality of men in the world against the absolute and universal consciousness.[21] We have little choice left if we set the problem off as either the concrete human self, as being-in-the-world, or the transcendental and universal consciousness, which in this case is identified with the ego *cogito*. By opting for the absolute consciousness we save the idea of a universal knowledge and of a rigorous science, but we lose the right to proclaim our own individual views; on the other hand, by opting for the self-as-being-in-the-world, who is conditioned by his facticity at least to some extent, we lose the right of striving for a transcendental philosophy. It is hard to see how one could transcend one's naturality in Merleau-Ponty's existentialist phenomenology. In the *Crisis*, Husserl more than once conceded that totality and perfection can only be an ideal for the philosopher or for the scientist, the goal toward which one's search tends;[22] however, lack of totality and perfection is not necessarily ambiguity, as Merleau-Ponty would want it.[23] In fact, it seems virtually impossible to accept the ambiguity as what-pertains-to the human condition, and yet to continue at the same time to tend toward the universal and the absolute as to one's goal. But we must ask whether, in positing the alternative between the human self, who is in the world, and the transcendental self as absolute subjectivity, something of the problem does not remain concealed. It is quite clear that Merleau-Ponty does not want to defend the irreducibility of man as a psychophysical unity but rather as a concrete agent in the world. However, this concrete agent is grounded in the body, or, ultimately, in the unity of the physical and psychical, that is, the psychophysical unity that Husserl had overcome through the reduction. I certainly do not deny that Merleau-Ponty struggled to go beyond the traditional concept of body, and neither do I deny that his treatment of the body is both interesting and sometimes illuminating. However, his phenomenology of the body willingly comes to terms with the scientific concept of body, which is bound, at least to some extent, to physical laws. This recognition of the fact that the human self is bound to at least some physical laws, which in turn determine one's psychical life, sounds both sensible and reasonable.

Another kind of literature, which at least some of us would nevertheless consider reasonable and sensible, though in a different sense, does not recoil from claiming that one could overcome the physical laws of one's body. Teresa of Avila tells us in the *Interior Castle* that while one's body and one's thought may be engaged in the struggle against naturality and sensitivity, the soul may be all absorbed in God.[24] What is striking in this statement is the fact that thought (also called imagination) is put *au pair* with body at the level of the natural condition. Clearly, what Teresa here calls the soul cannot be identified with what we usually call the psyche (or with what Husserl calls *die Seele*) — rather the psyche could be identified with what she calls the thought. The same self can think at a natural level, where this self struggles within the psychophysical order to which it also belongs, and it can also experience another type of thinking at another level, at which it transcends the psychophysical order. It is not that I want to naively identify what Husserl calls my de facto transcendental ego with what Teresa calls the soul or, better, the highest part of the soul. I simply use Teresa's description to highlight the problem of a transcendental ego, which while conceived as an individual ego and therefore not identified with absolute subjectivity, is at the same time conceived of as being "above" the natural order of things. Of course, I am not trying to claim that such a description of the transcendental ego could be attributed to Husserl with some degree of probability; it would be equally difficult to prove, or even just give good evidence, that Husserl attempted somehow to achieve such a conception of the transcendental ego. My only claim in respect to Husserl's position is that, while he holds to the concept of absolute ultimate subjectivity, he also strives for the concreteness and nonremoteness of the transcendental ego. Moreover, even more important, he reaffirms in more than one way the personal and individual character of the ego, as when in the *Cartesian Meditations* he calls it my de facto ego,[25] and when, in the *Crisis*, he states that the philosopher through the epoché can enjoy a new way of thinking and of theorizing, not because he has lost anything which belongs to the world but because he "simply forbids himself — as a philosopher, in the uniqueness of his direction of interest — to continue the whole natural performance of his world-life."[26]

By claiming that the philosopher forbids himself to indulge in the natural attitude, Husserl is stating that the initiative and the resolve of the will, on the part of the philosopher who performs the transcendental reduction, is a necessary condition for such a performance. Statements of the sort are not infrequent in the *Cartesian Meditations* and they are more frequent in the *Crisis*. The whole concept of the high task of the philosopher as a "functionary of humanity" would make no sense if the individual and personal ego of the philosopher should be suspended; and Husserl's complaint that so many practicing philosophers prefer the comfortable ease of the natural attitude would make even less sense.[27] If truth and insight could only be attained by the absolute and universal consciousness, then the individual and personal subject could not take even the initial step of the epoché by which he could transpose himself to the transcendental level and perform as an absolute consciousness. It is indeed hard to accept that man as psychophysical unity, which is in fact no more than the result of a certain attitude, could have the power to hold back the performance of the absolute consciousness. And ultimately, the whole conception of an absolute consciousness was bound to change and develop from the one of *Ideas I* once Husserl's concept of reason acquired more and more, as Landgrebe said, the significance of an open historical process.[28] There could hardly be any doubt nowadays that Husserl's ego is quite set apart from the *cogito* of Descartes. (Although Descartes' *cogito* had a broader breadth than Locke saw in it and it was not restricted to a mere rationalistic "I think," it was still a psychological ego in the natural sense, that is, it presupposed a separation between body and mind.) The activity of the Husserlian ego is an encompassing, living, accomplishing which extends its roots in what we could call the "primordial life" of the ego, where it uncovers itself as ego source. Does the ego lose its personal identity in such a discovery? The considerations made above seem to suggest that Husserl was at a later stage reluctant to dispose with the personal and individual character of the ego as he appeared to have done in "The Fundamental Consideration," when the phenomenon of reason had not yet shown its historicity to him in full. Why then would Husserl still insist on the absolute and universal character of subjectivity? One possible explanation is that the universality and absoluteness of consciousness was guarantee against the relapse into naturalism by which the ego is identified with a psyche and a body in the natural, taken-for-granted sense. Perhaps another more fundamental reason for the unachieved clarification of the transcendental ego as personal and individual resides in the difficulty of keeping such an ego wholly above the natural attitude without identifying it with the only and one absolute spirit. Therefore, the task remains for us to achieve a more satisfactory description of the transcendental ego as my de facto ego while holding on to its transcendental character.

MacQuarie University, Sydney, Australia

NOTES

1 *Analecta Husserliana*, vol. 7 (1978), pp. 23–36.
2 E. Husserl, *Cartesianische Meditationen, Husserliana*, vol. 1.
3 T. De Boer, *The Development of Husserl's Thought*, trans. by T. Plantinga (The Hague, 1978).
4 E. Husserl, *Die Krisis der Europäischen Wissenschaften und die Traszendentale Phänomenologie, Husserliana*, vol. 6.
5 E. Husserl, *Ideen zu Einer Reinen Phänomenologie und Phanomenologischen Philosophie, Husserliana*, vol. 3, pt. 1, 'Die Phaenomenologische Fundamentalbetrachtung,' pp. 56–132.
6 E. Husserl, *Formale und Transzendentale Logik, Husserliana*, vol. 17.
7 E. Husserl, *Ideen*, bk. 3, *Husserliana*, 5: 140.
8 T. De Boer, *The Development of Husserl's Thought*, pp. 505–6.
9 Ibid., p. 506.
10 E. Husserl, *Ideen*, bk. 1, *Husserliana*, vol. 3, pt. 1, p. 119; English translation by W. R. Boyce Gibson (London, 1958), p. 167.
11 T. De Boer, *The Development of Husserl's Thought*, p. 459.
12 E. Husserl, *Cartesianische Meditationen, Husserliana*, 1: 48–65; English translation by D. Cairns (The Hague, 1970), pp. 7–26.
13 Ibid., *Husserliana*, 1: 64; English translation, p. 25.
14 Ibid., *Husserliana*, 1: 64; English translation, p. 25.
15 See also L. Landgrebe, 'The Problem of Passive Constitution,' in *Analecta Husserliana*, 7: 34. (Here Landgrebe suggests an absolute principle of individuation.)
16 E. Husserl, *Cartesianische Meditationen, Husserliana*, 1: 99–121; English translation, pp. 65–88.
17 Ibid., *Husserliana*, 1: 103–6; English translation, pp. 69–72.
18 E. Husserl, *Die Krisis, Husserliana*, 6: 156ff. (par. 43ff.); English translation by D. Carr (Evanston, 1970), pp. 154ff.
19 E. Husserl, *Cartesianische Meditationen, Husserliana*, 1: 117ff. English translation, pp. 83–84.

[20] L. Landgrebe, 'Husserl's Departure from Cartesianism,' in *The Phenomenology of Husserl*, ed. R. O. Elveton (Chicago, 1970).

[21] M. Merleau-Ponty, *Phénoménologie de la Perception* (Paris, 1945); 'Le Philosophe et son ombre,' in *Edmund Husserl 1859–1959*, Phaenomenologica, no. 4 (The Hague 1959), p. 195–220.

[22] E. Husserl, *Die Krisis, Husserliana*, vol. 6; see particularly p. 171: "Das Ergibt Eine Verständlichkeit, Die (Was Freilich Ein Idealfall ist) . . . "; English translation, p. 168.

[23] M. Merleau-Ponty, *Phénoménologie de la Perception*. 'Les Sciences de l'Homme et la Phénoménologie,' from the series *Courses de Sorbonne* (Paris, 1961); English translation by J. Wild (Evanston, 1964). See also A. De Waelhens, *Une philosophie de l'ambiguité* (Louvain, 1970).

[24] "Yet the soul may perhaps be wholly united with Him in the Mansions very near His presence, while thought remains in the outskirts of the castle, suffering the assaults of a thousand wild and venomous creatures and from this suffering winning merit" (Saint Teresa of Avila, *The Interior Castle*, trans. E. Allison Peers [London, 1974], p. 34).

[25] See *Husserliana*, 1: 117; English translation, p. 83.

[26] *Husserliana*, 6: 154–55; see also p. 147; English translation, pp. 152, and 145 ("In this total change of interest, carried out with a new consistency founded on a particular resolve of the will").

[27] See *Die Krisis, Husserliana*, 6: 158; English translation, p. 155.

[28] L. Landgrebe, 'La Phénoménologie de Husserl est-elle une Philosophie Transcendentale?' *Les Etudes Philosophiques* 9 (1954): 315–23.

MARY-ROSE BARRAL

BODY–SOUL–CONSCIOUSNESS INTEGRATION

Whether man's consciousness arises from his bodily condition or is an active entity in itself vis-à-vis the world and the body are questions of crucial importance to which many, and sometimes contradictory answers have been given. Science itself has contributed to both the confusion and the attempted elucidation of the problem.

Between the Platonic notion of soul as unique source of life and activity guiding the body, and the contemporary theories of body–soul–consciousness relation in the world, there are many hypotheses to examine and difficulties to overcome, in order to shed some light on the subject. One of these theories, based on the phenomenological method of Husserl, and developed by Merleau-Ponty in particular, provides new insights into the nature of man, his consciousness and his relations to the world of subjects and of things in the world.

The investigation of perception, the analysis of the human organic structure and of the intersubjective relation among subjects contribute to the elaboration of a theory of the body–soul–consciousness dialectic which may prove true to experience. Man's body is seen as the focal point of his consciousness vis-à-vis himself, the other, and the world, which is the field of all perceptive experience. The relation of body and soul is the necessary foundation of consciousness; man is in relation with others and with the world even before reflective thought arises: at the root of it all there is already a "lived experience" wherein soul and body are one.

This paper is intended to raise many questions and promote further discussion on the nature of the soul and on its relation to the "flesh" in every facet of human experience and personal expression. It also aims to show through reflection on Merleau-Ponty's existential interpretation of man's experience, how, within the body–soul–consciousness dialectic the integration of the human subject may come about and/or be lost depending on the conditions of the body in its relations to the soul and to others in the world.

Contemporary philosophy has been concerned with man's relation to the world as the necessary counterpart of consciousness. Whether man's

A-T. Tymieniecka (ed.), Analecta Husserliana, Vol. XVI, 119–125.

consciousness arises from his bodily condition, or is an entity in itself vis-à-vis the world and the body are questions which still baffle philosophers and to which many contradictory answers have been given. Science itself has contributed to both the confusion and the attempted elucidation of the problem. We have come a long way from the position of Plato, for whom the soul inhabited the body and was the sole source of consciousness and intellectual activity. We also question the view that the soul is closely related to the body, yet an entity in its own right, independent of the body in its existence, though not in its operations. This Aristotelian-Scholastic position is somewhat closer to the contemporary notion of body—soul—consciousness relation, but still challenged on the concept of a spiritual form.

How do contemporary philosophers differ in this respect? How does consciousness appear to them within the framework body-soul-world? Is there a general agreement of the notion of man's nature or are there only diverging views which make it impossible to formulate a definitive philosophy of man in the world?

No doubt, there are mostly diverging views and we are not close to a synthesis in the matter of man's being in the world. However, some contemporary philosophers have dealt with the problem in a more scientific way, and have come up with appreciably coherent answers. Phenomenologists have studied the question from a different point of view; among them, J. P. Sartre, M. Merleau-Ponty, Gabriel Marcel have investigated the body-soul relation and the emerging of consciousness in man in the world, existentially, rather than conceptually. In fact, Merleau-Ponty says: "Contemporary philosophy consists not in stringing concepts together, but in describing the mingling of consciousness with the world, its involvement in a body, and its coexistence with others . . . . "[1] The approach of the phenomenologist is clearly different from that of the traditional philosopher; it purposes to be a study without presuppositions, a search for the real nature of the being in question which is presumed to reveal itself to the inquiring consciousness. But here is the difficulty: if the real world reveals itself to consciousness, is not consciousness itself taken for granted when, in fact, it is precisely the 'why' and 'how' of consciousness which is in question? Not only the nature of consciousness, but also the notion of soul and its relation to body need to be investigated. In other words, the beginning of the philosophical inquiry is here in question.

Perhaps one should first consider the visible, the material, that is, the body in its behavior. This is where science and philosophy have a common point of departure. For Merleau-Ponty, as well as for Sartre, the beginning is the material bodily reality which appears to my consciousness, and from which I am to derive the truth about man and his world.

Merleau-Ponty first considers the fact of man's "lived experience" previous to all reflection and such that vital activities are happening at a pre-conscious and pre-personal level. In fact, much of his philosophy is aimed at discovering this profound stratum of human life from which, he thinks, the roots of conscious life arise. He is aware of the level of his own consciousness and of the cultural development at which he begins his investigation. For this reason, he aims at rediscovering "and understanding the relation of consciousness and nature: organic, psychological or even social."[2] The study of man's behavior will reveal to him the true nature of man as a conscious being aware of himself, of the world, and of others. It is his contention that "Our own body is in the world as the heart is in the organism: it keeps the visible spectacle constantly alive, it breaths life into it and sustains it inwardly, and with it forms a system."[3] But how does he come to the discovery of consciousness or of the soul by merely observing the behavioral aspects of the body? The burden of the investigation is this: is human behavior merely the response to certain stimuli at the physical level, or is it the carrying out of intentional acts? Is man's activity merely a sequence of organic functions or is it a structured pursuit arising from a conscious subject? In spite of the fact that Merleau-Ponty does base his investigation on the findings of science (e.g., physics, biology, psychology), he comes to the conclusion that both the purely physical and the purely biological approaches to man are wanting, and psychological theories need revision. He maintains that there is indeed a physical organism which receives impulses and reacts to them, but that this is not all there is to man's life. He conceives of reality as of a threefold dialectic: the physical, the vital, and the mental.

Physical nature in man is not subordinated to a vital principle, the organism does not conspire to actualize an idea, and the mental is not a motor principle *in* the body; but what we call nature is already consciousness of nature, what we call life is already consciousness of life, and what we call mental is still an object vis-à-vis consciousness.[4]

What is most significant is that these three orders are each a new structuration of the preceding one. Therefore, the physical is always present; consciousness is rooted in the materiality of the subordinated dialectics. There is no "pure consciousness" as spectator of the world. In another phase of his investigation, Merleau-Ponty reinterprets the notion of form: it is not a static essence, but one which will be defined according to the level of organization of the organism in question. Form or structure is operative at different levels: the syncretic form remains engaged in matter and is exemplified by those animal organisms which are tied to the instinctual and are, so to speak, imprisoned in the framework of the natural conditions. The revocable form represents a higher type of organism which may in some cases have a certain independence from matter, is capable of adaptation, and may even exhibit reactions which resemble what we call "ends". The symbolic form is that proper to man, wherein the "sign" becomes "symbol". This form is distinguished by the emergence of novelty, creativity, in the ends of conduct as well as in the human activity itself, activity which is from within, a quality arising from its own interior structure — not merely from an exterior impulse. This dimension is lacking to the animal, whereas man can express himself in a behavior which assesses values and truth, and can equate the intention and that which is intended.

But Merleau-Ponty has not thereby described three types of soul: he has merely pointed out that there are different degrees of integration of organisms, each carrying with it a certain level of operations and at one point at least, a certain degree of consciousness. Actually, he does not describe how consciousness arises in nature — or in man — but discovers *that it is*, because *man is*.

The notion of soul in Merleau-Ponty follows his study of consciousness. He does not accept traditional theories of soul. Since he has made clear that physical, vital, and mental are degrees of being of the individual, it follows that soul and body are not at all distinguished or distinguishable; there are different degrees of integration in man. In the total integration, soul and body are one. They might be somewhat distinguishable if the integration were to fail in some respect. But even the language "body and soul" would not be meaningful except to describe the condition of total disintegration, death, wherein body has lost its meaning. If one accepts *structure* as the fundamental reality, then it becomes clear that body and soul are a unity and their distinction can be understood in relation to the degree of integration of the individual. Although Merleau-Ponty denies that the soul can rule the body or that the body can influence the soul, he nevertheless admits that the integration of the individual is never absolute — in fact it constantly breaks down; "All integration presupposes the normal functioning of subordinated formations, which always demand their due."[5] But are not fatigue, hunger, etc. influences of the body on the soul? The difference is in this: that it is not the body but the dialectical relation of the various levels of structuration which fails. The notion of soul is relativized. Soul is not a substance, which one could always identify, even admitting it to be in a dialectical relation with the other structures. On the contrary, soul arises, so to speak, from the other structures: this soul is developmental, never the same, therefore it cannot be defined once and for all. In a passage which lends itself to a variety of interpretations, Merleau-Ponty describes the body-soul dialectic thus:

There is the body as mass of chemical components in interaction, the body as dialectic of living being and its biological milieu, and the body as dialectic of the social subject and his group; even all our habits are an impalpable body for the ego of each moment. Each of these degrees is soul with respect to the preceding one, body with respect to the following one. The body in general is an ensemble of paths already traced, of powers already constituted; the body is the acquired dialectical soil upon which a higher "formation" is accomplished, and the soul is the meaning which is then established.[6]

There is then no single relation between body and soul; better said, matter and spirit are not two substances which can be subsumed into a synthesis. What occurs here is the most complex dialectical interaction, wherein the relation is never constant, never predictable, never definable. The intricacy of the dialectic of *body and soul* at the different levels is a relation which *is not*, but *becomes* at each new moment, at each new stage; to study man and to attempt a philosophy is to engage into a constant beginning, because man is not a fixed nature, but a processual being.

This tantalizing study of the body-soul relation defines Merleau-Ponty as a quasi-materialist, since soul seems to be arising from the structures of the material organism. When he says the *soul* is the *meaning* which is attained, what does he signify? Can matter give meaning to itself? Or is consciousness a product of matter? Yet Merleau-Ponty also refers to spirit, and definitely rejects a mechanistic theory of man. But how can one reconcile the idea of a developmental soul with spirit — traditionally understood as a superior entity responsible for the intellectual and spiritual activities of man? Perhaps it is time to use new concepts to describe the human condition, and to truly understand how development is not synonymous with mechanism. If soul is a form, then it must — even in the language of the Scholastics — "inform the organism" in its totality; nothing prevents this form from being nonmaterial. Thus it may well be true that

... Neither scientific thematization nor objective thought can discover a single bodily function strictly independent of existential structures, or conversely a single 'spiritual' act which does not rest on a bodily infrastructure.[7]

However, these infrastructures are not autonomous in a solipsistic sense. Man's body is in contact with other bodies in the world; and this world is the ground or the foundation of all conscious life. How does the interrelation of man to man and man to world attain?

"Between my consciousness and my body, as I experience it, between this phenomenal body of mine and that of another as I see it from the outside, there exists an internal relation which causes the other to appear as the completion of the system."[8] But the system is not complete without the world, which is the generalized field of all my experiences, of all my perceptions — for my participation in the world and my relation to others is founded on perception. I do not primarily communicate through thought, but I exhibit my being and come to know the being of the other through perception. I come to an intersubjective relation with the other because I understand his bodily behavior: knowing that my consciousness has a body, I can readily assume that the other's body has a consciousness. But the relation is not that easy to attain, since I am an autonomous subject and the presence of the other is in a sense a menace to me. Sartre stresses this aspect of consciousness and comes to the conclusion that the other by his presence is depriving me of my reality, because he intrudes into my world and even deprives me of my subjectivity if by his gaze he makes me into an object.

It is striking how from the same basic notion of consciousness of man in the world, different conclusions can be reached. For Merleau-Ponty, the presence of the Other in my world comes to consolidate my reality, because the object I perceive and the other also perceives becomes an intersubjective object, just as his perception of my body confirms me as a being in the world. In a sense there is a loss of my own private world, but only for the enrichment of a common one which results from the coalescing of different views of the world. There are difficulties, no doubt, in this meeting of consciousnesses. The self is always holding on to its own world view and to its own being; another always appears as alien; however, experience also shows that much positive exchange between the self and the other goes to effectuate a subjective and intersubjective existence.

The need for an exchange even at the pre-personal level is absolute. Consciousness could not arise in a vacuum or in isolation. Both the world and the other are necessary factors for this fundamental human development. G. Marcel goes even a step further. He maintains that the relation of consciousness at the personal level must have an added dimension: besides the acknowledgment of the other in consciousness, the recognition of an Absolute *Thou* in the transcendental order is required for man to be fully man, so that not only is there a conscious exchange in the world, but there is also a spiritual dimension in man which transcends the world. Yet, even for Marcel, existentially the body is the root of being and of relation, so much so that he can say, "I am my body," rather than "I have a body," thus proclaiming the unity of man's being.

This discussion has not taken into account the human relation which attains at the intellectual level — the Cogito. Deliberately, the study has been limited to the role of consciousness at the pre-personal level or at the level of perception not yet transformed into reflective thought. The study of this aspect of human experience has been conducted by phenomenologists — particularly by Merleau-Ponty — in a strictly philosophical and scientific manner so that the findings can have a certain validity even from a scientific point of view. However, the whole theme, being treated phenomenologically, does not have the strength of a demonstration — which, at any rate, would be difficult to achieve when dealing which the human experience at the level of incipient consciousness. Phenomenology is a description of phenomena — in this case, the phenomenon of human consciousness in the world. The phenomenologist does not build a system, does not define and categorize. In presenting the results of his investigation, he can only hope that the reader — or hearer — may undergo a similar experience and discover for himself that same reality as manifested in consciousness.

Johns Hopkins University

NOTES

[1] *Sense and Non-Sense*, translated from the French by Hubert and Patricia Dreyfus (Evanston: Northwestern University Press, 1964), p. 59.
[2] *The Structure of Behavior*, translated from the French by Alden L. Fisher (Boston: Mass., Beacon Press, 1963), p. 3.
[3] *Phenomenology of Perception*, translated from the French by Colin Smith (New York: The Humanities Press, 1962), p. 203.
[4] *The Structure of Behavior*, p. 184.
[5] Ibid., p. 210.
[6] Ibid.
[7] M. Merleau-Ponty, *Phenomenology of Perception*, p. 431.
[8] Ibid., p. 352.

PART III

THE SOUL—BODY TERRITORY

ROMANO ROMANI

NATURAL MAN AND HIS SOUL

I

If I had to give a brief and synthetic description of every condition of human inadequacy in a situation, I should express myself by saying that there is uneasiness in the temporal dimension. The present becomes confused with the past, the future flattens on the present, past and present are fleeing toward the future. Our uneasiness makes us believe that our happiness or unhappiness is a mode of being of time and prevents us from seeing that time is one of our modes of existing, that is, of being in *relation with*. The more the contrasts between happiness and unhappiness, love and hate, life and death, become acute in a given man, in a given society, in a given epoch, the more will the problem of time become the central problem of that man, that society, that epoch. Temporality makes us all partial, and every mode of being finds in temporality the most profound evidence of its own partiality. Certainly, partiality is space-temporal, but temporally it imposes itself at emotional levels that are unattainable when the spatial representation is separated from the temporal one: the uneasiness in the temporal dimension manifests itself, first and foremost, as a scission of space and time. Space and time develop as two elements related in and by the possibility that living is becoming: the distinction between them is at a breaking point that time and time again cedes to the pressure of happening, subsequently to become transformed into a new and different relation of becoming.

II

We are living in an epoch of psychology. The human soul has never before been taken into such careful consideration, studied and "analyzed" as is the case in this historical period of skepticism vis-à-vis every religion and crisis of every form of religiousness. Never before have people been led to believe that the human soul is affected by a soothing and yet incurable and even mortal illness. And the anatomists of the soul, although denying the existence of neuroses, yet prescribe a quite incredible quantity of psychic drugs to the restless people who consult them. The existence of the soul is made evident

129

A-T. Tymieniecka (ed.), Analecta Husserliana, Vol. XVI, 129–151.
Copyright © 1983 by D. Reidel Publishing Company.

by suffering just as in the case of any other sick organ. The pomp and splendor of psychology, therefore, are not celebrated in an epoch of happiness and peace, of serenity and richness of human relations, but rather against the background of unhappiness and fear, of violent passions and irreducible contrasts, wars of domination, and political and economic oppressions. It is not the mode of being of psychology that determines the mode of being of the contemporary soul: quite the contrary, psychology presents itself as the systematic form of the manifestation of this soul in uneasiness and suffering. This constitutes both the validity and the limit of the present-day *science of the soul*. The systematic character of psychology derives from its functioning as an administrative science in which the apodicticity and absoluteness of the cognitive foundation is not a *theoretical problem* but rather an *organizational fact*. In a psychology of this kind one can conceive of an increase of cognitions, but not a change of theoretical outlook, a different manner in which knowledge grows, a different relationship it may bear to the concrete subjective life of individuals and of society. A psychologist may conceive many different ways of theorizing in psychology, but only a single way of "knowing" and administrating the souls, that of the organization to which he belongs. As a *natural* science, indeed, psychology consists of the psychologist's feeling of belonging to the organization in which he works as the *administrator of souls*: this organization constitutes the insurmoutable limit of all the theoretical and practical operations of its members.

III

The transcendentality of thought is one thing, the incorporeity of the soul is quite another. Without this distinction it would be quite impossible to arrive at a philosophically useful reading of Husserl; but this distinction, which pervades the whole of Husserl's work as a problem to be formulated, is fraught with such a density of historico-theoretical connections as to be altogether discouraging. Here I shall therefore limit myself to some brief reflections that I nevertheless consider to be very important.

The conditions of subjective possibility that psychology may become a science are the same as those of any other science, but the object with which it concerns itself has a distressing characteristic: the impossibility of being perceived with the corporeal senses, and this no matter how perfect they might be and quite independently of any technical equipment (even the most advanced) they might call to their aid.

In this respect Pavlov cannot be less "spiritualist" than Jung. In the logic of the natural sciences the fact that the soul cannot be physically perceived is a problem that is inherent in the specificity of the method of psychology, that is, simply relative to the object and therefore such as not to call into question the logico-philosophical principles of the natural sciences themselves. But if the physical imperceptibility of the soul is responsible for its noncorporeity as far as the contemporary psychologist is concerned, for Homer, Virgil, and Dante, as also for a certain magico-religious thought (of which documents can easily be found in both the historical and the anthropological dimension of cultural experience), it is the noncorporeity of the soul that decides its physical imperceptibility:[1] with respect to this magico-religious conception, the only additional thing that is offered by the animist conception of contemporary psychology is the determinism of the cognitive methods bound up with the administrative needs peculiar of the present social organization. In it, in fact, the scission of body and soul not only remains and becomes aggravated, but the prohibition of all critical inquiry into the historico-theoretical conditions of this scission is rendered peremptory not by a theoretically possible scientific legality (in becoming), but rather by an institutionally necessary scientific legalism (factual, become). In the scientific institution, thus, the method of psychology has an effective stronghold, and of this stronghold it becomes the organ of social control.

IV

The existence of psychology as a science, however, cannot be reduced to its administrative function: in its administrative function, rather, there emerges with ever greater clarity the crisis of the scientific method that justifies it. Here we have the paradox that (precisely on the basis of psychology) opens the road to the problem of knowledge. And therefore psychologism is merely shown up and not instituted by the coming into being of psychology as a specific sector of the natural sciences: rather, it presents itself as an obstacle on the road of scientific progress in psychology, where the theoretical effort, far from causing an advancement of knowledge, runs the risk of becoming shipwrecked in a babel of languages that are mutually impenetrable, each incompatible with the other.

Philosophy begins at the point where there ends the sign of certainty that wants to seem the unity of knowledge. And it would therefore seem to be the task of philosophers to recompose this unity, to restore to men certainties more solid than those that have gone into crisis, and philosophy is even said to be in search of a certainty that will never become involved in a crisis. The thought of Husserl undoubtedly forms part of this tradition that sees philosophy as the unifying element and the founding activity of knowledge. Nevertheless, Husserl is not a systematic thinker, in him the founding intention does not coincide with the system, the unity of knowledge does not consist of a given (factual) knowledge, but rather of a possible (transcendental, critical) knowledge.

V

Fifty years of philosophical reflection had not passed in vain for Husserl if the dark and tragic years of Nazism permitted him to establish a limpid relationship between "the crisis of European humanity and philosophy" and "the crisis of the European sciences and psychology."

The spreading of the irrational with its terrifying violence does not call from the philosopher any useless moral condemnations, but rather induces him to serenely repropose rationality with everything vital, pacific, and projectual that rationality involves. His warning is calm and lucid: "*Blosse Tatsachenwissenschaften machen blosse Tatsachenmenschen.*"[2] In its form of "Ausdruck der radikalen Lebenskrisis des europäischen Menschentums,"[3] the crisis of the natural sciences commences with their expansion and then with an ever-increasing insufficiency of *critical rationality* (an ever-more ponderous and violent intervention of *factual reason*) in making the idea of the expansion of their use coincide with the idea of progress *sic et simpliciter*: "Die Ausschliesslichkeit in welcher sich in der zweiten Hälfte des 19. Jahrhunderts die ganze Weltanschauung des modernen Menschen von den positiven Wissenschaften bestimmen und von der ihr verdankten 'prosperity' blenden liess, bedeutete ein gleichgültiges Sichabkehren von den Fragen, die für ein echtes Menschentum die entscheidenden sind."[4] It is not a few problems but rather a manner of problematizing that recedes before the splendid patina of success that envelops the *positive* sciences. It is this problematizing, which has a history of its own in the complex and articulate process by which philosophical thought (or a philosophical thought) becomes specific and original, that decides whether man has the chance of being authentically human, that is, of going beyond the factuality of a mode of being enclosed in the "natural" definitiveness of a space and a time that only in relation to this possibility of going beyond are turned into the determinateness of historical time and geographical space.

The fact that the positive sciences, and physics first among them, leave the ambit of philosophy immediately appears in the modern era as a stepping out of the historical relativity of the theories.[5] The accumulation of observations of facts, the modalities of development by means of which they become imprisoned in mathematically expressed laws, seemed to provide a solid alternative to the successive verification of philosophical conceptions that had always been believed to be definitive, but subsequently — each in its turn — were thrown into crisis and left behind. It is nevertheless useful to distinguish the aversion for the authority of the philosophers from the aversion for the nondefinitiveness of metaphysics: the critique of the uncertainty of the various metaphysics, their relativity stigmatized as unfoundedness, is peculiar to the refounders of metaphysics but not to the founders of science — and this by virtue of the fact that a metaphysical conception was the premise of the founders of science, a metaphysical conception being here understood as a mode of being in relation with society and the world that more or less clearly corresponds to an idea of them that grew up together with mythico-religious language and/or was elaborated by philosophical thought. The problem of the metaphysical premises of the positive sciences is not the same thing as the problem of the relationship between the positive method of the sciences and the *metaphysical sensitivity* of the philosophers and the scientists. It is functional for positivistic psychologism to speak of "metaphysical sensitivity" in order to detach a "scientific" theory from the metaphysical premises that permit it to be critically considered in the logico-transcendental dimension of thought. And yet, is not this psychologism, in turn, merely a determined aspect of a metaphysical attitude, but one that is far more difficult to pinpoint inasmuch as its prejudicial refusal of the legitimacy of science (which it considers to be metaphysical) places it outside the reach of critical inquiry?

VI

In the experimental sciences the perception of the senses is a *positive criterion of truth* not because it is an element of the objective experience of the external world, but rather because it is the sign of a single and unknown objective reality. The great esteem in which Galilean experimentalism seems to hold the experience of the senses is nothing but the *reduction of this experience to a function of scientific theory*. The occurrence of an event in known or prearranged conditions is recognized as objectively true only in relation (positively or negatively) to a theory recognized as scientific, and this inasmuch as it is only through such theories that some light (however pale and uncertain) is thrown on the single factual reality. The abstractness of the positive sciences increasingly imposes itself as a situation where the complex abstractness of the facts replaces the simple evidence of things.

On the unsurveyed and uninquired assumption of the unity of reality, indeed, the facts become counterposed to the theories in such a way that neither the former nor the latter can be taken as a certain criterion of truth: in vain the so-called philosophies of science seek to present the disintegration of the criteria of truth as logical tolerance. On the contrary, the more theoretical uncertainty invades the field of scientific operation, the more the empty legalism of the scientific institution (of which the epistemologists have been elevated to the rank of poet laureates) behaves as if the best critique of science were the complete justification of its present mode of being.

The separateness of the necessity of the real and the possibility of the existing[6] is not a peculiar and exclusive feature of Western philosophy: but the history of Western thought consists of the effort of giving account, based on the awareness of the peculiar dimension of possibility, of the necessity of the real. While the thought of a given philosopher in its systematic unity endeavored to found the necessity of the real, philosophical culture quenched and attenuated this pretension by shrouding it in a problematical activity constitutive of a critical historicity rather than of a compresence (ingenuous but nevertheless tolerant) of doctrines. But this navigation in the possibility of the existing nevertheless constitutes the preoccupation of all those philosophers who continue (have continued) to regard the foundation of the necessity of the real as the foremost aspiration of their own philosophizing: all the more so as the spirit, denier of the necessary real as the scientific aspiration of philosophy, perorates the cause of skepticism and therefore also the negation of the possibility of knowing (which comprises knowledge as a possibility). And in this it would seem that skepticism and stoicism converge into an indissoluble thematic relationship when they consider a critique not as a transformative aspect, but rather as an element that destroys the possibility of knowing (or destroys a knowledge thought possible and actual before the critique). When one looks not only at the contents but also at the history of Western philosophy, one becomes inclined to think that the only thing capable of bringing it to the end of its road would be a particular form of thought that, without repudiating its skeptical spirit, succeeded in realizing its systematic tendency: it is to this task that Hegel dedicated the efforts of his exceptional genius. But what is the sense of this great effort and, above all, what was its outcome?[7]

It is in its factual givenness that the experimental method of Galilean physics imposes itself as the universal method of the sciences: the positivist attitude is not a kind of empirical concreteness, a closeness of science to the most immediately sensitive and common mode of considering the things around us. Rather, historically it comes into being first as the philosophical adjustment (and subsequently as the logical, sociological, and psychological justification) of the Galilean method as the legal form of science and of "doing" science in Western society. Within the positivist attitude, indeed, positivism and neo-positivism distinguish themselves by virtue of their historical function, which was that of bringing philosophical culture to full awareness of the character of a historico-social movement of thought that was associated with the Galilean experimental method. It is the level of this awareness that governs the possibility of a philosophical critique of the foundations of the positive sciences. When it came into being, the Galilean experimental method immediately began to struggle for its factual ambit of scientificness against the religious authorities: this struggle did not take place at the level of ideas but rather at the level of facts, and the ground for the factual open-mindedness of Galilean physics had already been prepared by the political thought of Machiavelli. As far as Machiavelli is concerned, the de facto exercise of political power precedes each and every religious, ethical, or philosophical justification of that power: from Machiavelli onward (and therefore long before the Reformation and the wars between the Catholics and the Protestants) European Renaissance thought ceases to have its factual ambit of scientific legality in religion, and in this manner there comes into being the space for a possible legality of the sciences aware of their own purely factual validity. For centuries the authoritarian defense of *true Christianity* and the spreading of *authentic Christianity* by means of wars had reduced religion to a simple form of knowledge (since faith consisted of believing in a particular teaching) and truth to a simple authority and, consequently, to a pure fact. The bare brutality of the facts drives the critical spirit, humiliated and impoverished by the *cruda realitas* that surrounds it, to seek the law that governs the succession of these facts and to abide by this law: and the facts *verify* the scientificness of a theory inasmuch as they are a measure of the truth, and the truths that people seek are factual truths. Indeed, factuality and truth coincide.

The law underlying the succession of events does not change its grey remoteness from the quality of things, from the adjustment of things to the mathematical unidimensionality of quantity: and the quantitative unidimensionality of the measurable phenomena becomes possible only as a result of the historical need of adjusting to the law of happening. Husserl's research into the constitutive premises of the Galilean mathematization of the universe (part 2 of *Crisis*), as a critical activity aimed at calling into question the reduction of the foundation of scientific legality to the mere use of the mathematical instrument, admirably highlights the exquisitely logico-transcendental origin of this instrument, but fails to inquire into the metaphysical premises that make its legalistic absolutization possible. Galilean science is not the simple mutilation of the Cartesian metaphysical project; between Descartes and Galileo there are more distant and profound bonds, but also more distant and profound divergences.

The Neoplatonism and the neo-Pythagorism of the Renaissance period are quite incomprehensible, at least as regards their outcome and their historical influence, if one overlooks the fact that in them there comes into being a concept whose complex vicissitudes are representative of the subsequent developments of philosophical thought: I am referring to the concept of the absolute.

In its originary purity, from which every definite quality attains its *form*, the *archè* of the Greeks, including the Platonic world of ideas and the Aristotelean prime mover, is the vital generative principle of the universe. Becoming is made necessary by a vital principle that *actively negates* (denies) the dissolution of dead inertia. And the finiteness of Parmenidean being is a conclusive *form* of life that abhors the infinity of space and the consequent linearity of time inasmuch as forms appear to be in contrast with the space-temporality of the entia, with their moving *freely* in space and their *free* becoming in time. The Parmidean being is space-temporally circular inasmuch as it is space-temporally free, is space-temporally finite inasmuch as it is space temporally alive. But, having denied the separate appearance of the infinite multiplicity of the spatial forms in the inexorably factual laws of linear time (the contrary of the freedom of the finite becoming of living beings), the mythical Parmenidean disposition of knowing for the purpose of founding would remain empty and suspended in the bewilderment of a disorientated existence (for every philosopher pays for the eruptive and destabilizing vitality of his thought with an insidious desire for quiet and security), without conferring upon the being those attributes of unity, immutability, and immobility that, making it the object of a definitive knowledge and therefore "naturally" unique and separate, also render it — in homogeneity with infinite space and linear time — the very contrary of freedom and of life. For Parmenides, therefore, the universality of truth always coincides with the uniqueness of the true: the mythico-naturalistic scission-contraposition of cultures, religions, philosophies, and views of the world that differ from each other consists precisely of this making the universality of truth coincide with the uniqueness of the true. The contrast between the ingenuity of the mythico-natural attitude and the profundity of critical thought turns the Parmenidean aporia into something shrill and strident, so strident as to become explicit awareness of a cultural uneasiness widely diffused in Greece at the time of the crisis of the myth and, much later, also in the West, contradictorily both in expansion and in crisis. The fact that our culture characterizes itself as Western was made possible by the uneasiness that the explosion of the explicit awareness of the Parmenidean aporia has provoked.

In the course of the history of our civilization the Parmenidean aporia has become a paradox far more enigmatic than those that were so dear to Zeno of Elea. The development of the conceptual elaboration, in fact, throws ever sharper light on the mythico-natural attitude but at the same time also inures it, and introduces into Western civilization more scission and conflict, both within and without. Given the uneasiness provoked by the inadequacy of the rationality of our civilization, the weaker spirits thus put the blame on rationality, while the obtuse reject the patient and vital critique of its forms. The former content themselves with an empty aestheticism, the latter agitate themselves in an irresponsible and barbarically menacing triumphalism.

The ingenuous endeavor to make the universality of truth coincide with the uniqueness and the unity of a true content of verbal communication is a consequence of the equally ingenuous illusion that truth, which moves freely in the formal life of language, can be imprisoned in a definitive communicated content. In this way one ends up by mistaking every paralysis of cognitive activity for *certainty of knowledge*: this equivocation peculiar to the mythical attitude of *natural man* gives rise to the the fictitious reasonableness of violence, to force gaining the upper hand, and to rationality having neither force nor a form of its own. The negation of the mythical ingenuity consists of gradually overcoming the attitude that determines it. The result of this overcoming process will not be the falling of a veil, but rather that transcendental thought will assume the fullness of its form. In fact, the equivocation, the illusion, and the ingenuity of the mythico-natural attitude do not constitute a content of error as compared with a content of truth, but rather *the insufficiency of the truth with respect to the overcoming of a de facto situation, with respect to a factuality to be overcome*: here we have a deficiency of vital form, because I here understand truth as the socially universal form of human living, and transcendental thought in its formal fullness as the condition that enables each man and all men to live humanly.

The great adventure of Western thought did not have in Parmenides the

origin of a presumed decadence or decline, but rather the starting point of
a new phase in the slow and far more ancient trend of human thought toward
the attainment of its own vital fullness. As regards this trend and the effort
of overcoming the mythico-natural attitude that it implies, Parmenides
represents the opening of a powerful discussion that, always at the level of
awareness that the various epochs and societies permit, involves the mode of
being and the destiny of the whole of mankind. The acquisitions that have
been made in the course of this discussion do not by themselves represent
anything superior, either quantitatively or qualitatively, to the acquisitions
of content of other philosophical or religious cultures. Rather, it is precisely
a presumption of this kind associated with the triumphalist barbarisms that
runs the risk of destroying (and in many cases have already done so) the
wealth of truths offered by the variety and profound complexity of human
cultures: this risk does not derive from the philosophical critique that has
turned European civilization into a pole of universal attraction, but simply
from the facts that have made it dominant.

Parmenidean being negates the infinite multiplicity of appearance, while
cognitively it affirms its own finite unity: the relatedness in the vital negation
of possibility and necessity ($\tau\grave{o}$ $\gamma\grave{a}\rho$ $a\vec{v}\tau\grave{o}$ $\nu o\epsilon\hat{\iota}\nu$ $\epsilon\sigma\tau\acute{\iota}\nu$ $\tau\epsilon$ $\kappa a\grave{\iota}$ $\epsilon\vec{\iota}\nu a\iota$)[8] becomes
a scission of appearance and being in the mythico-natural disposition for
seeking in knowledge the acquisition of a definitive content of truth: a
thought that, coinciding with being, denies the being of not-being does not
have the same dynamics as a thought that, coinciding with being, seeks
simply to reduce this coincidence to pure knowledge. In the vital negative
perspective, like life-giving water from a limpid spring, possibility-necessity
flows forth from every living self to the other-from-oneself: as content of the
unique truth, on the other hand, being reduces itself to a pure necessity that
paralyzes knowledge in a vitreous transparency and excludes possibility,
becoming from the real and cognitive sphere, excludes life in its ever-different
and passionate vicissitudes. In this way mythico-natural man *believes* himself
to remain firmly anchored in life without death, whereas each passing day he
really *chooses* death without life.

The unity of Parmenidean being implies the scission of being and appearance, of a universe of pure necessity and a universe of pure possibility, unity
and multiplicity, physicalness and psychicalness. But Parmenides did not
invent this scission, for it is already present in the mythical world of the
scission of the sacred and the profane: Parmenides deduced the critique of
the sacred in Greek culture from Xenophanes and turned this critique into a
problem that is no longer a simply religious one, but rather the starting point

for a new way of philosophizing, of conceiving philosophy. The ambiguity and the richness of this mode of philosophizing are wholly lost upon anyone who fails to see it in relation to the ambiguity of Greek myth in particular and of myth in general. When it is seen in the light of this connection with myth, on the other hand, Western philosophy as a whole (and at its origins the Eleatic school represented but one of its problematic articulations, albeit an important one) can be seen as a powerful effort to issue from the natural dimension of mythical language, but without losing — be it noted — the profundity and the richness, the vital pregnancy that language assumes in the modes and the forms of the myth.

In the natural dimension of mythical language, indeed, the possibility of making an overall and universal distinction between being and appearing, between physicalness and psychicalness, immobile unity of necessity and inarrestable fluidity of the infinite multiplicity of possibility, comes into being in (and not from) the factuality of a previous scission with respect to which human language-thought is transcendental. It is this scission that myth unveils (but at the same time adumbrates and covers) in delineating the separation of the sacred and the profane: it unveils inasmuch as it is in myth that the sacred has a statute of its own, a governing law, and also limits that are projected into accurately measured spaces and times; it adumbrates and covers inasmuch, in the various forms of myth to which the transcendental possibility of language has given rise, there does not emerge a unique validity of significant contents, and the only thing that appears as universal — in the various forms of its living — is precisely that transcendental possibility that turns a tragically ontological scission into a logical distinction. The scission that thought-language limits, and also transcendentally negates and transforms into logical distinction, constitutes the factual origin of the dualism that dominates Western philosophy, just as it does all the human civilizations. Western philosophy, rather, has as its origins the problematization of the scission that has permitted, and still permits, the thematization of the dualism into critical thought, that is, into a new evolutionary process of thought that opens the road to the overcoming not of one or several dualist or monist conceptions, but to the overcoming of the existential scission that man, held fast as he is on the sandbanks of the mythico-natural attitude, does not even dream of winning.

It is truly astonishing how great an influence the mythico-natural heritage of Parmenidean thought has exerted on Western philosophy and how, even today, one "returns to Parmenides" to underscore one's own participation in this tradition of the insufficient utilization of Parmenidean thinking. It is true that Plato and Aristoteles make use of the Parmenidean critique, but they consider as part and parcel of the thought of Parmenides everything in his work that contradicts this critique. The Parmenidean contradiction thus develops into the contradiction of Western metaphysics, while in the background there appears the Socratic "know that thou doest not know" as an isolated, continuous, and heartfelt appeal to bring the whole of this contradiction back into discussion.

The Parmenidean approach has two levels, i.e., a logico-critical one and a logico-mythical one.

The Parmenidean distinction between *alethèia* and *doxa* belongs to the logico-critical level. Here we are concerned with two words or, better, with two modes of living that produce the same awareness of one's own universe of life. Awarenesses fractionated (and also fractionating) into the infinite multiplicity of the world of opinions produces the people who speak of the not-being that is and of the being that is not. Inasmuch as one denies the being of not-being, the fractionation that is, the factual scission, the dualism and the dualisms "thinking and being" and "knowing and being" become one and the same thing; thinking is a form of living (production of transcendental forms and transcendentality of forms of life), continuous, uninterrupted production of truth that is life and of life that is truth. The continuity-activity, which is also determinateness of a form of thought and of thought as such, denies the formless infinity of something that is "contrary" to it in just the same way as an individual, social, or specific form of life, and indeed living in the determinateness of its being and giving form, denies actively and continuously the formless infinity of not having form and of not being continuity-activity. And even though the multiplicity of the forms of life bears witness to the discontinuity of the existence of each of these forms and also to the existence of the discontinuity, the discontinuity is denied as *not-being* inasmuch as *being a form* of life is impossible without its being continuity and therefore originariness, and being a form of thought is impossible without its coinciding with the originariness of the continuity and therefore without its having an origin in the originary, in life itself. In other words, everything is vitally and noetically impossible that, not having its origin in life, opposes itself to life as its contrary, the opposite of life as being and of being as life. But it is precisely by actively denying (negating) its contrary that *life manifests itself* as a form of possibility, makes death exist as an element of living, and in human thought makes it possible to conceive of impossibility as that on which possibility confers transcendental form by denying it. And all this because being is to be considered not only as it is, but also as it manifests itself; and yet the manifesting of being has nothing to do with its being given in knowledge as an object (as even Parmenides is led to think with his mythico-natural attitude), but rather has to do with its becoming possible and therefore with its being that is, at one and the same time, originariness of life and possibility of living. In its logico-critical aspect, therefore, the thought of Parmenides opens the road not to knowledge of life, but rather to a new awareness (developed only in and from the Socratic "know that thou doest not know") of the mode in which thought is in the process of becoming and becomes a form of being and, consequently, the negation of the being of not-being.

Just as in the unfolding of its possible forms life gives form to the impossibility of death (which it continuously overcomes), so thought in its being-becoming is a transcendental possibility (transcendental form of the possibility that life is) and in the transcendental possibility of its becoming gives the form of thought to the discontinuity-impossibility (death) beyond which it continually flourishes. This form, the not-being that is not, in the natural situation of mythical man constitutes the content of a necessary, absolute knowledge that breaks up the universe and hides the possibility of the true.

At the logico-mythical level, on the other hand, Parmenides considers being as the necessary object of unique knowledge, and unique knowledge as necessary to the unique object: in this way there completely disappears the dimension of the possibility and he affirms precisely that not-being *is*. The thought of Parmenides is made "necessarily" aporetic by the mythico-natural attitude. And it is this aporeticity that makes one think of Plato, as well as Aristotle, and to hold that in some way or other not-being is and thus turns the noetico-vital, critical negation of the being of not-being into an impenetrable mystery.

Isolated from the critical metaphor, indeed, the coincidence of thinking with being becomes idealism: it is not thought that is a transcendental form of life, but life that is a corruptible form of thought. In both Plato and Aristotle the affirmation of the being of not-being assumes the spatial image of a scission of the base from the lofty, a world of the immobile incorruptibility of the forms and a world of the corruptible mobility of matter. In both Plato and Aristotle the scission of body and soul, of consciousness and life, becomes fixed in the Western tradition of thought at the level of a more or less refined reelaboration and purification of the myths and the mystery practices that formed a part of Greek culture. The specific task that the philosophic reflection of Plato and Aristotle assumed in that culture so richly endowed with myths was rather that of pushing being as the object of necessary knowledge and necessary object of knowledge into the far distance, into the empirical, into the immobility beyond the heavens of the fixed stars. In being removed and petrified into the immobility of the necessary, however, being assumes a prevalently legislative function and loses its prevalently generative one: in other words, it loses its vital aspect as creator of ends and origin of liberty-possibility and assumes the aspect of the impenetrable origin of the emanation of laws. The problem of this double face of being, however, does not admit of either a historical or a hermeneutical illustration: it is rooted in the sacral form of the myth and the mythical form of sacrality and will continue to be so rooted until the logico-historical and logico-existential situation will not evolve toward a logically and biologically fuller form of the human.

In Greek philosophy the search for the *archè* presents itself as a vital need expressed in the terms of a cognitive problem. To deny death as principle and the principle of death is already a noetic activity in the purely mythical dimension of language, but the rational depth of philosophical research no longer requires that direct experience should reveal the presence of what is sought and the absence of what is refused: in the internal rigor of thought philosophers seek guarantees concerning the vital order of the world. In the absence of this evolution from the fluid loquaciousness of the myth to the toilsome and fatigued asperity of philosophical critique (no matter how slow and uncertain this evolution may be in actual practice) it would not be possible to highlight the condition in which the sacred in myth is at one and the same time the fount of life and the harbinger of death.

The scission of the sacred and the profane, inside and outside, reality and appearance, heaven and earth, *soul and body*, is the sign of the *hybris* in human being and thought is the road for overcoming this *hybris*: this is more or less what philosophizing must have meant to the Greeks. The fact that Greek philosophy subsequently reproposed the dualism of the myth as an attained truth is not a failure, but only the inevitability of certain roads having to be traveled right to their end.

VII

As far as Parmenides is concerned, the possibility that being may give itself in knowledge sanctions the impossibility of it giving itself in becoming, which latter must therefore be considered appearance and "not-being." What makes it impossible for being to become is, as far as Parmenides is concerned, the birth and death that becoming implies and which would be in contradiction with the fact that being must originarily (and therefore also definitively) have been living being. But the only way in which being can recognize itself as living is precisely that of *becoming*, at least as transcendental thought, the negation of the being of not-being or, in other words, of giving form (at least in thought) to the relativity of a contrary of itself: the impossibility of the disappearance of a content of truth is in contrast with the life (truth-possibility) of thought, just as the impossibility of the death of a singular being is in contrast with the possibility of being of being, that is, in contrast with living manifesting itself as the originariness of the possible and as the possibility of the originary.

On the logico-critical level, therefore, Parmenides discovered the originari-ness-universality of being as the originariness and universality of life, but on the logico-mythical level he only succeeded in thinking of being as of an ens that does not die; in this way he replaced the universality of being by the particularity of an ens. Given its particularity of an ens that does not die, being is no longer originariness-possibility (universality) of life, but necessity-impossibility of the absoluteness of a particular form of life. Platonic idealism and Aristotelian teleologism confer this chrism of abso-luteness (scission of being and not-being and consequent emergence in se and per se of not-being as an absolute) upon the vital form of transcendental thought. In separating this vital form as an absolute from all the others, both Plato and Aristotle remain impotent (as men immersed in the scission of the myth) vis-à-vis the task of rediscovering in the logico-transcendental dimension the fullness of that new modality of participation in the liberty-possibility of the universal that man had begun to lose and has continued to lose more and more in the dimension of concrete biological immediateness.

The participation in the liberty-possibility of universal being, therefore, can develop only as the transcendental negation of belonging to the unity of a single whole, that is, of the elevation of the particular to the universal that − like a curse and malediction − marks the split or separated condition of natural man. From time to time, therefore, in mythico-natural individuals, as also in societies, this lack of participation in the universality of being bursts out into the pretended totality (infinite extension in space and in time) of a particular mode of thinking and of individual and social being. Conformity with this one and only mode of thinking and of individual and social being constitutes for natural man a substitute for participation in the universality of being in the same proportion in which this participation is lacking; for this reason, therefore, the mythico-natural societies are armed against each other and tend to reduce the multiplicity of the modes of social living to a single one, just as they reduce to one the modes of thinking and even the modes of living and dying of their single members. For this reason, too, natural men are in a state of enmity with each other and succeed in maintaining relations with each other only (or predominantly) by means of common objects of faith (credence or belief) or common objects of interest (survival), and the unifying role of these objects is far more important than any faith and any interest. These objects of faith and of interest constitute the *reason* (ratio) of every mythico-natural society, a *factual reason* that must not be confused with the *transcendental rationality* that went into producing the rational form of the objects in question.

As far as its contents are concerned, a myth is an object of faith (credence), but these contents are enclosed within that transcendental form of rationality that is found to be more or less wanting or inadequate in development as compared with the need for transferring into the logico-transcendental dimension that tantum of participation in the universal that an individual or a human society has lost in the dimension of immediate biological concreteness. The contents of the myth, therefore, constitute the improper unifying element of the mythico-natural human communities, while the transcendental rationality that has elaborated these contents (and continues to elaborate and delimit them) constitutes the socializing element that attenuates the destructive consequences brought about by the infinity (nonfiniteness) of the split unification into a new possibility of participation in vital liberty. It is in the area of the sacred that mythico-natural man attains the unity of his own social community, always provided that he maintains the sacred within its logico-spatial limits. Outside and in the absence of sacrality, there is only chaos and scission in mythico-natural society, nothing but incoherent multiplicity; but even the sacred has at its center an abyss that calls for great vigilance, because it readily trespasses into something that natural man fears even more than individual death itself. Not-being that is not nevertheless hangs over both the sacred and the profane like a menacing shadow, makes itself felt in both absolute unification and in absolute scission, because the one just like the other is nothing but an aspect of man's incapacity of attaining his own vital finiteness, of participating in the universal, of really attaining the human.

VIII

Even though it remained within the bounds of the mythical attitude, Greek thought received and adopted the critical example of Parmenides and Socrates; both Plato and Aristotle worked on problematically open systems, and the absolutization of the transcendental was anything but obvious for them (both of them saw the being of not-being as the impossibility of thinking of not-being *independently* of being) and necessity descends (derives) from being as a principle of vital activity, the *anànche* is the contrary of the *mòira*. Possibility, nevertheless, remains the contrary of necessity and even if fate (inasmuch as it involves something that is ineluctable and passive) is the contrary of the necessity that being emanates as a form of the vital activity of necessity, fate itself is but an aspect.

In searching for the *archè*, the Greek philosophers began to overcome and also to relive the contradictoriness of the myth which, even though in its rational form is the product of the activity of transcendental thought, yet has a *sacral content* because the men who think it and live it "mythically" glimpse in it and seek in it (but also fear it and imprison it) the split unification, the self-contradictoriness of the impossible, the substitution of the particular for the universal, the not-being of which they say that it is, the void that within the fragile and concrete mobility of their own corporeal form, amid the changing of the seasons and the birth and the perishing of the entia all around, they believe to be indestructible and divine. Even the infinite space of Leucippus and Democritus with its infinite number of impalpable and invisible atoms in motion has an absolute of its own that regulates it and a rule of its own that is the absolute, and the inattainability (or otherwise) of knowledge is a question that concerns this sacred-absolute in its *existence for man*, as is demonstrated by the century-old dispute between skeptical religiosity and stoic religiosity.

In the fifteenth century Nicolaus of Cusa arrived at a new clarity in his awareness that the absolute as given in its elusiveness and provisoriousness is the condition that makes the attitude of mythical man toward the world a purely and simply cognitive (and not critical) one, just as it makes the soul of natural man split both in knowledge and from knowledge. Nicolaus arrives at the extreme consequences of the mythico-natural attitude, but nevertheless shares it and therefore once again encloses the absolute in the crystal of a suffered and profound rational activity: as far as Nicolaus is concerned, there is no form of continuity whatsoever between God and the world, but only scission, a scission that with respect to the world implies a single and absolute law of necessity and with respect to God a unique and absolute freedom from the law. The very atoms of Democritus were in their own way a living thing (each atom is in itself "continuous" and therefore live) whose law of

necessity corresponded only partially to the attitude of natural man toward
the absolute and must therefore be considered rather as the relationship
that the mythical need of arriving at a unique truth, of giving itself a unique
reason for what is happening, still maintained with the sacred. For Cusanus,
indeed, becoming is directly and completely identical with happening, happening with becoming (they are but two aspects of the selfsame imperfection
of the universe) and what is happening is a law that has been established from
all eternity: everything has already happened in God's decision and man
knows but a few glimpses of this unique and indivisible reality that transcends
the universe; the imperfection and fragmentary nature of impotent human
knowledge is but the mirror image of that spatiality to which man belongs
and in whose existence, which is already by itself a subjection to an obligating
law, all becoming resolves itself. The absolute unknowability and transcendentality of God guarantees, as far as Nicolaus of Cusa is concerned, the
knowability-rationality of the world, in the sense that he identifies the
rationality of becoming not with the possibility-liberty of living, but rather
with the inevitable necessity of dying; for Cusanus, therefore, all life is
outside the universe and in God and all death is in the universe and outside
God, but the necessity that dominates the universe, the death that holds it
in its grip, is a divine law.

Even though it has now emerged in all its impenetrable extraneousness,
the absolute still has two faces: the one that is turned toward man inspires
terror, while the other, which radiates the warm light of life, is turned elsewhere, just like the other face of the moon.[9]

IX

The scission that we find in Cusanus between God and the universe is the
same as that between *res cogitans* and *res extensa* in Descartes, and I think
that it was the condition that brought about the birth of the experimental
conception of Galilean science. This attitude, the consequence of a peak
in the awareness of how far the living of mythico-natural man is removed
from participation in the universality of being, sees scientific activity of
an experimental character, i.e., in keeping with the *truth of the facts*, as
a means of conquering some autonomous space (no matter how small it
may be) for the human hope of achieving liberation through knowledge; but
experimental science has as its foundation the principle of the necessity
of the laws of nature (becoming conceived as a succession of events or

happenings) that denies all possible liberty to each and every nature singular being, man included.

To conquer a space of liberty in the world of necessity, but at the same time to recognize this world as the only one that is really given, existing, and truly dominated by the inexorable and unique law of happening — this was the mythically self-contradictory intention that the founders of the modern science of nature were obliged to pursue by the violent crisis that European humanity was then passing through as regards both its religio-philosophical language and its political, economic, and civil life.

The Cartesian *cogito* does not therefore lie at the origin of the scission between *res cogitans* and *res extensa*: quite the contrary, this scission already existed in Cusanus, and the problem that Descartes had to face was that of calling back into discussion the necessity to which the world is subject and discussing it with the liberty of man as the starting point or, putting this into more historical terms, making the liberty of man independent of the necessity that dominates the world. The *cogito* is only seemingly the premise or condition of the scission between *res cogitans* and *res extensa* and the bridge that Descartes sought was not the one that leads from the liberty-being of the "I" to the necessity-being of the world, but rather that which leads from the liberty-being of God to the liberty-being of the human I: having accepted the geometric necessity of the world, Descartes had no choice other than to think of his own I as the image and likeness of God, that is, unextended like the God of Cusanus. From the inextension of God Cusanus had deduced the necessity that dominates — as divine law — the becoming of the extended universe. Descartes, on the other hand, started from the necessity that as law dominates the becoming of the spatial universe (*res extensa*) and deduced from it the nonextension of the I, which is because it thinks and in being (that is, participating in the universality of being), is also free and not subject to that becoming as necessity that he considered to dominate the spatial universe. As regards the deontological demonstration of the existence of God, it is almost as if Descartes were saying to Cusanus: "It is quite true that God is infinitely more perfect than man, but the human I is unextended like God." Descartes does not really offer a deontological demonstration of the existence of God at all, but rather a deontological demonstration of the existence of the human thought-being.

Like the God of Cusanus, however, the Cartesian I has the qualities and defects of an absolute: in the mythico-natural attitude, indeed, the *cogito* fascinates on account of the definitiveness of knowledge, which establishes a point of arrival in the fundamentality of the self-knowledge of the I and

not a starting point in the rediscovery in new terms of the coincidence of
thinking and being in the vital becoming of the critical process of transcendental thought.

The Cartesian mechanism is something different from Galilean experimentalism, since Descartes filters the Cusanian determinism through the *cogito*,
while Galileo abides by it quite rigorously: this does not in any way detract
from the importance of the Cartesian contribution to the early methodological
orientation and, above all, the subsequent theoretical development of the
natural sciences. But the naturalness of the sciences has caused the *cogito*
to be considered in its mythical aspect as a point of arrival of knowledge
and has thus frozen, as it were, Descartes' attempted critique into a role of
system foundation.

The scission between *res cogitans* and *res extensa*, which implicitly already
appears in Cusanus' "learned ignorance" but explicitly only in Descartes as
the new form of the scission "soul—body," lies at the basis of the contraposition of idealist humanism (*transzendentaler Subjektivismus*) and scientific
antihumanism (*physikalistischer Objektivismus*). This contraposition does not
constitute a more or less mistaken theoretical presupposition or practico-
existential prejudice of an individual philosophical conception, but rather a
latent degree of theoretical awareness of a historical, social, and individual
situation of scission, and the Cartesian *cogito* was one of the first and most
powerful attempts to overcome this scission through the transcendental
negation.

But the Cartesian *cogito*, even though it rediscovers and reopens the critical
horizon that the logico-critical level of Parmenides' thinking had already
adumbrated at the beginning of Western philosophy, does not establish any
scientism, just as it does not establish any idealism; on the contrary, at the
birth of the natural sciences it limits itself to indicating this critical horizon
as the indispensable precondition of European scientific humanism. To
describe the many ways in which it was illusorily thought possible to overcome the uneasiness deriving from the scission of *res cogitans* and *res extensa*
by means of a unification based on some form of "pure knowledge," i.e.,
autonomous knowledge and therefore once again *separated* knowledge
and, consequently, sign and repetion of the scission, would be to tell the
long and complex history of philosophical thought that accompanies and
determines the development of the natural sciences up to our own days. In
this history the birth in philosophical method of the concept of critique,
accompanied as it is by a new way of considering the transcendentality of

thought, represents one of the most interesting chapters and its decisive turning point is associated with the name of Kant.

At this point we should set out on a new line of reasoning, but these notes of mine have already come close to the end of what little they set out to say. In the Cartesian *cogito* the scission between *res cogitans* and *res extensa* presents itself as a reduction of the participation of man in the universality of being, in thinking: this *cogito* is a kind of last beach of natural man imprisoned by the impossibility, by the absolute that he is seeking as a liberation and which he suffers like a form of slavery. Upon a critique the absolute imposes the discipline of a definitive result, a knowledge, be it sought or found, that is definitive and *positive* (factual). In reproposing a scission between phenomenon and noumenon Kant's thinking reestablishes science by the side of metaphysics (critique tends toward the absolute unity of a definitive knowledge), the mortal body by the side of the immortal soul, the ideal unity of the world by the side of the ontological necessity of God. The equilibrium with which we are here concerned is extremely fragile, exposed to the upsets and upheavals that the first gusts of idealist wind did not fail to provoke.

The metaphysical attitude constitutes a horizon of possibility for social and individual human life, and in the absence of this horizon the various societies and individual existences precipitate into the chaos of scissioned unification, into the identification-confusion of scission and unity that emerges when the orderly and vital harmonization in the transcendentality of possible thinking (in the possible transcendentality of thought) of unity and multiplicity comes to lack. For this reason the situation at the level of philosophical thought (where philosophy has already distinguished itself from religion) and the development of mythico-religious language-thought, which historically preceded philosophy and continues to accompany it on its road, is always closely connected with the concrete vital existence, with the concrete "problems of life" of societies and individuals. The impossibility of overcoming the scission, the absolute in the progress and the development of the natural sciences, all involves a continuous sharpening of the already existing contrasts and of the already obvious social injustices, while this selfsame progress of the natural sciences is itself turned into war and social injustice. Like society, individual human beings become ever more lacerated and intolerant: the powerlessness of thought entrusts to the factual power of the sciences the organizational task of normalization that in all the mythico-natural societies compensates the vital-productive deficiencies of transcendental thought.

X

There is uneasiness in time, I said at the beginning of these notes, both in the premodern circular time and in the linear time of the modern age, I would now add. Perhaps there is uneasiness in the time that can only be represented spatially and which in some way or other, as Saint Augustine thought, is a distension of the soul. In fact, the being form of each one of us and of man (but not only of man) has a space-temporal *possibility* of manifesting itself (a manifestation that is inseparable from being, just as being is inseparable from being possible) that nevertheless cannot be reduced either to time alone or to space alone, but whose space and whose time become comprehensible and understandable only if one starts from the being that originarily is life of the form.

University of Siena

Translated by Herbert Garrett

NOTES

[1] In magico-religious thought, in other words, the soul is something that "materially" exists (air, light, animal, etc.); we are therefore concerned with a physical entity separated from the body and not recognizable as belonging to that corporeal form of which it is the soul.

[2] E. Husserl, *Krisis der europäischen Wissenschaften und die transzendentale Phäno-menologie, Husserliana*, vol. 6 (The Hague: Martinus Nijhoff, 1962) p. 4. Italics mine.

[3] Ibid., in the title of the first part.

[4] Ibid., pp. 3, 4.

[5] It is not that the natural sciences do not have a history, but theirs is nevertheless the history of a knowledge whose certainty is based on a factuality estranged from historical time: the more often the "repetition" of the facts confirms a theory, the truer is that theory. In changing the connotations of the Newtonian mode of conceiving reality, the turning point represented by Einstein has thrown this mode of considering the natural sciences into crisis and now the "philosophers of science" are in search of another concept of historicity that will nevertheless respect the "naturalness" of scientific knowledge.

[6] In Indian thought, for example, there is a sharp contraposition of the becoming that appears in our world of multiplicity and the being that is in absolute unity; the multiplicity is then considered as the fragmentation of the unique living singular being, the universe being anatomically described by the sum total of its part. The philosophico-religious text of the Upanishad begins to speak of the spatial and temporal world in anatomical terms, extension is truly a machine. Let me briefly quote from the beginning of this much cited text: "The aurora is the head of the sacrifical horse, the sun its eye,

the wind its breath, the fire Vwisvānara its wide open jaws, the year its being. The sky is the back of the sacrifical horse, the atmosphere its stomach, the earth its abdomen, the cardinal points the flanks, the intermediate points the ribs, the seasons the members, the months and the half-months the articulations, the days and the nights the legs, the constellations the bones, the clouds the flesh, the sand the nourishment, the rivers the intestines, the mountains the liver and the lungs, the plants and the trees the skin; the sun that rises is its anterior half, the sun that sets the posterior one; when it opens its mouth there is lightning, when it snorts there is thunder, and when it urinates there is rain; its neighing, indeed, is the very voice itself [*Vāc*, creating Word]" (*Upanissad* [Turin: Ronconi, 1977], p. 35). As can be seen, here the mythico-natural identification of participation in the universality of being with belonging to (or membership of) the totality of the one, of a unity, implies that intellectual knowledge by scission, anatomy (the universe is known through the sacrificial animal, through the totemic sacrifice). But the same sacrifice that makes possible this separated or "scissioned" knowledge is also at the basis of the unification in the ens (sacrificial animal) of being. Given this separation of necessity of the real and possibility of the existing, becoming seems impossible in an abstract sense and being in a concrete one. Both these impossibilities are self-contradictory inasmuch as thought is nevertheless in a process of becoming and the world gives itself concretely to the concrete living ens that is man. Unification and scission are therefore two aspects of the lacerating remoteness of mythico-natural man from the universality of being, as I shall endeavor to show below.

7 This question remains without an answer and is not taken up again in the text, at least not directly; thus this reference to Hegel might seem unjustified. I would ask my indignant readers to reflect about the underground streams that flow through a theoretical discourse, especially when it is concise and tense: ask yourselves (but do not ask me) why I do not continue to talk about Hegel, ask yourselves (but do not ask me) why I do not speak of Heidegger. I call them underground streams, but the soil beneath which they are flowing is really altogether transparent.

8 Parmenides, ed. Diels, fragment 3: "dasselbe ist Denken und Sein." Thought is here considered in its transcendental power (potential) of being-becoming form: not idealism, therefore, but simply awareness of the vital quality of thought (thinking) in its emergence as transcendental form.

9 The work of Nicolaus of Cusa (Cusanus) is extensive, and a careful analysis of it would call for a long and detailed treatment of its own. To choose a single quotation would be nothing but a wrong perpetrated against so important and profound an author. A thorough discussion of this important philosopher must therefore be postponed and attempted within the context of a more detailed exposition of the problems here treated.

CLYDE PAX

FINITUDE AS CLUE TO EMBODIMENT

The reflections in the following paper are centered around what I take to be a central motivation in the work of Edmund Husserl, namely, the desire to uncover and clarify a transcendental basis for human communication. My thesis is that the study of the finite character of consciousness in the thought of Husserl can lead toward an understanding of both consciousness and of body which is helpful in creating the possibility for a genuine and grounded communication. In order to pursue this thesis I wish to consider three questions. The first question is whether and in what ways transcendental consciousness in Husserl can be properly thought of as finite consciousness? Second, how does the finitude of consciousness signal embodiment? Third, what is the meaning of body such that it can be the embodiment of transcendental consciousness? All three questions are posed in the framework of phenomenology understood as a practical task for transcendental consciousness rather than as a purely theoretical undertaking.

I

In the beginning of his London lectures Husserl writes: "The wonder of all wonders is the pure ego and pure subjectivity."[1] This wonder was, of course, for Husserl not only a wonder but an immense problem at the heart of all his work; it was a wonder which was also a "dark corner haunted by the spectres of solipsism and, perhaps, of psychologism, of relativism."[2] Husserl proceeds to say, however, that the true philosopher will, instead of running away, prefer to fill this dark corner with light; he thereby indicates an acutely practical task for the philosopher.

It is perhaps well to tarry a bit on the threat of solipsism and relativism in order to keep fresh our appreciation of the efforts Husserl made to find a transcendental basis for communication, and thus to avoid an easy relapse into an uncritical acceptance of factual and naturalistic communication, a move which at least in Husserl's view would simply cover up the problem.

Repeatedly, in numerous passages, Husserl insists upon the centrality of the experience of consciousness: "There is no conceivable place where the life of consciousness is broken through, or could be broken through. . . . First of

A-T. Tymieniecka (ed.), Analecta Husserliana, Vol. XVI, 153–161.
Copyright © 1983 by D. Reidel Publishing Company.

all, before everything else conceivable, I am."[3] And in the beginning of the fifth of the *Cartesian Meditations* he writes:

Imperturbably I must hold fast to the insight that every sense that any existent whatever has or can have for me — in respect of its "what" and its "it exists and actually is" — is a sense *in* and *arising from* my intentional life, becoming clarified and uncovered for me in consequence of my life's constitutive syntheses, in systems of harmonious verification. (sec. 43)

A like degree of originality and centrality is attributed to consciousness by Descartes, of course, in his making the mind's self-experience the beginning of truth and the criteria for further truths. Similarly, when in the introduction to *Being and Nothingness* Sartre describes consciousness as consciousness through and through and limited only by itself he is giving a centrality to consciousness such that it is original and free from causal dependence upon anything outside itself.[4]

It would seem that one of the experiential roots which makes consciousness into this wonder of wonders and which at the same time gives rise to the threat of solipsism is the ability of consciousness to direct its gaze upon itself. Whenever consciousness directs its gaze upon itself it finds itself always already present and thus unable to grasp anything more original than itself. At the same time, however, consciousness finds within itself no satisfactory sense of self-origination of self-justification and thus becomes both a wonder and a problem to itself.

In the history of thought various routes out of the dilemma have been proposed. One attempted route has been to make consciousness absolute and infinite; a second route has been to see consciousness as originarily negative; and a third has been to make consciousness one of the entities of an assumed natural world. That Husserl did not wish to take the first route is evident not only from explicit statements, e.g., in section 21 of *Ideen I*, but also from his view of philosophy as a universal and rigorous discipline available to all who could and would take sufficient effort and care to go to the facts themselves. The second route would, I believe, appear to Husserl not as the pursuit of sense but as an absurd affirmation of countersense. The third route appears to Husserl as the unfortunate lapse of Descartes who by making mind into a *substantia cogitans* "became the father of transcendental realism, an absurd position" (*C.M.*, I., #10). The motivation of Husserl is toward a universal *philosophia perennis*, toward clarity, and toward overcoming what appears to him as a critical condition of European science. The seriousness with which he takes the threat of solipsism indicates not only his vision of the need to seek a transcendental basis rather than a naturalistic basis for human communication, but indicates as well that the route to such a basis must be from an explication of the life of transcendental consciousness and only from there.

To begin to move along this route I wish to begin with a consideration of the finite nature of consciousness. Furthermore, I find it helpful to begin not with the controversial fourth and fifth of the *Cartesian Meditations* (which, after all, represent a conclusion to a long effort of reflection) but with a consideration of the central eidetic structure of consciousness called synthesis, and of its various manifestations as association, inner-time consciousness, intentionality as "I can," and passive genesis, especially as Husserl gives to this latter a primacy over active genesis. Unlike Descartes, who discovers the finitude of consciousness in the experience of factual error, Husserl, I believe, discovers the finitude in the very structures of consciousness in its transcendental life. Unlike Sartre who sees consciousness as a negativity or at least as the root of negativity, Husserl sees consciousness as belonging to a realm of sense, a realm which, in fact, exists in and for itself.

The finitude of consciousness for Husserl lies in the fact that consciousness appears as Ego, as transcendental Ego. Consciousness is revealed as finite precisely because and insofar as its own eidetic structure makes it necessary that it appear as Ego. Although Sartre laments the awkwardness of section 61 of the *Ideen I* wherein Husserl distinguishes between immanent and transcendent essences, the distinction is crucial to Husserl's understanding.[5] In speaking of the residue of the transcendental epoché Husserl notes that while the empirical ego and its world are suspended, and while all the *cogitationes* even when considered transcendentally appear to be perishable, the transcendental Ego itself is revealed as abiding and necessary. He writes in section 57: "But in contrast [to all the cogitationes] the Pure Ego appears to be *necessary* in principle, and as that which remains absolutely self-identical in all real and possible changes of experience, it can *in no sense* be reckoned *as a real part or phase* of the experiences themselves" (p. 156). This self-identical permanence is not, he tells us, that of a "stolid unshifting experience, of a 'fixed idea' but a flowing permanence which belongs to every experience that comes and flows past.

As a residuum of the phenomenological suspension the Pure Ego appears as a "*quite peculiar* transcendence — a transcendence *in immanence*" (p. 157). Such an essence, an essence immanent to consciousness, is, as Sartre rightly remarks, awkward. It is awkward because it places an opaqueness within the very life of consciousness understood as a total realm of all possible sense and experience. But this awkwardness is the awkwardness of finitude itself. It is

an awkwardness for critical thought but an awkwardness which is phenomenologically given. The finitude of consciousness is evidenced precisely by the
fact that the transcendental Ego cannot be suspended and thus the evidence
is in the transcendental move itself. Our task is not to avoid this phenomenologically given awkwardness, but to seek to illuminate it.

Husserl does precisely this, it seems to me, in uncovering synthesis as
the primal form of consciousness (C.M., #17) and identification as the
fundamental form of synthesis (C.M., #18). Because consciousness as a
totality is structured by the mode of synthesis, the transcendental Ego
exists in principle only by being in some way absent from the totality of
consciousness while "belonging inseparably together" with the whole of
consciousness. The existent and experiencing Transcendental Ego, even
though given with apodictic necessity, is like every other existent a "practical
idea" (C.M. IV, #41). It exists only in an inability to be unqualifiedly the
whole of consciousness; yet only because the whole is already given is the
Ego able to contemplate its individual acts or even its own presence as self-
identical Ego. It is an immanent noetic essence whose correlate can be the
whole of consciousness. Consciousness as thought consciousness, i.e., as
correlate of the experiencing Ego, can be consciousness through and through
but the experiencing consciousness exists only by being in some way other
than this whole, i.e., by being personal and limited Ego. The finitude of this
experiencing consciousness is obviously not the finitude of an object but a
finitude which exists as an openness, as an incompleteness which belongs,
at the very least, inseparably to other manifestations of consciousness.
Paradoxically, therefore, it is the very finitude of experiencing consciousness
which links it inseparably to the whole of consciousness that reveals the
experiencing consciousness not as a negativity but as a finite positivity. It is
not only consciousness that can in no way be broken through or left behind
but the finite mode of consciousness as well. This consciousness appears
phenomenologically as individual Ego and lives its life as a practical task
necessarily directed toward what is other than itself. It is irremediably finite
and irremediably dependent upon the other for the sense which accrues to it
in its every lighting glance.

This is made more clear by the primacy which Husserl gives to passive
genesis in relation to active genesis. In the fourth *Meditation* Husserl writes:
"In any case, anything built by activity necessarily presupposes, as the lowest
level, a passivity that gives something beforehand; and, when we trace anything built actively, we run into constitution by passive generation" (#38,
p. 78). And a little further on he continues, saying that here we

encounter eidetic laws governing a passive forming of perpetually new syntheses (a forming that, in part, lies prior to all activity and, in part, takes in all activity itself); we encounter a passive genesis of the manifold apperceptions, as products that persist in a habituality relating specifically to them. (p. 79)

What does this primacy of passive genesis signify? Most deeply it signifies that consciousness always *comes to* awareness as Ego consciousness on the basis of a before, and does so "owing to an essentially necessary genesis" (p. 79). It is in virtue of this primacy of passive genesis that the intentionality of consciousness can best be understood as an "I can"; furthermore, this "I can" is an "I am" that is never totally clear to itself but rather presents itself as a wonder to itself, and as a need to question its own being. As transcendental experience consciousness is not consciousness through and through (which remains a conceived correlate of experiencing consciousness) but consciousness interrogatively present to itself. This experienced finitude is not a finitude disclosed by factual error but disclosed as a positive eidetic structure experienced transcendentally. It is only on the basis of this interrogative self-experience that there is the possibility of a further reduction to a sphere of ownness and the uncovering of consciousness as alter ego. Without this interrogative self-experience there would be neither clue nor motivation to uncover consciousness as *There* as well as *Here*.

Husserl's insistence that association is an a priori structure of consciousness before it is an empirical law would seem to confirm further that experiencing consciousness is present to itself only as a limited belonging to what is other than itself. Similarly the all-embracing form of inner-time consciousness is known as all-embracing, that is, exists and has the meaning of all-embracingness only from a present which is finite by reason of belonging inseparably with but differentiated from a receding past and an anticipated future.

II

In what way or ways does this experienced finitude of consciousness signal, or give a clue for, embodiment?

An adequate consideration of this question can be undertaken only in the fuller consideration of the subsequent question of what body might mean as the embodiment of transcendental consciousness. For the moment allow me to make some very brief, and I fear somewhat enigmatic, suggestions.

The first and most important consideration is that the signaling is transcendental and from the eidetic structure of transcendental Ego. Therefore it is necessary from the beginning to keep in mind that we are seeking clues for the embodiment of transcendental consciousness and not clues for the existence or meaning of body as natural object.

Synthesis as the primal form of consciousness reveals that consciousness as individual but transcendental meditator exists always in a realm of belonging to what is other for its own necessary self-identity. The clarity, the apodictic evidence in which the transcendental Ego lives its life, is enveloped or embodied by an Otherness which makes possible its own lighting. This envelopment of its own clarity by what is other might appropriately be thought of as an embodiment. The enlivening character of this embodying denseness or opacity is more fully manifested in the description of synthesis as passive genesis. As noted above, Husserl tells us that this passive forming is a forming which in part takes place before all active forming but which also in part takes place *in* all activity. In the consideration of awareness in the broadest sense, this opacity appears as the need to wait for evidence both as a condition for all sense arising in and from consciousness and as a condition which remains *within* the lighting and "justifies" this lighting as meaningful. Association understood as an a priori synthetic structure which unites consciousness with consciousness not only within a single life but also which unites a single life with alter ego places each transcendental ego in a realm of a priori obligation of integrity to itself and of respect for others. It not only makes possible (and necessary) the recognition of the Other consciousness but places each transcendental Ego in a moral realm a priori, a realm of obligation to communication as truth-task and thus denies to each Ego the status of a *Pour-Soi*. Its way of living its own self-identity is to-be-with.

The all-embracing form of inner-time defines the transcendental Ego as a flowing identity but more importantly for our present considerations, as a flowing that has a directionality. Its ever greater and receding past horizon, precisely insofar as it is the horizon of the Ego itself, reveals the transcendental Ego as an Ego that is able to age.

Finally, intentionality as "I can" places the transcendental Ego in a realm of instrumentality. As individual Ego it exists not only in a realm which exists in and for itself but in a realm where because of its finitude the issue of means and the question of differentiation of evidences is relevant to its own ongoing self-presence.

III

What is the meaning of body such that body can be the embodiment of transcendental and not merely natural consciousness?

In the closing section of the fourth *Meditation* Husserl makes the claim that "Only someone who misunderstands either the deepest sense of intentional method, or that of the transcendental reduction, or perhaps both, can attempt to separate phenomenology from transcendental idealism" (*C.M.*, #41, p. 86). In the consideration of the embodiment of transcendental consciousness we must, therefore, take care to avoid falling victim to what Husserl calls "the inconsistency of a transcendental philosophy that stays within the natural realm (pp. 86–87). What, then, does the self-explication of the central structures of the transcendental Ego teach us about its embodiment? I think above all else the approach of Husserl teaches us that the understanding of body must be an understanding which is a part of a practical truth-task of the transcendental Ego itself. This means first of all that the "natural body" must be recognized as a theoretical construct that is not equivalent to the embodiment of consciousness and that insofar as we use the term human body to designate the embodiment of ourselves as conscious beings, human body is not the same as natural body. Furthermore, it would seem to follow that the pursuit of a natural science of the human body, however extensive and probing, does not, *of itself*, lead toward or offer any guarantee for the clarification of body as the embodiment of consciousness. Such a line of inquiry, taken uncritically, like the inquiry of a naturalistic psychology regarding consciousness, would seem to augment rather than alleviate the "crisis" which troubled Husserl throughout his work.

The remarks just made should not be understood as a denegration or rejection of the natural sciences of the body but as calling attention to the immense task, which is indeed already underway, of thinking anew the relationship between the natural sciences of the body and a phenomenological explication of embodiment.

A reflection, from the phenomenological side, on the intentionality of the "I can" might be helpful in pursuing this thinking. If the "I can" reveals the transcendental Ego as belonging, in part, to a realm of instrumentality, it reveals the Ego itself as belonging to a realm of vulnerability where failure, error, and injury are possible. It is this vulnerability, it would seem, which accounts for the very possibility of the transcendental Ego's uncritical forgetting of its own transcendental Being and a settling into a purely natural self-understanding. This negative dimension, however, does not exhaust the meaning of instrumentality for the Ego. The inherent and life-giving opacity which envelops the Ego as an "I can" which is structured primordially by passive genesis makes the existence and continual self-constitution of the Ego into a need for a discriminating listening, a listening which makes possible the constitution of ever-further harmonious wholes. On the one hand this gives a pragmatic dimension even to the meaning of theoretical truth about the human body. On the other hand, it cautions the Ego itself that it can live its life neither apart from a consideration of its genesis nor in hasty generalizations which would destroy the richness or perhaps even the possibility of its full concrete existence. The uncovering (and construction) of the appropriate, i.e., actually possible discrimination of evidences is of course an immense and, in principle, never completed task. This practical and discriminating listening to the embodying otherness cannot, in the present historical time, escape the task of forging the theoretical tools needed to integrate the positive sciences of the body with a phenomenological explication of consciousness.

Furthermore, if, as Merleau-Ponty has argued, transcendental subjectivity is always already intersubjectivity, not only must we take note of a common objectivity but also of a practical listening that is thoroughly political. It is perhaps not too extreme to wonder whether the embodiment of consciousness does not, on the international level, call for a careful but courageous pursuit of detente. On a more intimate level — but a level that is pervasively relevant and problematic — might not a concrete appreciation of embodiment insist upon a study of sexuality as common but also as importantly and interestingly differentiated into an embodiment that is sometimes female and sometimes male.

The embodiment of every actually experiencing consciousness in a temporality that is at once all-embracing and in some ways irreversibly directional invites, it seems to me, a necessary and reasonable preservation of the difference between the young and the mature that is far from easy to maintain. This temporal embodiment would seem to indicate also that the transcendental Ego must live its life and achieve the identity attuned to an ever growing and irretrievable past; transcendental Ego must, thus, live its life attuned to the possibility that its own on-rushing flow threatens its own extinction as self-identical. And insofar as association is an a priori law uniting consciousness with other consciousness, the Ego must reckon with the dying of others, others who by reason of their unique embodiment and their aging and vulnerability escape the bounds of my own practical sense-giving for and with them.

In conclusion I would like to recall the motivation which seemed to me to be the driving force of Husserl's work, namely, the striving toward a fundamental and reliable basis for human communication. In the course of these reflections I have argued that the synthetic structure of consciousness, in its various forms and as discovered in transcendental perception, reveals the transcendental Ego as radically finite and as able to be self-identically itself only in a practical willingness toward an abiding but ever-changing and concrete communication with what is other. Furthermore, I have argued that this other acts as an embodiment, in manifold ways, of the transcendental Ego. This embodiment is essential for its own lighting and its own continual life. At the same time, this embodiment arising from the essential finitude of the transcendental Ego denies to the Ego a meaning of truth which is total and complete vision. Instead it presents truth as a practical task which is characterized as deeply by faithfulness and courage as it is by vision.

College of the Holy Cross, Worcester, Mass.

NOTES

[1] Translation as given by Herbert Spiegelberg, *The Phenomenological Movement* (The Hague: Nijhoff, 1960), Vol. I, p. 87, Note 2.

[2] Husserl, Edmund, *Formal and Transcendental Logic*, translated by Dorion Cairns (The Hague: Nijhoff, 1969), p. 237.

[3] Ibid., pp. 236–37.

[4] Sartre, Jean-Paul, *Being and Nothingness*, translated by Hazel E. Barnes (New York: Philosophical Library, 1956), p. lv.

[5] Sartre, *The Transcendence of the Ego*, translated by Forrest Williams and Robert Kirkpatrick (New York: Noonday Press, 1957), p. 51.

MARIA DA PENHA VILLELA-PETIT

TOPOÏ OF THE BODY AND THE SOUL IN HUSSERLIAN PHENOMENOLOGY

As the title suggests, the question posed here is the following: What are the problematical *topoï* in which the theme of the body and the soul arises in Husserlian phenomenology?

It is necessary, first, to situate this theme in relation to the transcendental reduction. By saying this we implicitly assume that we take into consideration the historical and philosophical horizon against which the necessity of this reduction stands out, namely, the horizon opened by the Cartesian "ego cogito". This reduction reveals that the Cartesian approach obliterated the very sense of its discovery because it was unable to apprehend the transcendental motive which it contained.[1] The transcendental reduction discloses the halfway point at which Descartes stopped, for, having failed to distinguish the egological transcendental sphere from the empirical self, he identified the ego with the individual soul conceived as a thinking substance, to which he opposed the universe of material things, of extended objects, including everything that can be called a "body."

In following in Descartes' tracks, Husserl's transcendental reduction holds the Cartesian dualism in suspension, thereby dismissing its metaphysical theses concerning the soul and the body. According to these theses, which Husserl explicitly denounced in the *Crisis*:

Body and soul thus signified two real strata in this experiential world which are integrally and really connected similarly to, and in the same sense as, two pieces of a body. Thus, concretely, one is external to the other, is distinct from it, and is merely related to it in a regulated way.[2]

But if the reduction implies the critical suspension of metaphysical dualism, under which angle will it permit us to recover the problematic of the soul and the body, if it is also true that this reduction does not spare man, but places him in parentheses, as a factual reality, as an empirical subject? This question leads us to recall something which is obvious for the phenomenologist, namely, that the reduction does not eliminate anything since, if it suspends all the doxa of reality, it is in order to open to investigation the field of "realities" such as they appear to consciousness, and this with a view to revealing the essential meaning of their being for us.

163

A-T. Tymieniecka (ed.), Analecta Husserliana, Vol. XVI, 163–171.
Copyright © 1983 by D. Reidel Publishing Company.

Yet if the field of consciousness is the terrain of phenomenological investigation, are there grounds for suspecting that, by its method, it gives a
privileged place to the sphere of psychic life? This suspicion could be backed
by the parallel which Husserl stated between transcendental phenomenology
and psychological phenomenology.

In other words, if by "soul" (*Seele*) Husserl designates the psychological
consciousness, not only in its intentional life but also in its egological unity,[3]
could we not deduce from the methodological privilege of consciousness the
ontological privilege of the "psyche," that is, of the soul? The counterpart
of this privilege could then be expressed in the following terms: within
the transcendental reduction, the body can only appear as a unity of meaning
for consciousness. Consequently, the first question which arises is, what
status is left to the body after the reduction? Might Husserl not have sub-
stituted for the Cartesian metaphysical dualism a phenomenological dualism
which, being more subtle, is also more tenacious?

Let us not be too quick to answer this question, which is closely akin to
the objection Heidegger made to all philosophies of consciousness: "if we
posit an 'I' or subject as that which is proximally [immediately] given, we
shall completely miss the phenomenal content [*Bestand*] of *Dasein*."[4] In
other words, is there not an ontological priority of man's concrete totality
which determines as such his modality of being in the world, and which the
reduction − which leads to a transcendental egology (the content of which is
that of the "psyche") − might keep us from recognizing?

To raise such an objection here may prevent us from recognizing the
recurrence and impact of the question of the body in Husserlian phenomenology. Is it not at the crossroads of all its important questions? Besides, the
body reaches its transcendental place when phenomenology moves on to the
constitutive tasks in order to fulfill the requirement of elucidation which it
poses vis-à-vis the units of meaning which present themselves as the noematic
poles of intentional life.

Among the constitutive tasks, the first − the elucidation of the constitution of a nature for and by the subject − reveals that the perceived points
back, by the way it is given, to a feeling and moving body. Thus the body
here is not a contingent body as in Descartes, for whom the union of soul and
body, even though real, was not at all necessary. According to Descartes, it
was possible, even while feeling light and heat (as psychic events), to imagine
oneself bodiless, from which he concluded that the tie of the soul and the
body could only be contingent.

Others may be more competent than I to interpret this imaginary suppression of the body, which Descartes places in the "dream category" in the First Meditation. Let us merely add that if the argument of body contingency seemed useful for him to prove the immortality of the soul (being moreover linked to his other metaphysical thesis), it proceeded from an obvious phenomenological mistake; and, we might say, was opposed to biblical anthropology, which Descartes, as a believer, did not mean to oppose.

Now, in Husserl, on the contrary, and this point is decisive, the "constitution" or perception of a nature implies the subjectivity of the living body (*Leib*). *Subjektleib* is the expression forged by Husserl to designate the body, such as it reveals itself through the description of the phenomenal appearing of nature — a strong expression that changes the place of the question of the body, insofar as, before Husserl, for a philosophy of subjectivity, the body could only be conceived in terms of an object.

This subject-body is that which things indicate as a zero point (*Nullpunkt*), that is, as the absolute here (*hic*) to which the phenomenal appearing of things, as always already oriented, points back. Indeed, perceived things display their aspects only in accordance with a mode of orientation and distance, relating to the subject's own movement, of which the body is the unique starting point. Or as Husserl expresses it in *Ideen II*:

Und auch die Auszeichnung hängt offenbar hiermit zusammen, dass der Leib zum Träger der Orientierungspunkte Null wird, des Hier und Jetzt, von dem aus das reine Ich den Raum und die ganze Sinnenwelt anschaut. So hat also jedes Ding, das erscheint, eo ipso Orientierungsbeziehung zum Leib, und nicht nur das wirklich erscheinende, sondern jedes Ding, das soll erscheinen können.[5]

From this phenomenological assignation of the body as the zero point of all perceiving and moving (perceiving and moving being moreover inseparable) there results that, as such, our own living body is not a physical part of nature which consciousness could represent to itself as an object, even if it can become so through an act of objectivation. This amounts to saying that the recognition of the body as *Nullpunkt*, that is, as a nonobject, frees it from the chain of physical causality, and reveals it as being that which is presupposed by all object representations or all causal connection of objects which could be held valid for a subject.

I will later have to ask whether Husserl himself went to the end of the consequence of this identification of the body as subject-body. In any case, the way in which the body reveals itself through the elucidation of the constitutive intentionalities of the physical thing is a sufficient proof of the novelty of the question of the body in Husserl and of how it transgresses Cartesianism. To emphasize this point, it is useful to evoke the surprising passage from the Second Meditation, in which Descartes states his astonishment concerning the body:

For the power of self-movement, and the further powers of sensation and consciousness [de sentir et de penser], I judged not to belong in any way to the essence of body . . . ; indeed I marveled even that the were some bodies in which such faculties were found.[6]

Here the geometrical and mechanical representation of the physical thing has such an ascendency that it silences all true experience of the body as a living one. In Husserl, on the contrary, the subject, which moves and perceives a nature, is necessarily an incarnate subject, a subjective body. And this is not posed as the result of a logical inference, but as that which is revealed through and correlated to the very appearing of nature.

However, the place of the body in Husserlian phenomenology is not only the appearing of the physical world; the body is implied also in the encounter of an other as alter ego, thus being central to the constitution of an intersubjectivity. In other words, if an other can have the meaning of an alter ego, it is because, in his body, he appears to me as an other self. Thus, it is the gestures and corporeal attitudes of the other which express him as being the subject of an intentional life similar to mine. Thus the other is not inferred from his body as it enters my field of consciousness, but his body straight away presentifies him to me as an alter ego. The incarnation of the other discloses him as an other self.

Nevertheless, everyone is aware of the difficulties linked to the constitution of the alter ego in Husserlian phenomenology. Now, although we cannot here examine thoroughly the reasons for these difficulties, let us underline that they are in part due to the fact that Husserl's attempt to think the "constitution of the alter ego" refers almost exclusively to the perception of the other's body instead of taking into account the totality of "his" attitudes toward "me" and of "mine" toward "him," which include "speaking" to each other. In other words, the fact that the other speaks to me (or can speak) and I to him proves that from the beginning I see and treat the other (and, conversely, he treats me) as an alter ego.

Is this not what Wittgenstein expressed in his own way when, in his attempt to undo the "other-mind problem" as a metaphysical artifact, he wrote: "My attitude towards him is an attitude towards a soul. I am not of the *opinion* that he has a soul,"[7] by which he meant, against Cartesians, that the mind (consciousness) or the soul of others is not something which we can seriously doubt.

Let us consider more closely how Husserl places the question of "body" at the core of the intersubjective relationship, constitutively understood. The necessity of embodiment is not limited to the community of human subjects. This necessity is clearly asserted if other minds, whatever they are, are to communicate (or enter in connection) with us:

Die Möglichkeit bleibt offen, das immer neue Geister in diesem Zusammenhang treten: aber das müssen die *durch Leiber*, die durch mögliche Erscheinungen in unserem Bewustsein und durch entsprechende in dem ihrem vertreten sind.[8]

This impossibility of doing without the body does not necessarily mean that Husserl excludes on principle the relation of man to other than human minds. For, curiously, and probably without his knowledge, he skips modern times to find himself among certain medieval thinkers. Do we not read in Saint Bernard that "without the support of a body ... the celestial spirit is incapable of accomplishing its charitable role"?[9] In other words, according to Saint Bernard, in order to succor man, an angel must necessarily be embodied. Yet this assumption of embodiment raises here a certain number of questions: for example, would it not be possible to ask whether or not the spirit, in its sensible manifestation, could not occasionally do without a body (*Leib*), in order to soar with breath or vibrate with sound?

It is important to point out that neither the constitution of nature nor that of intersubjectivity exhausts the question of the body in Husserlian phenomenology. We are again brought back to the body through the constitution of all the ideal objects of culture and even through the constitution of mathematical idealities.

Of course, what Husserl called the "origin of geometry" implied a going beyond, a taking off from the sensible forms given in the *Lebenswelt*, a taking off which made these idealities as such, that is, independent of the positions and orientations of our own living body. Nevertheless, the theoretical practice they require, and without which they could not perdure or be reactivated in a process of traditionalization, has to have some physical inscription. Thus, if the originary (*ursprünglich*) appearance of the geometrical "ideality" occurred, according to Husserl, in the "space of consciousness of the first inventor's soul,"[10] the "objectity" thus produced became fully objective only when it was delivered to intersubjective reactualization by means of its consignment, that is, of its graphic inscription in the flesh of the world. And such an inscription presupposes the body which accomplishes it as much as the one which reads it. By this analysis, then, Husserl indicates that even in the highest intellectual processes, everything in the operativeness of human consciousness must have recourse to the body in one way or another.

We thus arrive at the following initial conclusion: as soon as it attempts the rigorous investigation of the constitution of the senses of the perceived, of the other-self, and of the ideal object, Husserlian phenomenology succeeds in revealing the essential role of the body and, consequently, in recognizing the human subject as a fully incarnate one.

Does this mean that Husserl was able to draw all the philosophical conclusions which this recognition and his own descriptions demanded? An attempt to answer to this question requires that we turn now to the other side of our theme, the soul, in order to undertake a crucial reevaluation of phenomenological reduction. The need for such a reevaluation has, moreover, been clearly posed by Merleau-Ponty:

> What Husserl's investigation brings to light is the corporeal infrastructure of our relation to things and to others, and it seems difficult to "constitute" these raw materials from the attitudes and the operations of consciousness, which fall within another realm, that of theoria and ideation. This internal difficulty of "constitutive phenomenology" questions the method of reduction.[11]

Indeed, if the body determines for the subject his absolute here and now, it follows that, as such, the body can never be absolutely reduced by consciousness. For how could an absolute here authorize any complete reflection? Yet, up to a certain point, Husserl did recognize, in the assignment of temporal horizons to consciousness, the impossibility of a reflexive totalization. But the here and now of my presence in the world prescribes more than a "problematic" of temporal horizons. They refer to a blind spot, a *punctum caecum* of consciousness. And perhaps the inevitable consequence of this is a certain mystification of consciousness in its encounter with the world and the other.

Now, it is striking that Husserl discloses the relation to the other as the necessary condition for a subject to accomplish a first objectivation of his own body. By this uncovering of the other as the mirror which gives to the subject the body as a body which he has, so allowing the subject not to coincide with his body, but to have it as a body, is considered in Husserl only from the point of view of the objectivation of nature. The other is thus considered as the condition allowing my body (*Leib*) to be at the same time a *Körper*, that is, a natural thing within nature; whereas one could expect a more careful consideration of the implications of this mediation through the other, with regard to the constitution of the subjectivity itself, and hence of self-reflection. (Here I see room for a fruitful confrontation of phenomenology with psychoanalysis.)

It is not of course a matter of simply dismissing reduction as a mere trap. It can remain a propaedeutical way insofar as it allows me to loosen the familiar ties of my empirical being to a world of meanings and opinions which pretend to be self-evident but are not clarified. Yet the reduction cannot claim to be radical. Instead, it must finally lead to the recognition of our *irreducible* attachment to Being, which is that of our human corporeality.

Consequently, we must still question the status of the soul, and where it finds its "place," if any point of view dividing or separating it from the body may disrupt our understanding of the being-there of man. And since the body (*Leib*) has not been enclosed within a somatology which would prevent us from surprising it in its relation to the world and to the other, has the soul to be enclosed within the limits of a psychology, of which the soul, according to Husserl in the *Crisis*, is indeed the theme? [12]

The objection here to Husserl is twofold. On the one hand, it concerns the fact that the soul, identified with the egological consciousness, finds itself a prisoner of the enclosure of consciousness. But is it not rather in the opening of our being to the world, to others, and to God — opening which implies more than the *of*ness of consciousness — that the soul points to itself? I would suggest here that a certain way of conceiving intentionality, based on consciousness alone, falls short of the ecstatic movement of the self in which the human being discovers that he possesses a soul, that is, an interiority which cannot be reduced to the limits of his psychological consciousness.

According to this approach, interiority is not so much what constitutes itself, in self-certainty, to the exclusion of exteriority, as the spiritual dimension and reverberation of what happens to us in terms of encounter, of *Erfahrung*. Whence we feel the appropriateness of this instant of hesitation which makes Heidegger ask in *Feldweg*: "Der Zuspruch des Feldwegges ist jetzt ganz deutlich. Spricht die Seele? Spricht die Welt? Spricht Gott?" [13]

On the other hand, we must ask whether the place given to the soul in Husserl takes sufficiently into account the intimate relation he himself made between the "I" and the deictic "here" in his fifth Cartesian meditation. Again, this intimate relation between "I" and "here" (the "here" of "my" body) renders impossible any reference or self-reference of the "I" to soul (or consciousness) alone but, on the contrary, demands that the inseparability of the soul and the body be thought in the "I." In other words, my soul alone is not I. Thus, insofar as the soul "inhabits" the body as the reserve, the inner source which keeps the gestures, feelings, and actions of man from exhausting themselves in their corporeal manifestations, it is doubtful that it could be considered separately, without the body, in a phenomenological psychology claiming to be an "egology." How is it, then, that Husserl has attempted to raise the soul, apart form the body, as the theme of a descriptive theoretical knowledge displaying itself within the field of an egological consciousness?

The soul demands rather to be recognized through the traces of what we might call the inscription of the spiritual in the sensible. And, in particular, through what is articulate at the juncture of body and soul, and which articulates this very juncture, that is, speech. Reciprocally, by recognizing speech here at the juncture of soul and body, we open the way for an understanding of its sometimes vivifying, sometimes deadly, effects, depending on whether it integrates or separates the body and the soul: words of life or of death. But a purely descriptive phenomenological psychology keeps such an understanding out of its scope. Furthermore, it puts aside the question of others' spiritual experiences, which are only attainable through the "stories" of their "souls." And if these experiences ask for a phenomenological elucidation of the specific intentionalities they carry, they cannot therefore be considered within the realm of a phenomenological psychology in a Husserlian sense.

In conclusion I would like to suggest that only by a longer detour through what I may venture to call an "hermeneutic phenomenology," going beyond the disciplinary frontiers which subtly tend to confirm dualism, would it be possible to fully acknowledge the unity of the soul and the body, to which speech and, more generally, every activity and work of man, bear witness.

Centre National de la Recherche Scientifique, Paris

NOTES

[1] E. Husserl, *Cartesianische Meditationen* (The Hague: M. Nijhoff, 1963), sec. 10, pp. 63–64; "Leider geht es so bei Descartes, mit der unscheinbaren, aber verhängnisvollen Wendung, die das ego zur substantia cogitans, zur abgetrennten menschlichen mens sive animus macht und zum Ausgangslied für Schlüsse nach dem Kausalprinzip, Kurzum der Wendung, durch die er zum Vater des ... widersinnigen transzendentalen Realismus geworden ist. ... Darin hat Descartes gefehlt, und so kommt es, dass er vor der grössten alle Entdeckungen steht, sie in gewisser Weise schon gemacht hat, und doch ihren eigentliche Sinn nicht erfasst, also den Sinn der transzendantalen Subjektivität. ..."

2 E. Husserl, *Die Krisis der Europäischen Wissenschaften und die transzendentale Phänomenologie* (The Hague: M. Nijhoff, 1962), sec. 62, p. 219; *The Crisis of European Sciences and Transcendental Phenomenology*, trans. David Carr (Evanston: Northwestern University Press, 1970), p. 215.

3 Cf. E. Husserl, *Phänomenologische Psychologie*, Husserliana 9 (The Hague: M. Nijhoff, 1962), p. 283: "Zu jeder Seele gehört nicht nur die Einheit ihres mannigfaltigen intentionalen Lebens mit all den von ihm als einem 'objektiv' gerichteten unabtrennbaren Sinneseinheiten. Unabtrennbar ist von diesem Leben das in ihm erlebende Ich-subjekt als der identische, alle Sonderintentionalitäten zentrierende 'Ichpol' und als Träger der ihm aus diesem Leben zuwachsenden Habitualitäten."

4 M. Heidegger, *Sein und Zeit*, 13th ed. (Tübingen: M. Niemeyer, 1976), sec. 10, p. 46; trans. J. Macquarrie and E. Robinson (New York: Harper and Row, 1962), p. 72.

5 E. Husserl, *Ideen II*, Husserliana 4 (The Hague: M. Nijhoff, 1953), p. 56.

6 Descartes, *Méditations* (Paris: Pléiade, 1958), p. 276; *Philosophical Writings*, trans. G. E. M. Anscombe and P. T. Geach (Edinburgh: Nelson, 1954), p. 68.

7 L. Wittgenstein, *Philosophical Investigations*, trans. G. E. M. Anscombe (Oxford: B. Blackwell, 1958), p. 178.

8 Husserl, *Ideen II*, sec. 18, p. 86.

9 St. Bernard, *Sermons sur le Cantique des Cantiques*, trans. Albert Béguin (Paris: Le Seuil, 1953), p. 113.

10 Husserl, *Krisis*, Beilage III, p. 369.

11 M. Merleau-Ponty, *Résumés de Cours-Collège de France 1952–1960* (Paris: Gallimard, 1968), p. 149.

12 Cf. Husserl, *Krisis*, p. 241; *The Crisis*, trans. D. Carr, p. 238: "Purely descriptive Psychology thematized Persons in the pure internal attitude of the Epochè, and this gives it its subject matter: the soul." See also *Krisis*, p. 242; *The Crisis*, p. 239: "For Psychology is, afterall, supposed to be the universal Science of souls, the parallel to the universal science of bodies; and just as the latter is from the start a science through a universal Epochè, through a habitual vocational attitude established in advance in order to investigate abstractly only the corporeal in its own essential interrelations, so also for Psychology."

13 M. Heidegger, *Der Feldweg* (Frankfurt: V. Klostermann, 1956), p. 7.

HIROTAKA TATEMATSU

HUSSERLS SICHT DES LEIB–SEELE PROBLEMS

1. VORWORT

In diesem Referat möchte ich hauptsächlich über das Problem der sogenannten Ichspaltung und über den Kerngedanken der Husserlschen Schichtentheorie sprechen. Nach meiner Meinung ist Husserls spezifische Schichtentheorie von großer Bedeutung für das Verständnis seiner Sicht der Leib-Seele-Geist-Beziehungen. Berücksichtigt man nur die europäischen Philosophien seit Descartes, so hat es auch im Hinblick auf das Leib-Seele-Problem oder das philosophische Menschenbild viele Polemiken und Streitigkeiten gegeben. Ich denke vor allem an einen Gegensatz zwischen der Leib-theorie innerhalb des cartesianischen Dualismus und der phänomenologischen Leib-theorie eines Merleau-Ponty oder Sartre. L. Binswanger z.B. versucht in seinem Vortrag 'Über Psychotherapie' (1934) seinen Zuhörern folgendes klarzumachen: "Sie müssen nicht nur wissen, daß der Mensch einen Leib 'besitzt' und wie dieser Leib beschaffen ist, sondern auch, daß er selber stets irgendwie Leib 'ist'." Bekanntlich behauptet Sartre im *L'être et le neant*: "J'existe mon corps" (p. 418) und "je suis mon corps dans la mesure où je suis" (p. 391). Solche Behauptungen wären für einen Descartes undenkbar.

Im Gesamtrahmen der reinen oder transzendentalen Phänomenologie Husserls, die man als Philosophie des transzendentalen Bewußtseins kennzeichnen kann, scheint mir das Leib-Problem keine zentrale, letztlich entscheidende Bedeutung zu haben. Und dennoch: besonders im zweiten Band seiner *Ideen* und in vielen anderen Manuskripten zur Problematik der Intersubjektivität beschäftigt er sich ziemlich eingehend mit diesem Problem. Aus seinen Erörterungen dazu können wir viele neuen Einsichten gewinnen. Die historische Bedeutung dieser Erörterungen liegt vor allem darin, daß sie den gegenwärtigen Diskussionen über das Leib-Seele-Problem eine neue Richtung gebahnt hat, die zu neuen Ufern führt und besonders von Phänomenologen im erweiterten Sinn vertreten ist.

Zweifellos ist das Leib-Seele-Problem für die gegenwärtige Philosophie eines der wichtigsten und interessantesten Forschungsthemen, weil wir gerade in dieser Krisenzeit der Menschheit ein neues Bild des Menschen suchen. Diese Sachlage ist auch in Japan nicht anders. Besonders seit fünf, sechs Jahren ist

A-T. Tymieniecka (ed.), Analecta Husserliana, Vol. XVI, 173–181.
Copyright © 1983 by D. Reidel Publishing Company.

auch bei uns das Leib-Seele-Problem zu einem sehr akuten Hauptthema
geworden und in jedem Jahr erscheinen nicht wenige Bücher und Aufsätze
zu diesem Thema.

2. ICHSPALTUNG UND BEGRIFFE DES ICH

Nun möchte ich zur Hauptsache kommen und zuerst über das Problem
der Ichspaltung sprechen. Husserls transzendentale Phänomenologie will
die ursprünglichste Seinsweise alles Seienden aufklären, und zwar durch
konstitutive, sinngebende Leistungen des transzendentalen Ich und seiner
Bewußtseinsakte. Auf dem Weg zur Lösung dieser Hauptaufgabe steht die
Methode der phänomenologischen Epoché, und wie jeder weiß, fordert
Husserl damit die totale Einklammerung oder das totale Außer-Geltung-
Setzen der natürlichen Einstellung, in der man bei jedem realen Seienden
naive Existenzialsetzung oder die sog. Generalthesis vollzieht. Dabei wird
auch die Existenz meines eigenen menschlichen Ich ein Gegenstand der
Epoché und mein natürliches Ich wird auf das reine oder transzendentale Ich
reduziert. Was folgt daraus? Eben nichts anders als die sog. Ichspaltung.

Was besagt hier "Ichspaltung"? Geht es dabei um eine substantielle
Unterscheidung verschiedener Ichs, die als völlig heterogene voneinander
unabhängig existieren könnten? Gewiß sagt Husserl einmal, "daß alles eigentlich 'Subjektive', Ichliche, auf der geistigen Seite (der im Leib zum Ausdruck
kommenden) liegt" (IV, 96).[1] Wir müssen aber bei diesem Satz darauf achten,
daß die geistige Seite 'im Leib' zum Ausdruck kommt. Das deutet schon an,
daß zwischen Geist und Leib irgendeine Verbindung besteht. Husserl sagt
an einer anderen Stelle sehr deutlich: "Ich als natürlich eingestelltes Ich bin
auch und immer transzendentales Ich" (I, 75). Wie schon aus diesem Satz
zu ersehen ist, bedeutet die phänomenologische Ichspaltung gar keinen
substantiellen Unterschied im cartesianischen Sinn und keine gegenseitige
Unabhängigkeit zwischen verschiedenen gespalteten Ichs. Vielmehr ist das
Ich in seiner ursprünglichen Konkretion immer und je schon eine Einheit.
Jedoch zeigt dieses Ich mit dem Wechsel von der natürlichen zur eidetischen
oder transzendentalen Einstellung jeweils einen anderen Aspekt. Und jeder
andere Aspekt des Ich enthüllt jeweils eine andere Phase des Menschseins, das
in sich selbst eine zusammengesetzte Schichtenstruktur hat. Vom gespalteten
Ich redet Husserl folgendermaßen: "Mein transzendentales Ich ist also evident
'verschieden' vom natürlichen Ich, aber keineswegs als ein zweites, als ein
davon getrenntes im natürlichen Wortsinn, wie umgekehrt auch keineswegs
ein im natürlichen Sinn damit verbundenes oder mit ihm verflochtenes"

(IX, 294). Aus dieser Art der Wortwahl, mit der feinere Nuancenunterschiede unterstrichen werden sollen, können wir ein typisches Charakteristikum der Husserlschen Ansicht zu diesem Problem herauslesen.

In der Tat spricht Husserl sehr oft von der Ichspaltung und dem äußeren Anschein nach kommen in seinen Werken sehr verschiedene Ichs vor: z.B. das reine Ich, das transzendentale, das reale, das empirische, das natürliche, das seelische, das geistige, das personale, das konkrete, das Menschen-Ich usw. Diese mit verschiedenen Adjektiven je anders gekennzeichneten Ichs sind aber, wie ich schon angedeutet habe, nicht einfach voneinander getrennt. All diese Ichs stellen in Wirklichkeit ein und dasselbe Ich dar. Trotzdem bietet dieses Ich im Gefolge der Veränderung der Einstellung und der Vertiefung der reflektiven Analyse jeweils eine andere Phase dar, und diese Phasen konvergieren dann wieder zu einem ursprungsgleichen, einheitlichen Ich. In diesem Sinn ist die hier fragliche Spaltung sozusagen bloß eine abstrakte Spaltung. Was bedeutet dann das Wort "abstrakt"? Um auf diese Frage zu antworten, möchte ich wieder einen Satz von Husserl zitieren. Der Zusammenhang ist zwar nicht ganz der unsrige, aber sein Begriff des Abstrakten kommt klar zur Sprache. Vom reinen Ich, das "in den Akten des vielgestaltigen vereinzelten oder durch es verknüpften cogito" fungiert, heißt es nämlich:

Hierbei ist das reine Ich einerseits zwar als das in ihnen funktionierende, sich durch sie hindurch auf Objekte beziehende von den Akten selbst zu unterscheiden; andererseits doch nur abstraktiv zu unterscheiden. Abstraktiv, sofern es als etwas von diesen Erlebnissen, als etwas von seinem 'Leben' Getrenntes nicht gedacht werden kann (IV, 99).

Auch vom transzendentalen Ich erläutert Husserl gleicherweise: Es

ist hervorgegangen aus der Einklammerung der gesamten objektiven Welt und aller sonstigen (auch idealen) Objektivitäten. Durch sie bin ich inne geworden meiner als des transzendentalen ego, das alles mir je Objektive in seinem konstitutiven Leben konstituiert . . . (I, 130).

Also erst durch methodischen Vollzug der phänomenologischen Epoché hinsichtlich alles Objektiven werden wir uns klar der eigenen konstitutiven Leistungen bewußt, die wir sonst nur unbewußt durchführen. Das dabei bewußt gewordene Subjekt, das als "Zentrum aller Intentionalitäten überhaupt" fungiert, will Husserl mit dem Namen des reinen oder transzendentalen Ich kennzeichnen (IV, 109f.). Nach ihm kann man auch sagen, "daß das 'reine Ich' eigentlich nur eine Identitätsform der lebendigen Akte ist" (XIII, 246). Und ein "lebendig fungierendes" Ich ist vielmehr ein in natürlicher Einstellung lebendes Ich, d.h. ein nicht auf sich und sein Erlebnis reflektierendes Ich,

ein schlechthin "erfahrendes, denkendes, wollendes Ich" (XIV, 57). Kurz
gesagt: Das reine Ich ist nicht anders als der Ichpol, der nur durch phänomenologische Reflexionen auf die gesamte Struktur intentionaler Erlebnisse als
Kern aller konstituierenden Leistungen und als Form der Identität abstrahiert
werden kann. Und dieser Ichpol wird sogar in jedem realen Ich gefunden.
Husserl bringt dies folgendermaßen zum Ausdruck: "Das reale Ich schließt
das reine Ich in der Art eines apperzeptiven Kerngehaltes ein" und "es gibt
soviele reine Ich als es reale Ich gibt" (IV, 110).

Aus den obigen Darlegungen wird es einsichtig, daß bei Husserl das Ich
eigentlich eine untrennbare Einheit darstellt – ganz gleich wie verschiedene
Aspekte es auch im Prozeß der reflektiven Analyse zeigen mag. So sagt
Husser: "Das Ich als 'psychisch' aufgefaßtes ist das seelische, das geisteswis-
senschaftlich aufgefaßte das personale Ich oder das geistige Individuum"
(IV, 143). Trotzdem sagt er merkwürdigerweise kurz vorher, "das seelische
Ich und das personale Ich" ist "in seinem Untergrund dasselbe" (IV, 141).

Welche Struktur und Seinsweise hat nun das primordiale Ich (oder der
konkrete, ganze Mensch), bevor es in seinen verschiedenen Phasen analysiert
wird? Wenn Husserl auf diese Frage antwortet, scheint er selbst hauptsächlich
in natürlicher Einstellung zu bleiben. Nach ihm ist in der normalen Ichrede
unter dem Ich "der ganze Mensch mit Leib und Seele" umspannt (IV, 94)
und "das menschliche (bzw. animalische) Subjekt" ist identisch mit der
"Einheit von Leib und Seele" (IV, 139). Gerade in diesem Sinn ist der Mensch
"das substantiell-reale Doppelwesen", das aus der realen Verknüpfung des
realen seelischen Subjekts mit dem jeweiligen Menschenleib entsteht (IV,
120). Das in seiner Primordialität erfaßte Ich hat somit in sich selbst eine
konkrete Realität und Individualität. Von diesem Ich, das in "der natürlichen
Einstellung, in der wir alle leben," vorgefunden wird, spricht Husserl darum
folgendermaßen:

'Ich', das bedeutet für jeden von uns etwas Verschiedenes, für jeden die ganz bestimmte
Person, die den bestimmten Eigennamen hat, . . . , die ihre Zustände hat, ihre Akte
vollzieht, ferner, die ihre Dispositionen hat, ihre angeborenen Anlagen, ihre erworbenen
Fähigkeiten und Fertigkeiten usw. (XIII, 112).

Dieses personale Ich wirkt mittels seines Leibes in die Außenwelt hinein und
leidet auch unter ihr (I, 128). Aber genauer gesagt, ist der Mensch nicht ein
bloßes Doppelwesen, sondern vielmehr ein dreifaches Wesen, nämlich "mit
Leib und Seele und personalem Ich" (bzw. Geist) (I, 129).

3. HUSSERLS SCHICHTENTHEORIE

Jetzt möchte ich näher auf das Problem der gegenseitigen Beziehungen von Leib, Seele und Geist eingehen. Der bemerkenswerte Grundzug der Husserlschen Sicht scheint in der spezifischen Art seiner Schichtentheorie zu liegen, wonach eine niedrigere Schicht die nächst höhere fundiert. Andererseits bleibt aber auch die umgekehrte Beziehung nicht außer Betracht, dergemäß die höhere Schicht über die niedrigere herrscht. Die Schichtentheorie als solche ist natürlich gar nicht neu. Aber das Charakteristische an der Husserlschen Schichtentheorie läßt sich darin erblicken, daß hier zwischen den einzelnen Schichten ein Beziehungsgefüge angesetzt und sie in ihrer gegenseitigen Verflechtung reflektiert werden. Leib, Seele und Geist stellen also nicht einfach drei ursprünglich isolierte, und erst nachträglich aufeinander gestapelte Schichten dar. Vielmehr wirken sie immer und je schon aufeinander ein, stellen also eine funktionale Ganzheit dar. Das zeigt sich auch an Husserls Sicht der Realität im ganzen.

Hier werden wieder drei Schichten unterschieden: die materielle oder physische, die animalische oder seelische und die geistige Realität. Dieser Unterscheidung entspricht partiell die Zweiteilung der Natur in die "materielle" und die "beseelte, d.h. im echten Sinn lebendige, animalische Natur" (IV, 27). Die Gegenstände, die zur Natur im letzteren Sinn gehören, nennt Husserl animalische Realitäten und charakterisiert sie als beseelte Leiber (IV, 32). Unter diesen drei Realitäten stellt die materielle die unterste Schicht dar. Sie liegt also "allen anderen Realitäten zugrunde, und somit kommt sicherlich der Phänomenologie der materiellen Natur eine ausgezeichnete Stelle zu" (III, 375). In Anbetracht der Tatsache, daß Husserl sich letztlich zum transzendentalen Idealismus und Subjektivismus bekennt, erscheint es immerhin sehr beachtenswert, daß der Phänomenologie der materiellen Realität als Fundamentalschicht eine so große Bedeutung zugemessen wird.

Diese materielle Realität (bzw. materielles Ding) wird aber weiter in zwei Schichten unterschieden: in die "bloß" materielle, seelenlose Realität und in die mit einer seelischen Seinsschicht verknüpfte, materielle Realität (IV, 92).

Auch der Leib des Menschen wird dementsprechend zuerst in eine bloß materielle oder physische Schicht (den materiellen Leib) und in eine beseelte, psychophysische Schicht (den beseelten Leib) aufgeteilt (IV, 32). "Der Leib tritt zugleich als Leib und als materielles Ding auf" (IV, 158). Der Leib ist also einerseits "physisches Ding, Materie, er hat seine Extension, in die seine realen Eigenschaften, die Farbigkeit, Glätte, Härte, Wärme und was dergleichen materielle Eigenschaften mehr sind, eingehen (IV, 145). Aber der

Leib im eigentlichen Sinn ist kein bloß materielles Ding, sondern eben ein beseelter Leib, eine animalische Realität, obwohl er in sich als Unterschicht einen materiellen Leib voraussetzt und von ihm "fundiert" ist. Erst als ein beseelter kann unser Leib empfinden, und Husserl nennt diesen "fungierenden Leib" (XIII, 57) einen "aesthesiologischen Leib" (IV, 284).

Husserls Leib-theorie bleibt aber nicht bei dieser Zweiteilung stehen. Er sieht weiter im Leib auch eine geistige Schicht. Dazu sagt er: "Der Leib ist als von mir frei beweglicher Leib eine geistige Realität, zur Idee seiner Realität gehört die Beziehung auf das Ich als Subjekt freier Bewegung" (IV, 283f.). Wenn Husserl von einer "zweiseitigen Realität" des Leibes spricht, meint er, ohne den bloß materiellen Leib einzubeziehen, nur den aesthesiologischen Leib einerseits und den frei beweglichen Leib andererseits, den er "Willensleib" nennt. Dabei ist selbstverständlich die aesthesiologische Schicht die Unterlage für die Schicht "Freibewegliches" (IV, 284). Hier haben wir ein typisches Beispiel für eine Eigenart der Husserlschen Schichtentheorie, die Husserl selbst mit den Begriffen "Überschiebung" oder "Verflechtung" kennzeichnet. – Nun möchte ich auch an Beziehungsgefüge von Seele und Geist analogische Verhältnisse aufzeigen.

Die Seele wird zunächst einmal in die Mittelschicht zwischen Leib und Geist eingeordnet. Seele und materielles Ding gehören eben "verschiedenen Seinsregionen" an. Aber diese Unterscheidung schließt keineswegs gegenseitige "Verflechtung und partielle Überschiebung" aus (III, 39). Darum sagt Husserl: "Die seelische Realität ist als Realität konstituiert nur durch die psychophysischen Abhängigkeiten" (IV, 138), und die psychophysische Realität ist eine "Doppelrealität" mit der fundierenden Stufe der physischen Realität (IV, 347). Gerade weil die Seele im materiellen Leib fundiert und insofern leiblich bedingt ist, kann sie sich ausdrücken in leiblichen Gesten, in Blick und Miene. Überdies ist die Seele auch "im Realitätskonnex mit dem Geist" und "geistig bedingt" (IV, 284). In Husserls Ausdrucksweise ist die Seele "die Einheit der auf den niederen sinnlichen aufgebauten (und selbst wieder in ihrer Art sich aufstufenden) 'geistigen Vermögen' und sie ist nichts weiter" (IV, 123).

Was ist nun der Geist? Merkwürdigerweise ist es mir kaum möglich, in Husserls Schriften eine klare und bündige Definition dieses Begriffs zu finden. Das heißt natürlich nicht, er habe keinerlei Begriffsbestimmung gegeben. Z.B. sagt er folgendermaßen: "Geist ist nicht ein abstraktes Ich der stellungnehmenden Akte, sondern er ist die volle Persönlichkeit, Ich-Mensch, der ich Stellung nehme, der ich denke, werte, handle, Werke vollbringe etc." (IV, 280). Dieser Geist bedingt einerseits die Seele und bewegt in seiner Freiheit den Leib, während er doch andererseits in seinen Akten von der Seele abhängig und dadurch auch vom Leib mittelbar bedingt ist. In diesem Sinn ist auch der Geist mit seinem Leib verbunden und insoweit "naturbedingt" oder "gehört" zur Natur, obwohl er selbst nicht Natur ist (IV, 283).

Die bisherigen Überlegungen möchte ich folgendermaßen zusammenfassen: Im Menschen bedingen und durchdringen die drei Schichten einander und in dieser gegenseitigen Überschiebung bilden sie die funktionale Ganzheit der einen Realität des Menschen. Nach Husserl ist das Wort "Mensch" doppeldeutig. Es bedeutet einerseits den Menschen "im Sinne der Natur", und andererseits den Menschen "als geistiges Reales und als Glied der Geisteswelt" (IV, 143). Deshalb kann der Mensch als ein Naturobjekt nur mit seinem materiellen Leib in räumlicher Lokalisation existieren und gleicherweise nur mit seinem beseelten Leib empfinden. Aber der Mensch als bloßes Naturobjekt ist noch keineswegs Ichsubjekt oder Person; er "impliziert" nur eine Person (IV, 287f.). Dagegen ist die Person oder der geistige Mensch zwar kein Bestandteil der Natur, aber sie drückt sich im Umgebungsobjekt des menschlichen Leibs aus und wirkt so auf ihre Umwelt ein. Trotzdem kann der Geist als ein eben nicht nur natürliches Wesen in der Transzendierung aller Räumlichkeit seine ihm spezifische Funktionen erfüllen. Darauf weist Husserl hin, wenn er sagt, die Kennzeichnung des Ich als "geistiges" bedeutet "das Ich, das eben seine Stätte nicht in der Leiblichkeit hat" (IV, 97).

4. EINIGE BEMERKUNGEN

Zum Schluß möchte ich im Zusammenhang mit den obigen Ausführungen einige persönliche Bemerkungen hinzufügen. Husserls Sicht des Leib-Seele-Problems hat zweifellos einen großen Beitrag zur Überwindung der Schwierigkeiten geleistet, die durch den cartesianischen Dualismus entstanden sind. Aber nach meiner Meinung wäre es besser gewesen, hätte Husserl seine Leibtheorie mehr in die Mitte seines philosophischen Systems gerückt. Hier scheint nämlich die Phänomenologie des reinen Bewußtseins eine all zu große Rolle zu spielen. Zur Gewinnung eines neuen, vollkommeneren Menschenbildes müssen wir aber außer dem Problem der Erkenntnis auch die Praxis, die Tat des Menschen noch mehr erhellen. Dabei ist auf das Problem des Leibs ein größeres Gewicht zu legen. Denn wie Husserl selbst schon erkannt hat, wirkt der Mensch mit seinem Leib auf die Umwelt ein und schafft in ihr seine eigenen Werke.

In der japanischen Philosophie ist seit alters her die Tendenz, Leib und Seele als eine letztlich untrennbare Einheit zu fassen, immer sehr stark gewesen. Im Zenbuddhismus z.B. gilt die Seele-Leib-Vereinheitlichung, auf Japanisch "Shin-shin-ichinyo" als der ideale Zustand. Das Wort "Ichinyo" besteht aus zwei Schriftzeichen: das erste bedeutet "Eins" und das zweite "Ununterschiedlichkeit". Das Wort "Ichinyo" bedeutet also "Eins im Ursprung bei aller Unterschiedlichkeit in der äußeren Erscheinung" oder "Entdifferenzierung". Der 1960 verstorbene japanische Philosoph T. Watsuji behauptet, daß den Japanern in alten Zeiten der Unterschied zwischen Leib und Seele wie auch zwischen Subjekt und Objekt unbekannt gewesen sei. Noch ein interessantes Beispiel: Prof. Ohno, ein bekannter, japanischer Linguist weist darauf hin, daß die alte japanische Sprache kein Wort für "Natur" hatte und sowohl das Wort wie auch den Begriff erst aus der chinesischen Sprache eingeführt hat. Nach seiner Meinung könnte der Grund darin liegen, daß die alten Japaner die Natur eben nicht als Widersacher — und damit auch nicht als Gegenstand erfahren haben.

Zu allerletzt möchte ich noch kurz auf den Begriff des Menschen eingehen. Husserl hat "das Prinzip apodiktischer Evidenz als philosophisches Leitprinzip" gewählt (VIII, 126) und das ist für ihn nichts anders als "die apodiktische Evidenz des Ich-bin" (VIII, 166). Darum betont er, daß "der Satz 'ich bin' das wahre Prinzip aller Prinzipien und der erste Satz aller wahren Philosophie sein muß" (VIII, 42). Das besagt, daß Husserl bewußt von solipsistischen Standpunkt ausgegangen ist und dann erst den Weg zur Intersubjektivität gesucht und dessen Möglichkeit theoretisch zu begründen versucht hat. Dazu hat bei ihm 'der Leib als Mittel zur Erkenntnis des anderen Ich' eine sehr wichtige Rolle gespielt. Immerhin manifestiert der alte Husserl, daß im Sinn des Menschen, der schon als einzelner den Sinn eines Gemeinschaftsgliedes mit sich führt, "ein Wechselseitig-für-einander-sein" liegt, daß also ich mit jedermann "als ein Mensch unter anderen Menschen" bin (I, 157f.).

Husserl ist, wenigstens theoretisch, den Weg von der einsamen transzendentalen Subjektivität zur Intersubjektivität gegangen. Im japanischen Denken ist die Intersubjektivität aber gerade der Ausgangspunkt. Das zeigt sich schon an dem japanischen Wort für "Mensch" nämlich: "Ningen". Dieses Wort ist eine Kombination von zwei chinesischen Schriftzeichen. Das erste Zeichen "Nin", das für sich allein "Hito" gelesen wird, bedeutet soviel wie "Person". Das zweite Zeichen "Gen", das allein meistens "aida" gelesen wird, bedeutet soviel wie "zwischen". Also charakerisiert das japanische Wort "Ningen" schon als solches den Menschen als "Mensch unter anderen Menschen". Die japanische und asiatische Weisheit birgt manchmal in sich solche tiefen Einsichten wie Kleinode. Aber es fehlt ihr leider oft an der Strenge wissenschaftlicher Analyse und Begründung. Diese Strenge möchte ich von der europäischen Philosophie, in meinem Fall von Husserls Phänomenologie erlernen, wo die Idee strenger Wissenschaftlichkeit bis zum äußersten durchgeführt ist.

Nanzan University

ANMERKUNG

[1] Betreffs der Quellenbelege: Die römische Zahl zeigt den betreffenden Band der *Husserliana*, die arabische die Seite des Bandes.

EVELYN M. BARKER

THE EGO–BODY SUBJECT AND THE STREAM OF EXPERIENCE IN HUSSERL

In Husserl's phenomenology, it is important that a conscious experience is not a discrete, self-contained event, but occurs in an ongoing *stream of experience*. Thus a central purpose of Husserl's phenomenology is to describe "conscious experiences in the concrete fullness and entirety with which they figure in their concrete context — *the stream of experience* and to which they are closely attached through their own proper essence." [1] In the main part of this paper I argue that Husserl's characterization of the stream of experience in *Ideas* and *Phenomenology of Internal Time-Consciousness* is defective insofar as it applies to the stream of experience of the pure ego — a stream which is a philosophical construction, abstraction or fiction — rather than to the stream of experience of the real ego-body subject, a real existent. This defect is significant philosophically, for the stream of experience represents the temporalization of the ego-body subject as it realizes itself in thought, feeling, and action. In the stream the various experiences of the ego are not merely unified in a serial progression; they are also *centralized* into an ego pole, as Husserl realized in his later *Universal Teleology*, which I shall discuss at the end of the paper.

I first want to show that the stream of experience Husserl describes is an abstract logical construct which does not fit the real stream of experience of an ego-body subject. We can locate the difficulties in Husserl's conception of the stream in his procedures in the bracketing of psychophysical reality, including that of the experiencing subject's body. Husserl's moves here are more radical than his own phenomenological method demands; indeed, a better phenomenological description of the stream of experience of an ego-body subject is possible utilizing many of Husserl's descriptions of conscious experience. Husserl's crucial move is to substitute the *postulated* stream of experience of a "pure ego" for the *lived* stream of experience of a "real ego-body-subject." He does so because he believes (1) that the phenomenological method requires that he deprive the experiences of the real ego-body subject of "intentional relation to man's ego and man's body"; and (2) that this purification will simplify without distorting these experiences. It is these two beliefs which I shall claim are mistaken.

The stream of experience which is the *concrete context* of full-blown

A-T. Tymieniecka (ed.), Analecta Husserliana, Vol. XVI, 183–191.
Copyright © 1983 by D. Reidel Publishing Company.

conscious experiences is the sequence of experiences which comprise the life history of a "self-identical real ego subject," the kind of consciousness "that is naturally apperceived, given as human and animal at once, in close connection with corporeality."[2] Husserl admits that "only through the empirical relation to the body does consciousness become real in a human and animal sense." Why then does he describe the stream of experience of the disembodied "pure ego," rather than that of the real ego-body subject? Husserl maintains, quite rightly, that consciousness cannot lose its essential nature by becoming linked with the body. (What this seems to mean, however, is that it cannot become the kind of thing appearing through perspectives, like a material object.) He holds that consciousness does become "other" in becoming *individualized* as a part of nature in its association with the body. Consequently he makes a sharp distinction between the "absolute essence" of the flux, its "flowing thisness," and the states of consciousness of the real ego subject, in which an ego's individual real properties are manifested, properties that rest on bodily foundations.

I want to take issue with Husserl's contention that consciousness somehow becomes "other" in becoming individualized by its association with the body. This claim contains two different assertions: (1) that consciousness becomes other through being individualized; and (2) that consciousness becomes other through being individualized through its association with the body. Now it is plausible to conceive an embodied consciousness without the particular kind of bodies that human beings in fact have; and it is conceivable that such a consciousness would not be centered in the experiences of an individual ego; it might, for example, have a group consciousness like that imagined for ants in an ant colony by E. B. White in *The Once and Future King*. In such a group consciousness there might be a number of different consciousnesses, each of which has experiences, but the locale of any particular experience in one stream or other would not matter. Thus there might be a number of different streams in each of which a hunger experience is occurring, and the common consciousness would simply register all these different experiences indifferently. But in the case of an individualized consciousness, preference is always given to the conscious experience in one's own stream. Thus, your neighbor at table may be much more hungry than you, but it is your own hunger that you feel inside you that takes your attention.

But whether a consciousness not necessarily linked to a body must be a nonindividualized consciousness is a completely different matter. It may even turn out to be an essential feature of consciousness that it be individualized — indeed one would suppose this to be the kind of thesis phenomenology might even prove or disprove. In any case, it is true that being individualised is a firmly established possibility of consciousness. Thus we must entertain the possibility that being individualized, and even being embodied, is at least an *intrinsic* feature of consciousness, and in no way alien to it. If so, Husserl should not dispense with the stream of experience of the real ego-body subject on the ground of its being individualized. Husserl's description of the stream of the pure ego ought to apply, in outline at least, to the flux of the ego-body subject, according to his own views.

For, although Husserl distinguishes the psychological point of view from the phenomenological, he implies that they cannot be in contradiction: the psychological sees each psychic event as an inner state of a man or animal; the phenomenological, the experience in its "intimate subjective flow" — that is, "purified of its individualized ego-body character." But the perceptions of the "pure ego" are dependent on the real ego subject, whose purified experiences are its data: As Husserl puts it, the "pure" experience "lies" in the experience of the real ego-body subject: "With its own essence it takes on the form of psychical subjectivity, and therewith the intentional relation to man's ego and man's body."[3] If so, the essential content of both the pure experience and the unreduced experience of the ego-body subject must be the same — there ought not to be anything in the experience of the pure ego which could not be in the experience of the individual ego-body subject. If the description of the pure experience is to be complete, it must contain all the significance attaching to that of the real ego subject, that is, whatever could be shared or entertained by *any other real ego subject.*

But is this the case? Does "purifying" the experience of intentional relation to man's ego and man's body remove essential content from or distort the experience of the real ego-body subject? I believe that there is a whole class of experiences in which the loss of this intentional relationship emasculates the meaning of the experience: namely, those in which there is essential mention of the *whole stream of experience* of the ego-body subject, of its *finite character*, and its *dependence on the body*. The problem here is that the stream of the pure ego is an "endless continuum of durations," one which extends infinitely in the three dimensions of past, present, and future, whereas the stream of experience of an individual ego subject, as a unity resting on bodily foundations, obviously has a beginning and end, roughly bounded by the birth and death of the body.

Furthermore, as surely as an ego-body subject is aware of other ego-body subjects, it must be aware that its own ego stream is finite, on Husserl's own showing. If, by means of empathy, the ego subject projects other ego streams from observing mind-suggestive bodily behavior, it is also made aware of the cessation of ego-indicating behavior on the death of the body. Correlatively, each ego subject must expect that its own ego stream ceases at the death of its own body. There are many kinds of universally shared conscious experiences in which this awareness is essential — the fear of one's own death or that of a loved one are obvious examples. We could take Husserl's own example of *joy*, the joy of a man at the birth of a son, or more generally, the desire for a child. In their "intimate subjective character" these are entwined with a consciousness of the coming into existence of a being intimately related to the ego-body subject, one who will be a surrogate for the self in the human world, one who will take the ego's place after death. Would the joy exist at all, would it have its special quality unless the subject was aware of its finiteness?

All that the phenomenological method itself requires, I believe, is that such experiences be "purified" of features depending on the particular character and circumstances of the real ego-body subject — for example, whether the person desiring a child is an old wise hereditary monarch or a young beautiful and foolish girl. What Husserl's method demands is a description that would be true of *any* individualized ego subject, rather than of just one or of an arbitrarily limited number of ego subjects. In effect, Husserl does not merely *suspend* the belief in consciousness existing in empirical relationship with the body; he *denies* or at least fails to acknowledge its so appearing.

This is, I submit, a serious defect: The ego subject's awareness that the continuance of the ego stream depends on the integrity of the body conditions the wholesale experiences of ego. The vulnerability of the body affects the way we perceive material things; it evokes distinct emotions about the fates of other egos, and colors an ego's basic orientation to its natural and human environment. The felt exigencies of the body require the ego both to act and suffer in the world: although awareness of bodily needs may often be at the periphery of consciousness, or experienced in the neutralized mode, it punctuates the consciousness in a regular predictable way which continually modifies the experiences of the ego. Husserl tends to stress, as experiences in which the ego attains heightened self-awareness, those in which it realizes its *freedom*, in decision, action, or resolve. But, equally, such heightened awareness occurs through sufferings visited upon the ego in its experiences of *necessity*, mediated by the body. Sex experiences, too, are ones of heightened ego awareness, extremely meaningful events in the stream of the ego-body subject. Thus, the experiences of an embodied and individualized consciousness has characteristics not parallelled in a "pure" nonindividualized disembodied consciousness.

Since Husserl's purified experiences have lost the characteristics of a psychic event in the world of nature, when strung together in the continuum of the stream of the pure ego they escape the boundaries imposed on the experience stream of the real ego-body subject in its felt relationships to its own body. As a result Husserl's conception of the stream of the pure ego as an "infite unity" cannot fit the stream of the ego-body subject from which it is derived. But both phenomenology and logic demand that they should at least be consistent.

I shall now attempt to show that Husserl's conception of the stream of the pure ego is a logical construction (even a fiction) based on his phenomenological description of the temporal quality of an experience, rather than a phenomenologically reduced reality. In *Ideas* (#81) he draws a sharp contrast between the transiency of the single experience and the endlessness of the flux in which that experience occurs:

Every single experience ... is necessarily one that endures; and with this duration it takes its place within an endless continuum of durations .... Every single experience can begin and end ... but the stream of experience cannot begin and end.[4]

He derives the infinity of the flux of the "pure consciousness" from the temporal quality of the single experience — its reaching forward and backward to preceding and succeeding experiences. Each immediate experience, whatever its content, has a temporal quality marking it as a "now" experience, which contrasts and connects it to an experience "just past" whose memory is retained as part of the "now" experience; beyond the "just past" experience, the "now" contains also a modified retention of what was "just past" for that "just past," and so on without stop. Similarly, each immediate "now" pushes forward toward a future "now" which will be the culmination, continuation, or cessation of the "now" experience, again without stop.

Because of this continuous interlacing of the time features of a single experience along the three dimensions of past, present, and future, Husserl goes on to characterize the flux of the "pure" ego as an "infinite, progressive unity." But I would point out that these temporal features of a single experience do not require the *infinite* extension of the stream but only its *indefinite* extension — or perhaps the *illusion* of infinite extension in the directions of past and future.

In *Phenomenology of Time Consciousness* the logical construct basis of the "absolute flux" is most apparent as Husserl argues the paradoxical thesis that the flux has no *duration* even though each single experience which composes the flux does have duration. In appendix 6 he offers the strictly logical argument that anything that has duration must be something which may carry over in time either changed or unchanged. Persistence of this kind does belong to each single experience that occurs in the flux, but the flux itself is just this series of happenings connected into a unity by our thinking. What Husserl seems to have in mind here is an important point concerning the logical relationship between the flux of the ego and its individual experiences. An experience of anger can, he would say, properly be said to endure — it announces itself typically in an irritation stirring in the background of consciousness, comes center stage as it develops into full-blow flury, and then subsides leaving behind a resentment which reverberates through subsequent experiences. But the flux of the angered ego is not an entity running alongside these anger states, but is constituted through such states, and has no separate existence from them.

When we say that an ego is angry, we mean that a stretch of the flux with the form of anger endures for a period at a given point in a temporal route following other experiences which modify it, and followed by other experiences which it modifies. We make reference to an *ego*, thus signaling that each experience "stands in need of completion" (cannot stand alone) in respect of some connected whole which in form and in kind is not something we are "free to choose, but are rather, bound to accept."[5] The specific connected whole of the stream of experience is the ego subject. The ego *is* simply the stream considered from the perspective of "formal" factors appearing in it which enable us to make motived connections between the contents of successive experiences in the stream. In the stream of the ego "given to anger" we will find numerous experiences with the form of anger conjoined with many different contents; whereas for the ego "slow to anger" we will find few anger incidents with a more limited selection of contents.

Husserl is expressing an important point about the logic of the ego, but it does not prove that the ego stream is without duration, or that it cannot be said properly to persist or alter. It only shows that the duration of the stream must be of a kind which can be understood as the resultant of the durations of the conscious experiences constituted in it. The latter point raises no problems; on the contrary, it enables us to fix the bounds of the ego stream in terms of conscious experiences roughly coincident with the birth and death of the body.

Husserl's refusal to grant *duration* to the stream is grounded on his failing to assign *individuality* to the ego stream; a move which then requires him to deny that the ego stream is a temporal object.[6] But by retaining the intentional ego-body relationship, we restore the individuality of the stream; the real ego-body flux, is, in Husserl's sense, a temporal object. At one point he admits some sense in the notion of a conscious experience as a perspective: each conscious experience is a temporal perspective variation of the flux, containing within it reference to other moments of the flux with which it has motived connections. His description of the relationship between the real ego-body subject and its concrete experiences is also somewhat analogous to that of the material object and its perspective variations. Furthermore, both change and persistence in the character of the ego may be explained in terms of what Husserl calls *longitudinal intentionality* in the stream of experience — motived connections of conscious experiences and changes in associated (bodily-mediated) experiences in the course of its life history.[7]

It would be a worthwhile philosophical project to undertake a phenomenological description of the stream of experience of a real ego-body subject, focusing on the intentional relationship to man's ego and man's body: such a study would trace the course of the ego in its bodily career as it develops and changes in its transactions with the human world and other ego-body subjects. The notion of the embodied and individualized ego-stream may be fruitful both in theoretical philosophy of mind and practical ethical and political studies. I would, however, like to conclude by discussing Husserl's own interesting use of the notion in the late work entitled *Universal Teleology*.[8] In *Time-Consciousness* he seems to dismiss the stream of experience as a mere analogy, not to be taken seriously from the philosophical point of view. But in this work Husserl makes a strikingly original application of it in his conception of a universal teleology. And now, it is not the stream of the pure ego, but that of the real individual ego-body subject. In fact, Husserl acknowledges that in the earlier work he failed to give the stream of experience the egological significance which rightly attaches to it; he now emphasizes that the ego itself is an evolving process, continually developing through its deeds as it temporalizes itself in the objective world. He even describes the universal teleology itself as "the totality of monads *streaming* to universal self-knowledge in an unending gradual process" (my italics).

The notion of the universe as a system of ends including beings of higher and lower grades, evolving as its members fulfill specific ends, is by no means new with Husserl. It is present in classical authors, in Hegel and in Kant. But in most philosophers, the individual human being participates consciously in the universal system of ends only as a rational being, one who acknowledges goals and principles that are universal and validated by reason. Such a being

sets aside his private aims as an ego-body subject; in so doing, the individual
ego negates its own individuality as an ego-body subject. But Husserl argues
that the primordial basis of man's participation in the universal teleology lies
not in the human as a rational being, but rather in his sexual experience as an
ego-body subject. Husserl arrives at this conclusion through phenomenological
reflection on the intentionality of the sexual impulse, and procreation: It
is through fulfillment as an individual ego-body subject in sexual-social life
that the human subject "awakens" to the realization that his own stream
of experience merges with, emerges from, and flows into an ongoing stream
which is the endless continuing stream of the whole community of human
monads.

In outline Husserl's thinking goes as follows. In its typical heterosexual
manifestation in marriage, the sexual impulse has for its object a human
ego-body subject, so that the individual is made aware of his (her) membership in a human species distinct from other species that coexist in the natural
world. Yet not just any other member of the human species will fulfill the
sexual need: it must be a human being of the opposite kind, one who takes a
complementary sexual and social role. Nor will just any other human of the
opposite sex suffice; it must be one to whom the individual ego subject is
specifically attracted, and one who has a corresponding sexual desire for the
ego subject. The sexual impulse thus involves the individual ego-body subject
directly in a complex individual-social-natural world, bringing awareness
of (1) one's species character within a universe containing other species,
(2) one's distinctive social roles within a human world, and (3) one's own
distinctive individuality in having and being the specific object of a particular
affection of an ego-body subject. The fulfillment of the sexual impulse
in intercourse cannot, he insists, be divided into two different acts in two
different streams; rather, it is *one experience* which is the *merging* of two
streams.

Furthermore, the ego-body subject comes to know the sex act as being
also an event occurring in the natural world, setting off a natural process
in which the female partner produces a child. Through recognition that his
own being as an ego-body subject is mirrored in a comparable sex impulse
in another ego-body subject, and that his own sex impulses are mirrored in
the streams of other ego-body subjects, each ego-body subject realizes a
common human bond uniting him not just to his own ancestors and progeny,
but to the whole of mankind, past and future. The intentionality of the
sex impulse thus contains and develops the sense of a biological and social
human community implied in the fulfilling of the sexual desire of a real

ego-body subject. In the attainment of metaphysical knowledge of the universal teleology, and in the fulfillment of man's role within it, Husserl does not negate the reality of the individual human ego-body subject, suspend or overlook it.

From ancient times, philosophers have taught that the human ego-body subject must free himself from the body, if he is to fulfill his moral destiny as a human being. But Husserl here points out how bodily mediated experiences are crucial in the spiritual development of the ego needed for the fulfillment of that goal.

University of Maryland, Baltimore County

NOTES

[1] *Ideas* (New York: Collier, 1972), #35, pp. 104–5.
[2] Ibid., #53.
[3] Ibid., p. 151.
[4] Ibid., p. 217.
[5] Ibid., p. 221.
[6] *Phenomenology of Internal Time-Consciousness* (Bloomington: Indiana University Press, 1969), p. 152.
[7] Ibid., pp. 106–9.
[8] *Teleologia Universale* (1933); see M. R. Barral, 'Teleology and Intersubjectivity in Husserl,' in *Analecta Husserliana* (1979), 9: 221–233.

AURELIO RIZZACASA

LIVED EXPERIENCE OF ONE'S BODY WITHIN ONE'S OWN EXPERIENCE

In Husserl's phenomenological vision the anthropological problem poses itself as the need for explicating the ulterior aspects of the subject-man as distinct from the bio-psychic moment of human individuality.[1] This is particularly important in bringing out the epistemological and philosophical limits of scientific psychology, especially now that this scientific psychology has — quite rightly — claimed and established its autonomy with respect to philosophical psychology.

In other words, one has to bear in mind that it is one thing to draw up theoretical and experimental models for the description of the psychic sphere of man, but quite another to concern oneself with the consciousness of man. Human consciousness, in fact, is intentionally open to both the ultimate essences of reality and to values, and strives as much for knowledge of being as it is ready to take operational decisions in full awareness and responsibility.

Seen in this light, therefore, the specific ambit of competence of scientific psychology is the field of psychic mechanisms, in the twofold sense of the mnestic-informative system and of the biological correlate of the psychic mechanisms themselves. The specific sector to be attributed to the world of philosophical ethics, on the other hand, concerns the moment of freedom, of obligation, and of behavioral decisions, moments that are collocated in the unique and unrepeatable situations of the existential experiences of the personal consciousness of each individual man.

It must be clear that, continuing along this line of argument, the long-standing and indeed classical aporias that oppose necessity and freedom, bio-psychic mechanistics and spiritual autonomy, will be overcome and left behind by a new vision of man. And the novelty consists of the fact that each aspect of the alternative, rather than representing a totalizing vision of the human being, constitutes only the consequential result deriving from a wider perspective approach to man as such.

We could thus say that man represents a nucleus of problems from which there emerge many fields of research, and each of these fields makes use of appropriate categorizations and appropriate linguistic assertions. The fields of research, in turn, establish a basic complementarity between them, they integrate, enrich, and valorize each other. Nevertheless, there may be

A-T. Tymieniecka (ed.), Analecta Husserliana, Vol. XVI, 193–200.
Copyright © 1983 *by D. Reidel Publishing Company.*

encroachments from one field into another as the research hypotheses are pursued and worked out further, and temptations to totalize one field at the expense of the others may come to the fore. When this happens, the complementarity of the different themes simply gives way to the contradictoriness of the solutions proposed in the various fields, and consequently theories will delineate themselves that afford a reductive treatment to the anthropological theme taken as a whole. In this connection one need only think of the materialist philosophic proposals emerging from premises of a bio-nature or, going to the other end of the scale, the idealist proposals deriving from a philosophic psychology that even denies the possibility of founding an experimental scientific psychology.

The considerations we have made so far find their epistemological support, at least on the linguistic level of the assertions that have been used, in what has become known as "Hume's law"; according to this well-known law, there exists an unbridgeable qualitative diversity between description and prescription, between scientific interpretation and ethical decision.

In the last resort, therefore, the problem is that of avoiding all totalitarian imposition of values in the name of a pseudo-justification of a scientific nature. This is particularly important in our own day and age, which is characterized not only by an increase in the prestige of scientific research, but also by a growing appreciation of the necessarily ideological nature of scientific knowledge itself.

If we now return to the central core of the theme that constituted our starting point, namely, the relationship between scientific psychology and philosophic psychology, we find that the considerations made above seek to recreate space of a philosophic nature (and significant from a theoretical point of view) within the bounds of the anthropologic problem. Space, moreover, that already delineates itself, at least in our opinion, in the scientific treatment of the manifold aspects of the anthropological problem itself. In fact, it is only too easy to note that, even when we have taken into consideration all the demands made on man that have been developed by the different sciences that concern themselves with man, we are still very far from having resolved every anthropological problem, and this is as true on the theoretical level as it is on the ethical one. In other words, there always exists a problematical ambit that is not resolved (and perhaps even cannot be resolved) at the level of the descriptive models of a theoretical nature proposed by scientific knowledge. In that case we could perhaps maintain that the dimension of human being unknown to science is a mysterious dimension that forms part of the *Lebenswelt* (to use an expression taken from Husserl's phenomenology) and not of the world of the interpretative assertions that can be proposed by science.

It is precisely at this level, therefore, that one has to look for the possibility of founding both the theoretical and the ethical dimension as philosophico-spiritual dimensions of the human person.

Even though there can be no doubt that one must not lose sight of the autonomy of the theoretical ambit with respect to the ethical one, the problem of a theoretical foundation of the anthropological thematics assumes particular importance in a line of interpretation of a phenomenological nature. In this connection it will therefore be helpful and desirable to reflect about some significant considerations made by Husserl and to keep them in mind, above all, in an interpretative view that aims at bringing out their methodological significance.

In fact, Husserl came onto the scene following Kant's attempt to realize the objective foundation of philosophical knowledge (inasmuch as it was scientific) in the ambit of the phenomenological world and Fichte's attempt to reestablish the doctrine of science in an idealistic manner and (basing himself also on the Cartesian idea of a universal philosophy and a universal science) dedicated himself to the project of establishing a scientific philosophy in a rigorous sense. This really constitutes the project that underlies the whole of Husserl's speculation.

As regards the theme with which we are here concerned, this consideration brings us back to the Husserlian problem of the phenomenological refoundation of a scientific psychology capable of overcoming the crisis of the objective sciences. Working to this end, Husserl also endeavored to provide an answer to the needs of a philosophic psychology that would make possible an effective understanding of man in his totality.

It was precisely in the light of this need for profundity and objectivity that Husserl inaugurated a philosophical interpretation of the anthropological thematics that was based on and elaborated from intuitive awareness of the experience of one's own body (*Erlebnis des eigenen Leibs*). As is well known, this interpretative line was subsequently to find its most significant developments in the phenomenologico-existential philosophies of Sartre and Merleau-Ponty on the one hand, and of Marcel and Levinas on the other.

In Husserl's perspective, therefore, the originary intuition of the *Erlebnis* of one's own body, rooted in the solipsistic matrix of the subject-man, constitutes the foundational moment of a reductive process of a phenomenological nature, a process that finds its tendential point of arrival in the objective essences of lived experience.

More particularly, a clear understanding of Husserl's notion of "own body" requires one to bear in mind two important conditions: (1) the overcoming of the naturalistic attitude in favor of the transcendental attitude, and (2) Husserl's attempt to provide a scientific foundation of psychology in a philosophic sense, but starting from a negative appreciation of psychology inasmuch as it is erected on a biological foundation of positivist derivation. Seen in this light, the own-body concept is therefore collocated somewhere between the concepts of "soma" and "psyche," and thus implies that in lived experience the "object body" should be overcome and left behind, giving way to a psychologic and spiritual category understood in the philosophic sense of "I."

From the systematic point of view, Husserl's starting point (in *Ideen*) is the distinction between *Körper* and *Leib*: the former is understood as the physical body, objectively capable of being delimited in space and time and immersed in causal connections, while the latter is interpreted as conscious corporeal experience.

Consequently, awareness of one's own body leads reflection into the ambit of the intuitive immediateness of lived experience. It follows from this that, between the horizon of animal nature and that of human nature, experience of the "I" draws its origin, its significance, and its positive valency, as it were, from the unavoidable reference to the *Erlebnis* of one's own body.

When analyzed in the light of phenomenology, this experience leads Husserl to the recovery — in the description of the *Erlebnisse* centered on one's own body — of the only authentic and (as one might say) nuclear elements of the subjectivity of man. In fact, although man is immersed in the world, he lives (experiences) himself in a privileged manner and at the same time opens himself to his similars by means of analogic experiences of an empathic nature founded on awareness of his extraneousness (otherness). Our own body, therefore, enables us to live our sphere of ownness[2] (*Sphärender Eigentlichkeit*) and to distinguish it from what is extraneous to us, even though it comes into relation with us.

Moreover, we must not forget that through the experience of our own body we discover ourselves to be spatially and temporally determined; and we localize the various cognitive experiences by means of the two sense-perceptive modalities of "contact knowledge" of the tangible, receptive type and "remote knowledge" of the audiovisual type. Last, our own body is also the privileged place of emotional experiences, as also of behavioral decisions. But the most important thing consists of the fact that through our body we have the possibility of living not only the representations of the exterior world, but also, and above all, the private representations of our own personal interiority.

Developing the own-body thematics, Husserl thus undoubtedly overcame the traditional problematics connected with the subject-object relationship or with the contradictory positions of realism and idealism, of solipsism and intersubjectivity. Moreover, and this is really the most important aspect, he furnished the starting point for a new approach in philosophy and in science; an approach that, rather than dwelling on the contradictory moment of the opposing alternatives, reemphasizes the continuity of the intentional processes of human consciousness.

In this connection we must not forget the positive consequences that the subsequently developed phenomenologico-existential philosophies were to derive from the theme of corporeity, nor the contributions later to be made by phenomenologically oriented psychology and psychiatry. The own-body concept as outlined by Husserl also makes it possible to over-come the traditional dualism between body and soul; this concept, in fact, constitutes a new approach capable of avoiding the reductive simplifications that privilege only one aspect of human being and therefore inevitably end up by forgetting the very complexity of man.

We are effectively concerned with an experience in which each man, using himself as a starting point, finds within himself the occasion for opening his own being onto the world and man himself. It is an opening that in the experience of one's own body highlights the initial moment of the cognitive itinerary for getting under way in a phenomenological manner a theoretical project that extends to the highest experiences of transcendental subjectivity.

Although for the sake of simplicity in analysis we have here separated the notion of the "own body" from the overall context of the arguments used by Husserl, this notion must nevertheless be seen in a wider and more articulated perspective. For in this wider perspective it constitutes but the starting point of a phenomenological process that leads from one's own body to the "I," then proceeds from this personal I to the transcendental ego, and eventually uses this latter as the foundation of the intersubjective relation.

This phenomenological process is obviously present in all the treatments of the anthropological theme undertaken by Husserl, but we find it expressed and summarized in a particularly significant manner in his *Cartesianische Meditationen*. At this point in our discussion we will refer specifically to this part of Husserl's arguments.

In the *Meditationen*, indeed, the *Erlebnis* of one's own body (already treated by Husserl in *Ideen*, where he puts forward the thematics we have

so far considered) is analyzed by means of a process of phenomenological
reduction that, in a series of successive epochés, proposes the passage from
the natural attitude to the transcendental attitude.

In this general framework, then, in this gradual opening of the solipsistic
matrix toward the intersubjective horizon, Husserl highlights the experiential
Erlebnisse that, starting from the psychophysical I, pass through the personal
I and attain to the "transcendental ego." This process is specified by means
of the concepts of ownness (*Eigenheit, Eigentlichkeit*) of *extraneousness*
and *otherness*, concepts that, using the sphere of our "own" as the starting
point, enable us to recover the horizon of the other in an intentional tension
protended toward transcendency, and this already within the sphere of
subjective experiences and both on the level of our body and on the level of
the personal "I."

In fact, we may here point out that, as far as Husserl is concerned, solipsism
constitutes an originary and indispensable experience, and it is within this
experience — and also for the purpose of breaking out of this experience
— that there is proposed the *Erlebnis* of the other already inherent in the
intentionality of the transcendental ego; and this *Erlebnis* assumes the con-
figuration, the specific dimension of *extraneousness*.

From the ontico-noematic point of view, therefore, experience of the other
proposes extraneousness as the intentional pole of conscious experiences,
of *Erlebnisse*; it is in these latter that the being-there of the other assumes
a place in the various levels of the egological horizon and thus becomes a
being-there-for-me. As we have already said, therefore, the extraneous is
already inherent or *innate in one's own* and can be experienced only *starting
from one's own*.

In Husserl's perspective, however, the theoretical horizon in which the
extraneous is recovered in one's own self, and also starting from one's own,
is provided by a series of epoché processes to which the experience of the
Erlebnisse is submitted. In this general framework, indeed, *ownness* is focused
into *one's own*, while extraneousness gives rise to the image of otherness in
the truest sense of the term. Thus the ego is integrated into the alter ego, and
is completed in it. At this point we must underscore that it is precisely in the
ambit of one's own that the lived experience of *one's own body* comes to be
privileged.

In the horizon of one's own, therefore, the process of phenomenological
reduction of the *Erlebnisse* (lived experiences), starting from the originary
experience of one's own body, gives rise to the *psychophysical I*, the *personal
I*, and the *transcendental ego*. And it is within the ambit of this latter, then,

and at a level already elaborated, that there are structured the experiences of otherness (which emerges from extraneousness) and of intersubjectivity as an intermonadic relationship.

It follows that the moment of one's own assumes a nuclear value within the transcendental ego; for in this moment there is structured both one's ownness (or one's peculiarity) and one's experience of othe other, this latter experience, as we have already seen, being rooted in extraneousness. In the transcendental ego, moreover, the *I-am* reveals itself as the originary givenness, so that Descartes' cogitative moment, the *I-think*, finds its theoretical and practical prolongation and enrichment, as it were, in the *I-can*.

Consequently, we are not here concerned with subordinating ethics to theoretical philosophy, but rather with underscoring the foundational importance of philosophical psychology. This latter, in fact, constitutes the only theoretical and nuclear element capable of conferring upon the anthropological problem a significance and ulterior aspects of a spiritual nature.

University of Siena

NOTES

[1] The views developed in this essay are based on the following of Husserl's texts: (1) *Ideen*, bk. 2, sec. 2, pars. 35–42 and bk. 3, chap. 1, par. 2 and app. 4; (2) *Cartesianische Meditationen*, pars. 42–46. For an interpretation of the concepts developed in *Ideen*, book 2, sections 2–3, the reader may compare the following parallel texts, which particularly concern themselves with the theme of the *Leib* (living body) in relation to *Einfühlung* (empathy). (1) *Zur Phänomenologie der Intersubjektivität I, Husserliana*, vol. 13: no. 3, *Die Einfühlung*, Texte aus dem Jahr 1909, Beilage XI (1914–15); no. 4, *Stufen der Einfühlung* (um 1910), Beilage XIII (1913), Beilage XIV (1913), Beilage XV (1913); no. 6, Aus den Vorlesungen *Grundprobleme der Phänomenologie*, Wintersemester 1910/II, and especially chap. 1, pars. 2–4; no. 7, *Wie konstituiert sich die Realität Mensch, wie gewinnt für mich mein Leib reale Einheit mit meinem Subiektiven und vorher bei Anderen?* (1914–15). (2) *Zur Phänomenologie der Intersubjektivität II, Husserliana*, vol. 14: *Texte aus dem Zusammenhang der Vorbereitung eines "Grosen Systematischen Werkes"* (1921–22), and especially nos. 1, 3, Beilage IX, no. 2, and Beilage XXX. Moreover, it will be useful to provide the reader with some references to analogous concepts developed by Husserl after his *Ideen* and thus to integrate the texts listed above. To this end the reader should consult *Zur Phänomenologie der Intersubjektivität III, Husserliana*, vol. 15: no. 16 (1931), Beilage XVI (1832), Beilage XVII (1932), no. 17 (1931), Beilage XVIII, no. 18 (1931).

[2] Here we come face to face with the problem of how to render the German *Eigenheit* and *Eigentlichkeit*. The translator's first instinct was "ownness" and "peculiarity," but

this distinction is far from incisive when one bears in mind that *Eigenheit* in colloquial acceptations is usually rendered as "property," "characteristic," etc. The matter is further complicated by the conventional Italian rendering of the two terms, namely, *appartenenza* and *appartentività* (both derived from the verb *appartenere*, "to belong to"), and which are of course the terms actually used in the text. They are probably best rendered in English as "belonging" and "belongingness," but even here the contentual difference between the two does not exactly spring to the eye. Without continuing the linguistic disquisition, which could be well nigh endless, we may note that English phenomenological literature does not seem to contain any trace of this dichotomy. The translator in any case does not feel qualified to propose an English distinction between the two terms, all the more as he is convinced that any attempt to do so in the present context would create more confusion than clarity. He has therefore opted for the simple "ownness" to represent them both! Strange but true, here at least it seems to make the meaning clear in spite of the reduction. The reader remains of course free to substitute any two terms *he* feels to be best suited to express the difference. (Translator's note.)

PART IV

SOUL AND BODY IN PHENOMENOLOGICAL PSYCHIATRY

LORENZO CALVI

LIVING BODY, FLESH, AND EVERYDAY BODY:
A CLINICAL-NOEMATIC REPORT

In my experience as a phenomenological psychiatrist a dialogue with the patient is not confined to the time of the clock. Its time extends in fact beyond immediately physical presence, becoming a time within which I await further donations of meaning, a monad with windows (E. Straus), divested of all my "worldly" involvements, remaining open to the own body in which those virtual movements arise, through which, on the basis of a transcendental constitution, an empathic consonance (*Einfuhlüng*) with the patient takes place. My intentional effort is to facilitate a network of relationships and meanings leading up to a self-givenness of the "patient" in his essential aspect, of fulfilled significance, without residuals, at once typical and normative. Such an "epiphany" is a unity resulting from cooperation between the meaningful forms of the patient's world and those of the psychiatrist. The field in which it manifests itself is a consciousness both of one and of the other.

Because of its metahistorical character, an anthropological epiphany persists until new donations of meaning give rise to another. Psychotherapeutic work then proceeds through a cooperative opposition between the two, or even more, epiphanies, a multi-dimensional space of possibilities emerging through this dialectic. In my view, Husserl's phenomenology makes it possible to conceive of a plural ego — one figuratively "cubist" — as the locus of primordial intersubjectivity, disclosed through suspension of that conventional world with its merely historical relations from which the patient suffers in his anxiety, and which the phenomenologist seeks to render transparent through "bracketing" (*epochē*). It is just because patient and psychiatrist confront one another within experiences which are different but comparable, namely, between that of anxiety and that of bracketing respectively (Calvi, 1963), that I am able to convey to the patient who has told me he cannot go on living, that he has the possibility of grasping his life as a multiplicity on the level of the transcendental constitution of the various aspects of his own self, both of those which manifest themselves in his different epiphanies, and those which are intuitively possible.

I will now describe some of the features of my experience in the process of encounter (*koinonia, Begegnung*) with a patient who manifested two

A-T. Tymieniecka (ed.), Analecta Husserliana, Vol. XVI, 203–210.
Copyright © 1983 *by D. Reidel Publishing Company.*

epiphanies of differing relations between his body and its flesh, the latter term understood here as material physical matter (Husserl's *materielles Leibding*).[1]

Through my clinical experience, understood in phenomenological perspective, I have come to conceive of the living body (*Leib*) as incarnating the ego in the field of transcendental constitution, as existential individuality. I believe that in some conditions the flesh may appear as transparent through the body to reveal its inanimate nature prior to individuation. The "everyday body" (*Körper*) is the living body as reduced by estrangement (*Entfremdung*) to become a "thing" for the biological sciences, as well as for the orders of economic production and societal arrangements.

Living body, flesh and everyday body are linked in a *vertical* synchrony, representing the consistency of the ego. They are transparent for one another, their mutual transparency the *condition de possibilité* (Ricoeur) of the metamorphosis of the bodily ego within diverse epiphanies. The *horizontal* synchrony of successive epiphanies may be termed the personal myth, in the sense that it is in the space of the myth that the metamorphosis as a whole occurs. To my way of thinking, the ego is plural and "cubist", insofar as it arises out of the dual synchrony within an encompassing intersubjective frame. And since all of this does not form a logical system, it must be approached dia-logically and phenomenologically.

In this presentation, which is an "intentional," or "noematic," case report, the issue is not one of the everyday body, because my patient had, strictly speaking, few biological problems. Nevertheless, like every hypochondriac, he believed that he had some, and so he often consulted a cardiologist about the supposed complaints of his everyday body.

Pietro has been suffering from a serious neurosis since adolescence. He relates something which happened when he was 17, at a time when his illness had not become fully manifest. During a visit to the "Brera," the famous art gallery in Milan, he was singularly affected by two paintings: Raphael's *Sposalizio della Vergine*, and Piero della Francesca's *Pala di Brera*. He experienced an irresistible impulse to compare the two works, and would switch frantically from one to the other, back and forth, realizing the phenomenal quality of "softness" in Raphael's painting and the phenomenal quality of "hardness" in that of Piero della Francesca.[2] During the following years Pietro developed his aesthetic sensibility; but though he acquired an excellent musical culture, he never became interested in painting. Consequently, he retained a deep impression of that adolescent episode, as he might not have had he sought to compare his intuitions with the critical literature on these two painters. Pietro has but a superficial knowledge of Raphael and Piero; he has never familiarized himself with their work nor with painting in general, which remains much like a foreign language for him. Probably it was because his intuition occurred in a sphere where Pietro had remained altogether naive that it initially excited my curiosity, setting up an intentional tension in need of time for a clarifying donation of meaning.

Pietro's experience (in the vitally intimate sense of *Erlebnis*) of the hard and the soft has so effloresced and ramified, that his existence repeatedly betrays a dialectic between these two poles, even though it is imperfect since the pole of softness always remains dominant. Diachronically it was noteworthy that already as a child Pietro had refused cream and cream cheese, which he described as "slimy." During his school years he had himself adopted a "slimy" attitude, making himself "very small and inoffensive," constantly wearing a honeyed expression on his face to secure the good graces of his teachers. Now at age thirty, disillusioned and critical, wrapped up as always in his anxiety and hypochondria, he lives out the epiphany of himself as "a shit." This is not a simile based on conscious comparison, but something immediate, intuitive, primary. I believe this epiphany to be a striking example of the flesh becoming evident through the body. When Pietro describes his symptoms, the appearance of the flesh through the body is partial; but when he relates his feelings, the transparence of the flesh is more nearly total. His body, now reduced to "a shit," is virtually filled up with flesh.

These are a few moments in a story replete with examples of the hard-soft antinomy, one defining a dynamic situation which, we can well imagine, is completely immobilizing. The antinomy of hard and soft affects the very quality of the "shitness," depending on the extent to which the flesh appears through the body. Its psychological corollary appears in the weakness, or at times in the intropunitiveness of Pietro. An example of this is his handwriting, with its "soft" fecality.

My mother wanted me to have an upright and incisive handwriting, a sharp, 'doric' style; you wouldn't believe how often she would slap me during my first year in school, or how many of my exercise books she tore up. At the end of the year my teacher gave me my report card with the words: 'You can bring this to your mother.' In the last grade of primary school I still got good grades in handwriting; but in the first year of secondary school I let myself go and began to scrawl. Now everybody tells me that I write in a filthy manner; my handwriting looks like written *cacca*.[3]

By contrast, as in the following example, where they are experienced as a penis or snake, the faeces are hard:

I feel the shit inside myself as something pressing, penetrating, as an inner penis; my bowels are a big long snake that moves and winds around the aorta (in his mistaken notion that the latter conveys blood to the heart), giving me a heart attack.

Not only does the long snake threaten to 'attack' his heart; it also robs him of nourishment, "that food that the heart has in fact to give to my 'body'."

After having discovered Pietro's internal space as a place where his body suffers assaults, threats and starvation, let us consider what his external space is like. I find three examples in my notes: Rome, the town where he lives, and his work environment.

Last year I went to Rome . . . that historical center . . . the narrow streets looked like bowels . . . I didn't like it, it is gloomy, oppressive . . . it is like baroque churches, like tombs, crypts . . . I prefer towns with wide, geometrical streets and orderly traffic. Also in this small town of ours there are alleyways in its old part which look like guts. (Note the stylistic contrast between "baroque churches" and the earlier "doric handwriting").

The factory where he works consists of separate buildings connected by underground passages which Pietro must use several times a day. These passages make him "melt," become "soft." Any space whatsoever turns out to be so threatening for him that for extended periods of time he will go to work only by car with his mother. Her presence, and subsequently mine as well, reduces the visceral quality of space, restoring its relational values.

When I look at myself in the mirror, says Pietro, I see myself withered and decayed; losing hair gives me the idea of decay; I also feel myself to be on the decline with women. Sometimes I feel better and then I feel myself I am young. I tell myself: how would I face up to reality if I had a decrepit body? When I am in good physical shape I can enter into new relationships.

The alternative possibilities of being either infirm or in firm shape can be seen as another parallel to the contrast between the soft and the hard. The hope of getting into good sharpe never leaves Pietro. In everything he discovers the counterpoint of soft on hard. He says, for example, "The eggs save themselves." That is to say, an egg is not repulsive because, even though it is soft, it assumes a shape when cooked. "It looks like a glans."

On the other hand we cannot simply say that a hard world surrounds a soft Pietro. His external space is also visceral, a mirror image of his internal space, even one with it, giving us that decayed thing which is Pietro, in the shape and figure of shit, as we have already seen. The peristaltic process is irrestible: "It is an imposition of reality; being pushed along frightens me because I am afraid I will not be able to manage it physically; it is like being on a train whose stops are not in my power to decide."

Because of the visceral space in which he lives, Pietro feels that the passage of time follows a rhythm that is alien to him.

I remember the clock in the corridor of my grammar school; all of a sudden the clock grew big, time stopped, and my heart began to throb wildly. Also, I now feel a discrepancy between actual time, which is quite slow, and time as I live it, which is very quick. When I was 17, and I played the piano, I speeded up the rhythm to 120 measures instead of 70. In Ravel's *Bolero* there is a place where one has the impression that the rhythm accelerates, while it actually remains the same; it is an invitation to let go and have an orgasm. My heart was throbbing at an excessive rate; my neurosis began when it occurred to me that my heart would soon wear out.

Pietro yearns for his heart to "float in his thorax as a whale in the ocean, finding its nourishment in the ocean." He longs for such a heart, powerful and invincible, with a wide and nourishing space at its disposal. But his body on the contrary remains compressed within a visceral space, pursued by peristaltic rhythms, starved by the long snake so that "it is always on the point of breaking down."

Here I interrupt my account of Pietro's narrative to observe that the decay of his body is located as much in his heart, even as the decay of his heart is coextensive with his body. If one were to consider the frequency with which he has asked for electrocardiograms, it would be necessary to focus on the centrality of the heart, with its symbolic significance, for him. Yet Pietro's eidetic intuition mediates an image of his body as nearly always falling apart or, less often, bursting. The alternative to decay is not well-being but an explosion, since if his body were to become hard, it would clash with a world also hard, with an inevitable explosive outcome. Pietro anxiously asks the cardiologist whether his coronary arteries are obstructed, and whether there is impending danger of a heart attack, because he knows he must submit to the doctor's language, and not question him about the extent of his decay nor his proximity to an explosion. It is not irrelevant that the semantics of hypochondria frequently show the language of the patient to be an involuntary caricature of that of the physician (Calvi, 1966).

For Pietro to be told that his electrocardiogram is normal is without consequence for the dialectic of the hard and the soft, the shaped and the decayed, since the former relates to the everyday body while the latter pertains to the counterpoint of living body and trans-appearing flesh. His repeated cardiac consultations are attempts at exorcism. But as self-submission to the "world" of the everyday body and that of medical science, it is also "treason" in the domain of the live body.

Emprisoned within a split which leaves no room for existential fulfillment, or its mundane expression as health, Pietro's existence is one of self-abandon, with the exception of rare explosive moments, to floating, not like a whale in the ocean, but as a center of decay in a putrescent mass. In fact, Pietro presents an image of himself whose margins are blurred, which does not stand out in any relief. Surrendered to his epiphany as "shit," he has more the character of a lump of matter evacuated for a brief and anonymous period than of a creature born for continuing personal evolution.

Pietro constitutes himself as a body only in order for the latter to return to its condition of flesh. Even the rhythm of the heartbeat assumes that of peristalsis; his body has become entirely visceral, so that his existence no longer alternates between the systole of tension and the diastole of relaxation, he is consumed in the spectacle of its self-consumption. His epiphany, described by Pietro himself as "shit," connotes an utterly fragile embodiment, one so precarious that death constantly arises from it.

Flesh, says Pietro, is an illness to me; some mornings I don't get up because I have the simple feeling that my spirit does not want to assume the burden of the flesh, does not want to hold it up. It's no use for my mind and me to deceive ourselves: the flesh senses death.

I believe that Pietro's eidetic vision is one of an extreme instance of transcendental constitution: that of the metamorphosis of the body from animate to inanimate nature.

Nevertheless Pietro has not entirely lost the prospect which was opened to him during his playing of Ravel's *Bolero*. Music is still able to dislodge the rhythm of his heart, free him from the long snake, from the visceral and faecal magma. He praises Beethoven's *Waldstein* sonata as "an outpouring of the heart, and an invitation to run."

He is fascinated by running athletes he sees on television and dwells with fascination on slow-motion sequences, seeking to grasp movement in its nascent form, virtually as kinetic intentionality. Dancing, whether watched, imagined, or mimed while listening to music, elicits an inner excitement. Music and dancing awaken moments of metamorphosis in Pietro, constituting himself as a running and dancing epiphany, one synchronous with the other one we already know, and even others not important here, all of them together giving his ego its thickness and manifold consistency, within the intersubjective atmosphere set up by his *koinonia* with the psychiatrist in what we call "psychotherapy."

The epiphany which runs and dances also includes that youth who ran back and forth between the painting of Raphael and the painting of Piero della Francesca. Pietro's excitement was that of his need to realize the significance of the two artists' tactile and motor dispositions for himself. His effort to discriminate between the hard and the soft represented not only an aesthetic concern, but also included a nascent "intentional criticism".

Mario Praz, the critic and writer, reports a conversation with Bernard Berenson, in which the latter lauded Walter Pater for "his attitude in the presence of masterpieces, expressed in something like a dance before an altar . . . ." Though his tone conveys a certain irony, and though I need this after my acquaintance with Pietro, and therefore not in a state of "worldly" irony but in an attitude of *epoché*, I believe I can also grasp Pater's astonishment, his enthusiasm, his devotion, in the presence of a work of art. I am sure that he was expressing this in his irresistible impulse to trace out a pattern of the painting in space, thus to retrace, reveal, and confirm the original irreal pattern as he had intuited it, from its inception in the artist's brain to the image of it he now bore within himself.

With due acknowledgement of differences, I believe that Pietro was also moved to and fro by an analogous aesthetic amazement, his sense of an unexpected consonance between his own bodily disharmony and that of the two paintings, one of which showed him his own softness, the other his deficient hardness. Here I shall not consider the problem of shifting from this psychological aesthetic level onto a purely aesthetic one; this move, says Formaggio, involves all the difficulty of Jacob's struggle with the angel. All I shall say here is that Pietro did indeed reach to the meaning of art in an uncommon way. From there a message came to him, brought to him on that one day by Piero della Francesca and later by Beethoven, a message altogether bodily, so that he now discovers his present and most reliable embodiment through the performance of music. It is in this region of the life-world that he receives a message of existential hope, "sitting by his window when evening comes" (Kafka).

Translated from the Italian by Erling Eng

210   LORENZO CALVI

NOTES

[1] Here, as in my previous publications, the word "body" corresponds to "Leib" in German, and the word "flesh" to Husserl's "Leibding." This explanation is necessary because Trotignon, for example, in this same volume, uses "corps" in the sense of "Körper" and "chair" (flesh) for "Leib."

[2] "Realizing", i.e., experiencing eidetically, the Husserlian "Wesensschau."

[3] *Cacca* = "Shit" in Italian. However, it is a mild word, usually employed by children, and is not taken as coarse.

BIBLIOGRAPHY

Calvi, L.: 1963, 'Sulla costituzione dell'oggetto fobico', *Psichiatria* 1, p. 38.

Calvi, L.: 1965, 'Réfléxion phénoménologique sur la mort intentionnelle', *Evolution psychiatrique* 1, p. 85.

Calvi, L.: 1966, 'La semantica intenzionale dell'ipocondria e la costituzione trascendentale del corpo proprio', *Psichiatria* 4, p. 65.

Calvi, L.: 1969, 'La fenomenologia del diabolico e la psichiatria antropologica', *Archivic psicologia neurologia e psichiatria* 30, p. 390.

Calvi, L.: 1980, 'La consistance corporelle chez l'hypocondriaque', in *Regard, accueil et présence* (Edited by G. Lantéri-Laura), Privat, Toulouse, p. 61.

Calvi, L.: 1981, 'Per una fenomenologia del sollievo', in *Anthropologia fenomenologica* (Edited by L. Calvi), Franco Angeli, Milano.

Eng, E.: 1976, 'Locating Erwin Straus', *Journal of Phenomenological Psychology*, Fall, 1976, p. 1.

Formaggio, D.: 1977, *Arte*, Isedi, Milano, p. 118.

Husserl, E.: 1965, *Ideen zu einer reinen Phänomenologie und phänomenologischen Philosophie* (Italian translation), Einaudi, Torino, p. 797.

Kafka, F.: 1958, *Eine kaiserliche Botschaft* (Italian translation), Frassinelli, Torino, p. 226.

Praz, M.: 1979, *La casa della vita*, Adelphi, Milano, p. 133.

BRUNO CALLIERI

THE EXPERIENCE OF SEXUAL *LEIB* IN THE TOXICOMANIAC: PHENOMENOLOGICAL PREMISES

It would be a naive and restrictive simplification to conceive human sexuality as a mixture of representations and reflexes rather than, basically, a global way of coexistence (of which the alter ego forms an integral part). In the conceptuality of empirical psychology, sexuality is intended to be understood as an independent reflected functional apparatus, in which context it is possible to recognize organs anatomically defined, sources of pleasure feelings, functions physiologically determined, an object of sexual desire, adequate sensorial and sensitive stimulations, and representations which can be the stimuli, according to the laws governing the association of ideas, of imagination or conditioned reflexes.

But we should note that the human body seems to be supported by a sexual scheme consisting of strictly individual, circumscribing erogenous areas which trace a sexual physiognomy and evoke the gesture of the lived body, either masculine or feminine, that is integrated in this emotional totality.

As a rule, the erotic perception has both spatial and temporal aspects; moreover, every person has the potentiality of planning (projecting) his own sexual world, sexually motivating his world, finding his place in erotism, holding or pursuing his scope to the utmost satisfaction in every possible manner, either by the most surprising sublimation or by extreme repression.

However, we can say that, in respect to the human presence, there is always an intention, a sexual effort, which is capable of evoking a cycle of motions and stasis, to be performed on the basis of a fixed model and finding in them its own satisfaction. There are, therefore, sexual components in every situation peculiar to normal human relations, in every aspect of the lived body (*Leib*).

This consonance of relations, or authentic way of projecting oneself, is common to normal sexuality: although it is possible only when *my* body is not seen as an impediment or explicit obstacle. It is necessary that I feel my body as belonging to me and at my complete disposal. Far from declining to mere instrument, it must become my own transcendency. At no time as perhaps in this situation of encounter, the body-which-I-am (*Leib*) and the body-which-I-have (*Körper*) tend so much to become insolvably united, beyond the alternative of for-oneself and in-oneself.[1]

A-T. Tymieniecka (ed.), Analecta Husserliana, Vol. XVI, 211–216.
Copyright © 1983 by D. Reidel Publishing Company.

The lived body, the *Leib*, by its exterior modulation, establishes and constitutes the erotic aspect of the perception, which makes possible an authentic communication with the body of other people. The latter is possible because my-body, in its external modulation, is much more than mere experience: it is, at the same time, present and participant, inherence and project (see Merleau-Ponty and de Waelhens).

The heroin toxicomaniac, it appears, lacks the capacity of projecting himself authentically, since he often considers his body merely as an obstacle, or rather as an uninterrupted source of urgent needs. He can remain a presence but will seldom become a participant.

The communication with other peoples' bodies cannot be based on the *Leib* (my body for another's), on my own *Leib*, but becomes restricted and entwined with the *Körper*, the body for-me. For this reason — illustrated by Merleau-Ponty — sexuality for many heroin dependents (the great majority, according to my own experience) tends to remain a mixture of representations and reflexes, and loses the possibility of being one of the fundamental ways through which, at a bodily self level, it express itself in an effective situation; such a heroin-dependent sexuality tends to remain within an autonomous circle (the "be done" cycle) and is no longer global and cognitive, as it should be, or closely connected with the whole being.

At present, even from an anthropological point of view, I agree with Freud's statement that the term "sexual" is not necessarily the equivalent of "genital," that sexual life is not the mere effect of genital and endocrine developments, and that libido cannot be reduced to instinct.

Sexual life is history; and man's sexual history gives us a clue to understanding his life, just because he projects his way of being-toward-the-world (being-at-the-world, W. A. Luijpen), toward others and his own future, through his sexuality.

Caught in the drug's conditioning vice, the heroin addict no longer possesses the capacity of projecting himself in a situation; he loses contact with history, acts only in the present moment, and lives exclusively for the next injection.

This is, perhaps, the fundamental characteristic of his *Dasein*, which radically weakens his projecting self in a sexual situation, his sexuality lived at a personal level; his lack of project, therefore, is linked to the weakening of his sexual *Leib*; the *Körper* prevails, i.e., the body-which-I-have, the immediate impulse (though weakened even in the neurobiological quality by the drug abuse); all these facts prevent encounters with others; the encounters are reduced to mere hazards with another body or, rather, with the body of someone else, to a simple instrument of one's own sexual satisfaction, an odd epiphenomenon of existence.

In my medical experience, I would say that the toxicomaniac, in the majority of cases, loses his metaphysic aspect of sexuality. The latter, i.e., the emersion of something beyond nature, must not, in my opinion, be localized at the same level of knowledge; it begins when, by an encounter, one opens himself to someone else (Buber).

What we are looking for is not just *a* body, as a body *enlightened* by the life of consciousness, i.e., a body consisting of an axiological behavior, the *very* axiological behavior.

The risk of considering the bodies of other people as objects within (during) sexual experience, a risk which Sartre accepts as necessary, has certainly to be faced by everyone of us, but it becomes particularly incumbent and imminent for the toxicomaniac, because the latter, owing to his "mania" or slavery, is no longer able (or at least not entirely) to reassume within his conscious and free intentions the sense arising from the bodily self. He is, therefore, unable to overcome the risk, to go beyond danger in an effective human situation.

Concentrating himself on drug keeps him from reaching, as Merleau-Ponty has explained, an osmosis between sexuality and existence, from living the sexual body in its existential i.e., anthropological (Buytendijk, 1961) aspect of *mit-einander-sein*.

I think it can be affirmed that a heroin maniac often loses the bi-personal constitutivity of sexual behavior. In his sexual life, which becomes progressively poorer, the heroin addict, who could live and manifest the extensive range of psychopathological disease in sexuality (from impotentia to ejaculatio praecox, from depersonalization to paraphilia), is no longer able to transcend the inner space of his being-for-himself, where sexual instinct might be localized.

The instinctual autonomous structure of vague (generic) and amorphous (von Gebsattel) sexual disposition cannot be estinguished within the love inhabited I-you encounter, the *Wirheit* (we-ness). This is perhaps one of the reasons why one tends to regain in surface by the practice of the swinging or couple love what is lost in deepness (the toxicomaniac is not the only one involved in this situation). This moral custom, examined thoroughly ten years ago (by John Webster, Sexe d'échange), tends to enrich interindividual relations and to go beyond old schemes "valid *only* in a different reality already *vanished*," beginning with wife swapping (the exchange of wives).

To understand such sexual practices it is necessary to overcome pseudo-cultural prejudices, presuppositions, phantomlike restrictions, or the narrow-mindedness rooted in conventional behavior. Group sex is certainly linked to the premise that "sentiment and sex are two different phenomena" (Webster). For the reasons mentioned above, the difference is often verifiable in many heroin maniacs, due either to bio-psychological or to cultural causes.

On the other hand, we know that in an inveterate heroin addict efficiency of sexual instinct and affection are, generally speaking, both subdued to extremely extended variations and reveal a remarkable tendency to decline and disappear; and with the latter the capacity of realizing the encounter, namely, the desire which "becomes personal," tends to reduce itself. The sexual *Leib* cannot be realized; therefore, *dual* reality cannot take place: the place for swinging remains, and more than ever it appears an authentic possibility of recovery.

In the sexual situation of a toxicomaniac we discover, in the first place, genuine uneasiness, often a real incapacity, in welcoming another person and relating to him according to his demands. Thus the other person is no longer perceived in the fulness of his being, but according to standards of utility and profit which do not coincide with his urgencies and leave him deprived of his preeminent connotation of alter ego, of his dialogical and interlocutory sense (aspect). Accordingly, the encounter capacity of the heroin toxicomaniac is reduced to the exclusive possession of the partner by the disposion to introjection, and to the removal of his paradigmatical dimension of being a person.

In many toxicomaniacs sexual *Leib* is, if we can put it in this way, "the side which takes," but never "being-with" and "mutual transparency." In a heroin maniac's sexual encounter the other person plays a peculiar role, which means that he is encountered not according to the offers, but to the urgent demands of the former.

The loss of oblativity in the erotic-sexual act is in this case particularly evident, as it is, almost at the same level, in the paraphilia. In the latter we remain within an existentially closed model, exclusively linked to the "bodily thing" (*Körperding*) as source of satisfaction for one's own momentary erotic pleasure.

As far as marijuana and the other hallucinogens are concerned, I cannot affirm that users of these drugs reveal such an anthropological structure decay. But it is almost certain that chronic cannabis intoxication, according to R. and G. Chopra, two Indian scholars, considerably weakens every physical activity, including sex.

Even if we occasionally doubt it, there is no evidence that cannabis stimulates sexual desire and strength (Benabud). Many smokers told me that marijuana increases their pleasure during the sexual act, by improving their "creativity"; the latter seems to me a genuine experience, not just an imaginary one, at least in the same way as pleasure is increased by aesthetic contemplation and particularly by music.

In the most serious cases of toxicomania, caused by concentrated essence of tetrohydrocannabynol, passivity, indolence, and lack of ambition is common. Sexuality is here subjected to a similar fate. (The same thing cannot be affirmed about sympathicomymetic amines [metedrine, sympamine, ritalin, maxiton, pervitin].)

Amphetamines act as stimulants and excitants, when chronically assimilated, even indirectly. They increase aggression by a considerable strengthening of this component of sexuality, which then becomes deeply modified through violence. In this case, it appears, the anthropological structure decay of sexuality raches its highest peak. The dual aspect, the we-ness (the *Wirheit*) are totally lost.

On the other hand, the sexual instinct toward aggression is sharpened by a substantial deformation of the *oblative* character, of the dialogue capacity and Binswanger's *modus amoris*.

My experince regarding politoxicomania is rather scarce; I have the impression, however, that within among various drug addictions, cocain and, above all, amphetamin are predominant, the latter being characterized by violent sexuality and an impulse toward destruction.[2] The use of amphetamine, besides the fact that it may evoke a paranoic attitude, implies a tendency to dissociate sentiment and sex. This is surprising, since such a dissociation has always been considered, today as much as ever, not only acceptable but even desirable. However, it does not mean that anthropological recovery of the human being's unity failed or that failure will be its fate.

Man's reflections about his own existence and fate, his irreducible singularity and interpersonal coexistence, incline us to think that it would be wiser neither to believe in dissociations nor to encourage them, even when pseudo-needs induced by modern society pressure us in that direction.[3]

We can undoubtely affirm, therefore, that nowadays a doctor must not neglect the phenomenological side of the problems if he does not want to demote the *Leib* which he should *encounters* to the *Körper* which he manipulates.

University of Rome

Translated by Paul Gabriel Weston

NOTES

[1] Sartre has introduced the fundamental distinction between the body for-me and the body for-someone else. Without his distinction, all problems referring to the body remain obscure and defenceless against the attacks of the positivists. This idea appears extremely fruitful.

[2] See F. Hacker, *Aggression* (Hamburg, 1973), and E. Fromm, *Anatomia della distruttività umana* (Milan, 1975).

[3] See L. Kofler, *Aggression und Gewissen* (Munich, 1973).

ERLING ENG

KINESTHESIAS AND HORIZONS IN PSYCHOSIS

This presentation grows out of an abiding interest in Husserlian phenomenology, and the wish to articulate my present understanding of psychosis. What I would like to do is relate my acquaintance with each to that of the other, believing that each may be of value for disclosing something of the other. Such an undertaking which starts from the psychotic condition as a constituted subjectivity is one of an hyletic, and not a pure, transcendental phenomenology (*Hua.*, 3: 214–15). There is of course an initial price to be paid for such an hyletic emphasis within an attitude oriented to a pure phenomenology. Yet if phenomenology is to realize its ties with philosophical praxis — and this involves its biological relevance in the broadest human sense — then it must be willing to take the risk of starting from below, again and again. Hence its necessarily fragmentary appearance in the context of classical philosophy. This also makes it possible for breaks in our ongoing experience to become occasions for developing enhanced intelligibility. Husserl, in a 1926–27 lecture, put it like this:

The world and we under abstraction from individual and collective death? The ideality — the open horizons, the indefinite possibility, that it can keep on in the same way, that I can keep right on unhindered. Horizons, on which I practically count, with counter-possibilities not currently indicated. But they are present — Horizon above all carries with it the real possibility of break, indeed it is real, the break is not merely possible in phantasy. Consequently a phenomenology of the irrationalities in contrast to, and correlative with, one of rationalities, of idealities.

But the irrationality as possibility remains itself subject to rational consideration. The relations between ideality and irrational facticity are also subject to laws of essential being. (*Hua.*, 14: 561)

Is not the possibility of "break" here also a reappearance of that death from which we and the world have first of all been abstracted? So that the "irrationalities" are now all of the instances in which this death reappears, if merely in the form of the jagged edges of the rupture. Thus deaths are correlative with lives, as irrationalities, idealities, are with rationalities. Something like this is also to be seen in psychosis, but where the irrationalities have gained the upper hand, in varying ways, over the rationalities, death over life.

217

A-T. Tymieniecka (ed.), Analecta Husserliana, Vol. XVI, 217–225.
Copyright © 1983 *by D. Reidel Publishing Company.*

To understand an acutely psychotic person, one who has suffered a "break," involves one in his death. Usual ways of initiating a relationship prove all but futile. As I come to realize the plight of a person who is no less detached from the world than I am from him, I can either try to withdraw, or accompany him through his underworld. There is a peculiar fascination in this, stemming from a recognition of the sameness and difference of our respective ways of experiencing. Thus its analogy with the sexual relation. There is sameness in the sense that my thoughts and perceptions are not immediate for the other. But the difference is that the psychotic individual has suffered an impairment of his ability to accomplish this indirectly, in mediating movement.

The way in which the existence of the world has become thoroughly problematical for the psychotic, without being thematized as a problem, can be viewed in the perspective of the epoché. If the Husserlian bracketing out of existence be viewed as a willful and partial experimental death for the sake of exposing fresh life beneath the patina of customariness, then conceivably psychotic remoteness could be seen as an involuntary and massive suffering of death toward the same end of life. But in psychosis the suspension of belief in the existence of the world is not voluntary but involuntary, and that makes all the difference. Husserl's recourse to the epoché, or suspension of belief in the absolute reality of the data of naive consciousness, is provoked by just those restrictions of perspective which thwart awareness of the life-world. Analogously, the psychotic, in his helplessly willed surrender of belief in the existence of the world, suffers an ascription of reality to the underworld of memories and kinesthesias, or passive movement tendencies. Having suffered a loss of mediating movement between the center and the circumference of his experiencing, he attempts to secure a world through immobility on the one hand, or through rearrangement of constantly shifting fragmentary viewpoints on the other. There is failure of progressive movement and, with that, of any future save repetition of what has already taken place. Only the intensity of desperation in the repetition tells of the continuing obscure presence of life at the heart of the encumbering deathfulness.

To institute understanding with one in this condition, and for myself, it is necessary in A. R. Bodenheimer's words, to "stand under" him, i.e., to make an effort to *accompany* his expressions, i.e., the kinesthesias of his self-presentation. This is to participate in his death. With Howard this began when, after a long period of apparently fruitless encounters, I yielded to his insistence to read a list of passages he had made from the Bible, largely from Daniel and Revelation. This despite misgivings, and some embarrassment on my part. But the images this reading evoked for me admitted me into the situation of his loss of world meaning. He had now become plausible to me, in his existence, as a person. Cocteau has said that the deepest dream of the poet is not to be famous, but to be believed. This was also true for Howard. But it was impossible as long as Howard could not grasp the words of the verses as also his own. As another patient said, "She [an imaginary persecutor] has the words coming out of me." Or "The words don't say what I mean." "It" speaks "like a tape recorder" said still another. "I," now but a compliant ear to an invisible source of commandments, can only try to smuggle messages out through mute signals secretly delivered.

Ordinarily we rely on conventional meanings of language, realizing it as a guarantee of shared understanding. But in psychosis it becomes a secret "code," not so much guaranteeing intersubjectivity as it does sinister meaning. Yet within the darkness there is still light. Traditionally implicit in the prophetic-apocalyptic words Howard gave me to read, and which I returned to him, the sense of "code" is that the individual vision of doom also contains the cipher of universal salvation. Here the position of life and death are reversed. Instead of life being merely the condition of death, death, now grasped as spiritual death, becomes the condition of enhanced life, no longer understood as merely natural, but also as spiritual. His definite though clouded sense of this possible transformation in the meaning of his experience often appeared in his assumption of the role of teacher toward me, a role I gradually grew to respect. I could also grasp the way in which Howard's apocalyptic pronouncements of an imminent end to the world were prefigured for him by the pollution of the Appalachian hills and streams of his boyhood. I could now believe him in the sense of biblical prophecy, whose doom holds a promise within and beyond any futurology, since it includes the possibilities of repentance and grace. Its anticipation of ruin depicts a moment in the transformation of the meanings of life and death in the course of human existence, individual and collective.

The path of psychotic discovery lies on the unillumined side of the phenomenological constitution. Phenomenology is a return to philosophy in the sense of Socrates' dictum that philosophy is learning how to die. The phenomenological epoché is now an experimental anticipation of one's own death, an initiation into an understanding of the experience of the transformation of the meanings of life and death in human existence, one suffered involuntarily, and hence merely partial — and to this extent self-destructive — by the actively psychotic individual. Moreover, to the extent that such a transformation of the meanings of life and death is always in part suffered, is always in part "process," it is to just such a degree, and in this sense, that it is possible to consider psychosis as the universal condition of mankind, inseparable from the realization of selfhood, a realization which proceeds from life in despite of experienced death, rather than from death in despite of experienced life.

The phenomenological epoché opens a Pandora's box of memory and imagination, exposing kinesthesias, or passive movement tendencies, beyond those to which Husserl most often refers. The epoché represents a self-abdication of my ordinary power over the world and myself through its suspension of my ordinary naive sense of the certainties of the world. In this way it is related to skepticism. If skepticism, like that of Descartes in the *Meditations*, serves to release phantasms of the past amid my ordinary future-oriented perception, Husserlian phenomenology also can allow those which have contributed in one way or another to my constitution of the life-world to reappear. In this sense it is something like a 'psychotogenic skepticism." Hence, for Husserl, the complementarity of "kinesthesias" and "horizons" in the constituted spatiality of the life-world (*Hua.*, 14: 534–61; 16: 328–29). Moreover, Husserl thematized the twofold character of horizon as "inner" and "outer." Although he did not do the same for kinesthesias, either as directed toward a focused future, or toward a fragmentary past, the very notion itself holds a duality of feeling and movement.

Retention and reproduction dominate the condition of psychosis. Its phenomena provide impressive evidence of how self-deception thrives on the naive acceptance of an uninterpreted past. (Thus its application by analogy to the cultural sphere.) A psychotic person may speak of being "roboted" or "hypnotized from underneath." The live body is submerged and strangely impersonal, as if speaking through a mask. The lost live body may be sought for within this dead body. This can be seen in self-mutilation of some psychotic patients, as if seeking the live body in watching the flow of its blood. Howard chortled at the passage from Revelation where "The blood came up to the horses' bridles." Another patient who cut her wrists stopped when she could not get the blood to flow as she wanted. Another patient was distressed because she had been "turned inside out," the inside body now on the outside, hiding the outside body from others.

As long as I believed that I was the unique privileged observer, Howard and I remained sealed off from one another. Only his insistence, in the face of my resistance, eventually compelled me to recognize that in some inexplicable way our positions had become reversed. He was the potential teacher, I the unwilling learner. I should also mention his daily encouragements, like

bringing iced tea or coffee, his finances permitting, and refusing my offers
to pay, camping outside my door. These attentions were offered with such
evident gratitude that it would have been ungrateful for me to have refused
them. At the same time he lived in fear of being duped, tricked, or betrayed
by others. His kinesthesias of withdrawal, and avoidance of others, whom he
saw as sinister, mocking, or seductive, conveyed a pervasive suspicion. On
entering my office he would place his hand on the small of his back, then on
the nape of his neck. I soon learned the futility of asking him about this,
since he either could not nor would not tell me, as if he simply supposed I
already knew. I satisfied myself by attributing some sense to them, like "Get
thee behind me, Satan." When leaving, he gave me a secret handshake with
which I complied. This implied a special understanding between us, as did
his concern about the possible theft of my notes of our contacts. Always
inquisitive when he found me at the typewriter, he expressed a wish that I
write about him. As our acquaintance developed, he discussed with me the
sign to be used the following day. This was a placement of one hand on some
part of the body, for example side, back, or neck. Outside my office when
others were around, he would sometimes use such a sign, though he never
required it from me. He seemed content to do the "signing," as if I were
somehow already included in his display. It also seemed that if at first his
signs were a one-man show, he subsequently included me, tangentially, in
their selection.

Eventually I discovered that Howard chose a particular chapter of the
Bible for each day. It was not long before he began to include me increasingly
in making his choice. He even came to ask me about the meaning of certain
verses, whereas formerly he had been the unquestioned interpreter. When I
explained a verse to him, I was sometimes surprised to also realize something
of myself, about him, and about our relation. After years of acquaintance
I found him readier to choose verses from the Psalms than from Revelation.
The horizons of the Psalms are those of the embracing high hills, in contrast
to the collapsing horizons of Revelation, with their falling angels, raining fire,
and earthly ruin. This turn of affairs coincided with an expression of his wish,
and readiness, to leave the hospital.

His shifting of preference from Revelation to Psalms conveys the promise
of reestablished horizons, at least in memory, from within the fallen ones of
despair. His earlier apocalyptic preoccupation included both the sense of
disillusion and realization. These characterized the prophetic visions of
transformation to which Howard was drawn. Our difficulties sprang from
the inaccessibility to himself and to me, of his kinesthesias and mine, our

virtually reflex accompaniments to particular horizons of expectation. Together we discussed the meaning of such words as "No man is a prophet in his own country," and "Cast not your pearls before swine." More attuned to one another, we no longer spent as much time on the imminent end of the world.

The phenomena of psychotic experience may not only be clarified by phenomenology, they may also in turn enhance our understanding of consciousness. I have referred to Howard's shift from solipsistic gestural display to shared verbal agreements. Viewing the body and world as partners in experience, whose relationship is represented in, through, and as consciousness, in psychosis this partnership has become so flawed that flight is attempted to one or the other pole, singly or alternately. Body and world alike are constitutions of coming into being and passing away, of the life and death of the body. In psychosis there is a loss of movement in this kind of exchange so that the difference between the live and dead body, between *Leib* and *Körper*, is no longer a rhythm of fulfillment, but an inner opposition, inseparable from the loss of rhythmic interchange with a world also of life and death. Now the dead body suffers the gaze of a live, but destructive, world to which it is exposed. Its display is experienced as a kind of displication of self and world, in contrast to ordinary explication of self and world. Hence too the moment of the obscene, as suffered and expressed in psychosis. Such displication tends to become progressively solipsistic and destructive.

Body and world may be seen as facing texts, each explicative to the other. Within the movement of mediation, there is a reciprocal enrichment of meaning. With impairment of mediation, as in psychosis, there is a relapse to a struggle between life and death as bodily, in whose terms events of the world are understood. World now becomes a projection of such a body, which is no longer a living mirror, but a magical projective one. That death previously indiscernible within the text of the body now appears in the form of mysterious influences and persecutions. Attempts to renew intersubjectivity take the form of a vain display of one's body and words, seeking reversal of death into life, through exposure of the dead body to the live world, or of the body to a dead world.

This polymorphous destructiveness may be understood if we consider the vital roots of the syntheses through which the constitutions of consciousness are effected. Vital initiatives, at bottom, necessarily encounter resistance from present constitutions. Thus kinesthesias involve opposition as well as apposition; horizons exclude as well as include. The moments of oppositionality in kinesthesias, and the moment of exclusion of the horizon, are inherent in constitution as a temporal process. In psychosis there is a break-down of the passive syntheses of experiencing. With this the paired moments of opposition and exclusion come to the fore. Action, no longer self-realizing, is now implicitly transgressive vis-à-vis an imprisoning horizon. With this, the domain of delusional conflict is constituted as the diabolic counterpart of those symbolic fulfillments through which freedom is realized through being in the world over and beyond the world (Binswanger). Just as the positionality of human uprightness involves a moment of opposition to gravity, so human consciousness holds a moment of oppositionality to the already constituted. To the extent of the failure of ordinary syntheses of experiencing, this antagonistic moment, with its constricting horizon, is thematized in terms of threat, destruction, and guilt. Hence the virtually erotic craving for suicide sometimes seen in acute psychosis.

Recognition of this moment of vital oppositionality to the world as already constituted in its own right, as we see it in psychosis, can extend the reach of Husserlian phenomenology. Such a moment, ordinarily suspended in more complex constitutions, is intrinsic to consciousness. It remains virtual throughout the entire range of human activities, capable of reemerging as a decomposition product of consciousness at every level. At the most complex levels of human consciousness, this *transgressive* moment may be seen in the lives of major innovators, whether of invention, scientific theory, art, literature, or music. It accounts for the way in which an otherwise inexplicable guilt often accompanies, or hinders, creative fulfillment, just as it possibly accounts for the phenomena of psychotic depression, and suicidality.

We may now recall Husserl's proposition that "The world is real only within the continually prescribed presumption that experience will continually go on in the same constitutive style" (*Hua.*, 17: 258). So that Ludwig Binswanger has referred to psychoses as "Modes of human existence in which the consequentiality or sequential nexus of experience is rendered doubtful, and therewith the possibility of leading's one life" (*Manie u. Melancholie*, 15ff.). With this in mind, what about the "break" to which Husserl has referred? Is not this "break" already prefigured in the recurrent vital upsurge of each moment of temporalization? Ordinarily the pristine nisus of each instant remains suspended within the ongoing synthetic activity of inner time-consciousness, with its play of protentive and retentive moments.

We may now reconsider the first sentence of our opening citation from Husserl. What does Husserl mean when he refers to experience as "abstracted from death"? Apparently a detachment of experience from its bodily pole. Is not this an attempt to exclude the "break" from experience, to conceive consciousness as streaming and continuous rather than as intermittent, or pulsing? But does not the latter feature return in the duality of moments in the horizons and kinesthesias, as well as in the possibility of a "break"?

There is this radical difference between the "break" of which Husserl speaks, and that of psychosis. The Husserlian "break" is experienced as an interruption of the ongoing rule of experience. But in psychosis it is suffered as an atmosphere of pervasive dread or incomprehension. The psychotic ekes out his remaining life within a "break" which is real only for others. If he could understand his transformed existence as an interruption of the ongoing rule of existence, it would already mean a resumption of the ongoing synthetic apperception: he would no longer be "psychotic." In this case we discover a limiting instance of Husserlian phenomenology. Nor is it clear how this kind of irrationality can be related to other rationalities through the discovery of *Wesengesetze*. One way of attempting this would be to understand psychosis in the framework of cosmic evolution. But that, by involving us in the meanings of life and death, would take us beyond Husserlian phenomenology.

But if this is so, it is also the case that we have forgotten the earlier mentioned distinction of Husserl between a pure and an hyletic phenomenology. What would a pure phenomenology, one corresponding to the hyletic phenomenology I have elaborated, starting from psychosis, be like? Just this: one which recognizes the experience of the other as already appearing within the bodily sphere, here taken in the sense of a primordial expanding spatiality of manifestation. Manifold bodies appear within sheer bodiliness, of, through, and in which world is manifested. Consciousness is a weave of bodiliness in its various modes, but which has reversed its direction in psychosis. Consciousness is now suffered within a particular body which, as already constituted and in relation to the world in open consciousness, is experienced as "dead." The live body (*Leib*) is discovered in and through the world opened by and open to consciousness (*koinos kosmos*); the instrumental body (*Körper*) is discovered by and in the domain of a singular world closed by and enclosed within consciousness (*idios kosmos*). An hyletic phenomenology starts from data understood primarily in terms of the instrumental body, whereas pure phenomenology starts from givens grasped in the shape of the live body. The constitutive movement in hyletic phenomenology has the form of a sheer ambivalence, a nonprogressive dialectic, whereas pure phenomenology discloses an open hierarchy of progressive self-clarifications. Psychosis, with its domination of consciousness by the instrumental body, presents a situation of sheer ambivalence that precludes stable and progressive constitutions, a situation all but excluded from live bodiliness. Phenomenological acquaintance with psychosis is able to demonstrate the latent presence of the *idios kosmos* as unacknowledged in the domain of the sheerly factual as defined under the sway of instrumental bodiliness. Moreover, scientific findings, when reduced to absurdity by their extrapolation into science fiction, provide additional evidence for the possible thesis, *in the foregoing sense*, that psychosis is the universal condition of mankind.

With regard to the problematic of body and soul in Husserl, we may now see how his shift of emphasis to the difference between the live and instrumental modes of bodily existence (*Leib/Körper*) is in response to the issue of the nature of fact in modern science, as this has become problematic for us in the consideration of ecological issues. If ecological be understood in a purely phenomenological sense, then the psychotic person, unclear about the relations between what is within and what is without the intervening boundary, or even the importance of the distinction, provides us with a valuable point of departure for reflections on the relations among body, consciousness, scientific fact, and world.

Veterans Administration Medical Center, Lexington, Kentucky

L. PARADISO and V. RAPISARDA

SELF-ACCEPTANCE: THE WAY OF LIVING WITH ONE'S BODY IN OBESITY AND MENTAL ANOREXIA

To speak of persons in psychiatry, only really makes sense if we restore to the word its etymological sense as used by the Latins: a mask.

A diaphragm, rigid, immutable, ambiguous and mysterious, on the one hand covers the face and on the other offers the stage and spectators a physiognomic characterization alluding to a way of being required by the role. It is opaque to the eyes of others and therefore protective; but the outward view is precious for the person living behind the wooden features. And from those features words come out, sentiments, infinite modes of being, of revealing the self and expressing intentions in the only possible "existential" way, among so many alternatives, in order to reach the epilogue.

It is not indeed person understood in the metaphysical sense of noumenon, an essence expressed a priori that then vanishes, if taken to extremes, in an impersonalized and hyperuranic ideal. Nor is it a positivistic vision of the problem where the person reduces himself to his personality, that is to say, to what appears superficially in his occasional and contingent way of revealing himself, in fact in practical psychiatry, as on the stage — that paradox of life — it is not possible to make such a distinction — fertile as it may be for further theoretical developments in philosophy — without incurring serious mythodological and interpretive errors. That would be an escape into the imaginary, the world of ideas so very far from the painfully incarnate world of the mentally ill.

But existence presupposes the necessity of putting ourselves in that determinate period of time and on that determinate stage with our own "persona" taking upon ourselves all the ambivalence contained in the term: with our will our conceptions and our prejudices but with our body as well. This body at one and the same time recognized as an immanent object and a transient subject. Two sides of a single existential reality where neither aspect can ever completely disappear, out in a Gestalt dimension takes on now the role of figure, now that of ground. Man understood in this way sees himself against reality neither as a mere abstract capacity, a cogito opposed to a cogitatum, as being in itself according to the idealist view, not simply as "being for its own sake" (to borrow the words of Sartre), as objectivity, that is, as a concrete entity conceiving external reality in an objective way,

227

A-T. Tymieniecka (ed.), Analecta Husserliana, Vol. XVI, 227—239.
Copyright © 1983 by D. Reidel Publishing Company.

experimenting with it and recognizing it as something other than itself as positivists would have it.

The body, then, is found to be that concrete place where one becomes aware of being in the world.

Man sees himself as a source of meaningfulness even before any abstract symbolizing, placing himself in the world through the body: he is (Leib) thanks to the body; he has (Körper) in the dialectical relationship between the ego and the world. The body is existentive possibility because it is the point of departure for all my directivity and for any intentions whatsoever outside myself (Graumann).

In this manifestation of the noumenon which is the expression of the personality through the historicizing of the body there lies the tension of being. In some measure being is appearing, offering oneself to others, allowing myself to be that image of me that others have made of me. And, in its turn, this image reverts to the subject in a game of recalling as in a line of mirrors, returns to its own foundations in some "basic precariousness" (Callieri) because it is a motive of absolutely personal differentiation and integration, not following any norms but the normativeness of an individual person.

And for this reason if the recall is a deformed image, it is an image perceived as a distortion of the being itself and thus not integrated at a personal level but rather rejected and denied because unacceptable.

How much weight those aesthetic and cultural canons have that prevail in a given society and even more in the different social strata of a certain period of time is so obvious as to make it not worthwhile to dwell upon. The recalls of the "other" at this impersonal level are of the *Mitsein* type (to use the term coined by Heidegger) and so superficial as not to make any significant impression except on the surface of the body and body image, but even then not enough to break the ego-body integration. The processes leading to the dissolution of this integration are, however, those applied when the recalls are of the Binswanger *Miteinandersein* type.

If we go back to the formation mechanisms of the body image, we will see that it was possible for it to be built up thanks to the experiences gained in an ambience full of highly significant recalls, that is to say, in the family.

Already in the first stage of interuterine life there is a progressive formation of body image thanks to the proprioceptive stimuli arising from the vestibuler apparatus, and the muscle and articulation receptors due to the first hand-to-mouth movements in the uterus. Later on these basic sensations take second place to the tactile impressions of sucking at the mother's breast with the mouth-to-breast movement.

When the child begins to suck its thumbs it begins to become aware of itself as a being having an identity of its own and the exploration of this identity is a source of pleasure over and above the symbiosis with the mother. Thanks to tactile stimulation, exploration of objects forms the basis of the postural model, even before that awareness of Euclidean space occupied by bodies having a substance of their own.

Surprisingly enough, acoustic, olfactory and optical stimuli do not play a primary role in the development of the body scheme; indeed they seem to be not fundamental compared with protopathic sensitivity that furthermore appears earlier.

We are not inclined to accept, just like that, Lacan's well-known theory of the "image in the mirror" according to which the child of about six months understands the unity of his figure as an organized whole. Clinical observation shows that children blind from birth have a perfect image of their own body; and in them there is no dissociation between the perceptive ego and the expressive ego.

Furthermore children who have a limb amputated before the age of five, thus considerably after the "triumphal assumption of the image" (Del Carlo Giannini), do not display the phenomenon of the ghost limb that, in order to appear, requires a complex nervous integration at various levels: cortical, spinal and peripheric (the telescope phenomenon of the ghost limb is a proof of this).

Anyone who has had at least a little experience of painting knows that the memory of a face or a body, or of the quality of an object, are not so much in the psychic memory of the eyes as in the sensation in the finger-tip pads that have touched and stroked and in that deformation that the tactile receptors have registered. This emotional somatic memory is a way of living the past in the present with a sensation that exists here and now outside the time of rational memory.

The problem of body image is rather bound to the development of logical primary and secondary cognitive processes. Due to these secondary processes the ego understands itself to be not just a body for itself having internal space and a contour limiting it from outside space, but also a body for other pepole, a body for dialectic use, the object of the looks and expectations of others, which is a way of saying to be there for others and not to be in some way always and in everything master of oneself.

That is why the image we have of ourselves in the mirror is never an objective or real one but always an idealized effective one. In fact if we protract our self-admiration beyond a certain time, as it were enchanted or mesmerized, we begin to perceive a strange sense of uneasiness and impersonalization that derives from our perceiving of ourself as an object.

On the other hand we sometimes make an effort to recognize ourself in some instantaneous photographs that catch an attitude we have never been able to verify in the mirror and almost tend to reject because they offer us an image that does not correspond to our ideal persona. Our self-awareness overrides the static one of an image. Thus we have the ego performance, but at the same time the manifestation of the ego: the ego of hopes, emotions and desires but also a hypothesized object of recalls, expectations and conflicts.

We pride ourselves as a body but we feel we are something more inexpressible than just this appearance and nevertheless "my face is not the image of me, it is me myself" (Chirpaz).

It is in this eccentric dislocation of the ego, in that loss of awareness of an ego laid bare and protected by the body that the experience is built up of a distorted perception of the body self that is transformed into a failure to accept the psychic ego. But if on the one hand this is a fracture, on the other it is salvation from anxiety. Anxiety is materialized, incorporeal and incarnate and allows the amputation of the *Körper* in order unconsciously to save the *Leib* from destruction and annihilation.

At this point we shall describe the history of Alice C. to illustrate what has been briefly outlined above concerning the concept of persona as seen from the psychopathological viewpoint of obese and anorexic subjects. Alice was a patient who asked us for help for her extreme state of physical debility due to mental anorexia that had begun in adolescence after an obese childhood.

She was the second child of a rather elderly middle-class couple who seemed to get on quite harmoniously together; perhpas they had cultivated this air of harmonious living in order to give an external appearance of success in this field rather than really having an understanding as a couple. They were both unsatisfied with each other but at the birth of their first child, a boy, and of the second, a girl, they seemed to have found in a neurotic way some compensation for their mutual dissatisfaction assuming the role of parents completely devoted to furthering the good of their children. They had a kind of competition to see who could make the bigger sacrifices, each of them wanting to capture the affection of the children and giving the children the role of making up for their own frustrations by coming up to their expectations. The poor state of health of the boy led the mother to pay more and more attention to him and he had established a relationship of dependence on her and learnt to monopolize her care and attention.

Alice felt unhappy with being thus rejected and so she tried to capture the attention of her parents by complaining periodically of abdominal pains the nature of which was never cleared up despite numerous visits to doctors. All the same she had on the whole managed to establish a good degree of balance in her ego, developing by reaction a certain spirit of autonomy and independence also in the matter of solving her little day-to-day problems; this earned her some gratification as her mother appreciated her for it: in a certain way she found compensation in the affection of her father, an apparently authoritative man but basically weak and insecure; he showed off this little girl, pretty and with good manners, as the best visiting card for his role of a good father of a solid middle-class family.

However, the first successes at school of Alice's brother called on to him the attention and pride of the father who began to project on to him, nursing in his heart all the satisfactions that would be coming to him from this son. Thus the brother came back into the limelight with his intelligence and powers of reflection and judgement, his father calculating that these qualities were a guarantee of success for the future career of his son, a career that could only be successful. Little Alice's progress went unobserved remaining always in the shadow of that of her brother who was held up to her as a model; so it seemed a foregone conclusion that by following in his footstops she also ought to earn some approval.

The girl was required to meet certain aesthetical norms and a code of good manners. They used to say: "she's a lovely little doll" but at the same time accuse her in public of looking at herself in the mirror thereby exposing her to the disapproval of everyone. Alice was convinced that to earn appreciation she had to become the same as her brother. She began to follow his taste for games and to prefer the boys as playmates, complainting that girls were affected, stupid and insupportable. She consciously avoided games considered more appropriate for a girl unless she managed to involve her brother in them too and she tended to prefer competitive games.

At the age of 7 she bgan to show a considerable increase of appetite and consequently the tendency to get fat which her parents considered worrying. As her body grew she acquired an extraordinary sensation of strength that allowed her to compete with the boys and so to be considered their equal; she felt that she could defend herself on her own and this was the source of her sense of security.

Her parents expressed their anxiety at the by now unpleasing and clumsy appearance of their daughter and threatened, without much conviction, to put her on a diet. They also complained that she could no longer get into the clothes her brother had outgrown because they were too small.

This sensation of physical expansion gave her the impression that she had at last achieved physical space in the home having a body of her own to dress in new clothes and having finally destroyed the image of only being a fragile, insignificant, passive and decorative little girl.

With the coming of puberty the different roles of being a boy or a girl began to be traced more clearly. The boy would continue studying and go on to the university, whereas the girl would receive an adequate education from certain tutors equipping her with all that a young lady of a certain class needs to know. Furthermore her brother was given freedom and privileges that she was not.

It then seemed to Alice that she had lost once more that equal footing she had built up so laboriously with her brother.

The brother was mixing in a socially acceptable group of young people and thereby earning more approval; whereas Alice's participation was prevented both from a sense of overprotection and a hidden doubt about the role that could be played in such a group by Alice, a fattish, ingenuous girl who ate too much.

The attitude of rejection of the parental figure, typical for all adolescents, was aggravated by the open hostility and lack of confidence on the part of Alice's father: she was followed whenever she left the house, her correspondence was controlled and she was closely questioned about the type of relation she was having with one person and another.

And Alice felt herself being sucked into the usual whirlpool: "Do what you're told! Don't do that! You must understand! You can't understand! Don't go there! ... Don't be ... !" And so she reacted. She felt herself to be just a body, sometimes a corpulent body, from which her soul had no hope of emerging, and at other times as the object of other people's desires as conjured up by her parents' fears. This body prevented her from recognizing and accepting herself as a person. She began to follow an ideal of being slim and by drastically cutting down on her food managed to lose 20 kg.

This rejection of food displayed an ambivalent attitude towards her parents: if on the one hand she was rejecting together with the food the parents who supplied her with it and on whom she was still financially dependent, on the other she was losing those sexual characteristics of femininity (amenorrhea, breasts disappearing, hips non-existent ...) almost as if to reassure her parents about her relations with boys whose company she preferred (finding them more like herself and more interesting). At the same time she was hardly ever aware of the romantic sentiments she aroused in them and if she got to know of them was very decided in pushing the boys away.

Secret eating occurred also in this case and we feel it is something that has never been sufficiently understood. It is our opinion that at least in this case, and perhaps not only in this one, the patient was compelled to secret eating by the genuine need to allay the pangs of hunger. But if food on the one hand represented a real threat to her being slim, on the other it had been selected as a weapon to use against her parents. In fact with it she had been able to capture their anxious attention on the state of her health (she said: "I had become delicate and liable to illness like my brother when he was younger"; (through food she thus rejected her parents and her financial dependence on them.) The sense of guilt generated by such a rejection could only be tolerated by secret access to food constituting almost a splitting of personality. In a kind of psychical lethary she said: "I used to get up in the night and walk around in the kitchen undecided, then Alice opened the door of the fridge . . . ."

It was when she had reached this stage of extreme emaciation and almost of dissociation that Alice came under our observation.

On the basis of the case we have just outlined, we do not feel able to dismiss the problem of obesity and anorexia simply by saying that they are due primarily to an unbalance in one of the cognitive phases of the body scheme development no matter how dynamically they may be understood and in process of continuous adjustment. Nor does it seem to us sufficiently complete, in order to understand the sense of mental anorexia, to revert to the hypothesis — true as it may be — that mental anorexia is the consequence of that reelaboration of and readapting to those internal relational circumstances that had previously acquired a state of balance. Such a phase that has been defined as "an identity crisis" cannot be understood only in the sense of a disorientation deriving from the need for endogenous reorganization.

Nor can we accept that the "crisis of opposition" that is always to be found in this evolution phase of life is, as some would have it, a way of giving priority to the endogenous proprioceptive messages over the external ones. It is almost a mechanistic way of looking at the problem, depriving the body of the person that lives in it.

The best understanding of the problems connected with obesity has been achieved by considering obesity as an equal and opposite condition of anorexia, as has been expressed by H. Bruch. And we shall continue here with a sequence of contrapositions.

The organization of hunger perception is the result of complex transactional relations within interpersonal relation frameworks. In fact it seems most probable that in the case of obesity there is a loss of the capacity to recognize any other stimulus as being different from that of hunger; or else, especially because of a defective transaction with the mother, there is the learning in the evolutive years to respond with hyperalimentation to different kinds of requirements or needs that needed their own appropriate responses.

In fact excessive eating can play a stabilizing role in a precarious psychical state deriving from a relationship between the child and a mother who has not the time for individualized interpersonal relationships. This could be one of the main reasons why obesity is found more often in the lower classes where the greater number of children, the authoritative character of the mothers and tendency of these mothers, on account of their cultural heritage, to give first place to the satisfying of hunger favor the tendency of meeting all needs with alimentary gratification.

Individuals who have achieved cognition of this kind are dominated by a strong sense of fatalism, weakness of character, a sense of interiority, a condition of dependence and by the idea of not possessing their own body, of not managing to dominate it nor being able to identify with it.

As H. Bruch observes:

The lack of responses regularly and consequently congruous to his need deprives the child in this phase of development of those essential bases on which to construct his own physical identity and the perceptive and conceptual awareness of his own functions.

Since the precocious experiences of the child will then condition his models of action in the years to come, that is why the obese child uncertain of recognizing his need will become a dependent subject, insecure, not master of his own body and not well identified from the sexual point of view. In him the desire to eat becomes the only possibility to find security.

Sometimes, on the other hand, the food is invested with an exaggerated emotional value or becomes a substitute for affection, security and gratifications.

Also the role of the father in a family with obese children seems to be not an indifferent one: sometimes he is a weak passive figure but at other times he is experienced as an oppressive figure, master of the child's life with his continual hypothesizing on the future and requiring the child to come up to his expectations while being very little concerned with the real needs of the child. But perhaps rather than the behavioral features of one of the patients, it is that complex of the family functions and the type of dynamic interaction of the members within the group to determine an alteration of the basic mechanisms in a logical integration of thought and behavior. The obese person, like the anorexic, is not only the passive victim but operates active transactions with members of the group, not only because his behavior appears adequate to the role that more or less consciously has been given him by his relatives (maintaining a matrimonial tie, compensation for frustrations, the possibility of social benefits, etc.) but also because with his continual eating or refusing to eat he arouses hostility and condemnation and thereby offers new, uncertain and elusive material to close once more the cycle of distorted and contradictory messages.

On superficial examination the family of the anorexic seems to be less undermined with these conflicts, but delving more deeply one notices that, perhaps on account of the more privileged social extraction, the expectations required of the child are more pressing, both those for success in his studies and those of a behavioral and aesthetic model; this gives the child even more the sense of not being master of his own life and of his own body.

This expropriation of one's own body leads to incorrect perceptions of its functions being established. The compulsion to eat food in the obese and the tendency to secret eating in the anorexic seem to indicate that they wish to dissociate their shape from what they eat and attribute other ends than plastic ones to the alimentary function. We are unable to share the theory that for the anorexic food is the only dialectical reality, but rather it is a means or a way to a relationship. On the other hand the processes of self-observation that are reported, wherever they are looked for, as being always based on the being used to one's own body, deny that there is any rejection of the dialectical intersubjective dimension.

In these psychopathological conditions the body itself is no longer a projection of the person and becomes something heavily objectified that is experienced as heavy, opaque and valueless in the anorexic, whereas for the defenseless being that the obese child feels himself to be it is protective, reassuring, almost a new uterus. This failure to make use of the body to follow through one's choices and the failure to enjoy it is still tied up with confusion about the sexual role in the wide sense of the word. In fact obese patients are frequently found to interpret the symmetrical blots of the Rorschach test indifferently as a man and a woman, whereas the normal interpretation is two men and two women. That is to say there is a tendency of the obese to go towards a sort of androgen, a man and women at the same time; but this is not experienced with that sense of fulness and exaltation as it was by the ancient Greeks, but rather as an attempt to make up for that sensation of solitude because they feel themselves to be undesirable. Once again the body is set up as a wall, an impenetrable barrier against the world of other people, sometimes excluding them but at other times protective and reassuring. Also on this occasion the obese is unhappy in every sense, he tends towards solitude, to shut himself off onanistically with the result of a progressive process of cutting himself off from the world.

This sexual ambivalence and rejection of sexuality is still clearer in the case of the anorexic where the rejection of food guarantees and, in the imagination of the adolescent, should guarantee a sylph-like figure that does not betray the incipient feminity. This presentation of oneself as an asexual being is the extreme attempt to idealize one's own person, to make it become essence, a categorial principle protected from any kind of worldly relationship and all the less carnalized in an instinctual relationship of the sexual kind.

We do not want to dwell here on the problem of rejection of the sex-connected role and thus on the introjection of model figures belonging to the same sex; this because, on the one hand, it would lead us on to interpretative ground and, on the other hand, we consider it to be a question of a more general character common to all adolescents.

How central the property of sexuality is for the anorexic is borne out by the fact that he nurses distorted and misinformed phantesies of oral fecundation; from these comes his fear almost that an abdomen that is not flat, or even a concave one, can give rise to the doubt of contamination in an instinctual kind of dual relationship. The anorexic is not so much concerned about being unpleasing to others so that they reject him as about satisfying himself and the idealized image he has of himself. This need to conciliate more and more his own apparent image with a rarified and elusive sentiment of himself, persistently "reasoning in itself" (Callieri), leads the subject not to take into serious consideration the worrying appraisals of his state of health and of his unpleasing appearance, but rather thay are gratifications confirming that he has reached that annihilation of his body that he was seeking.

One might well think that for the anorexic the alter ego references are void of meaning and dialectic; in reality, even with an apparent contradiction, the anorexic tries desperately to communicate almost that last SOS, that core of himself he has held he must and can save in the destructive direction his adolescence has taken. It should be pointed out that these relationships are limited to purely plural relationships and never cemented in one of a dual kind; at the same time it should be borne in mind that the illness occurs at an age when the sexual identity and psychological maturity are so uncertain and weak as not to allow the subject to risk committing his whole person in a relationship which is highly perturbing because so significant. This is particularly so in a subject as fragile as the anorexic even if he is equipped with apparent will power. It is, so to say, a mechanism of existential rationalization.

When the anorexic spies on his own figure secretly in the mirror he is not so much expressing the loss of self-presentation as the fear that his self-presentation is not sufficiently expressive of his self. We do not mean to say that anorexia is a kind of intellectual narcissism nor of a crystallization excluding others, but a desperate appeal to others to take notice of his soul and his sufferings also through a body that is sorrowfully mortified but finally "transparent".

This need to save the deepest psychical self from those equivocations into which others could fall is particularly exemplified in mental anorexia where the body is perceived as an incumbrance, something falsifying and obscuring and therefore repulsive and the object of rejection by the person tormented inside it.

This is not the case in the obese who submits to his body almost without any apparent wish to react. He is often accused of not having enough will-power to slim. But despite his grotesque efforts he does not manage to readapt to his own body; thereby he shows his inability to readapt to his profoundest ego.

This defeat where the *Körper* is concerned will then lead the obese to lack of confidence and a defeat in the matter of *Leib*, that justifies that kind of renouncing passivity of being in the world and offering others an image of oneself that is considered repulsive and falsifying, but at the same time it is an image he cannot free himself of. This escape into the imaginary or, better, into phantasy, because the body is not experienced as the person but almost as a ghost, is on the one hand the salvation and on the other the impoverishment, inflexibility and impossibility of the ego to express itself in the way it believes to be the only way.

In the anorexic and the obese the trend to internalize, not indeed a revealing and maturing trend towards a more authentic mode of being, represents an escape towards the imaginary. This trend is determined, conditioned, sclerotic, tending to the imaginary and thus unable to spread out towards the external.

It is not a question of body sensation or body function but of body significance and body relationship: body awareness and gestures are words expressed by the body. If the partial and anonymous sensations are integrated at the level of the body, then our body becomes that thing that sustains and forms us participating in our destiny, it is that "incarnate awareness" (Callieri) and "employed temporality" (Merleau-Ponty). On the other hand the ego needs this ambivalence and ambiguity because I am also my body (one sees dissociative syndromes caused by sensorial deprivation) and the only way I can recognize my body is to live it, to experience it as freedom but also as a part of my total being.

University of Catania

BIBLIOGRAPHY

Arieti, S.: *Manuale di psichiatria*, Boringhieri, Milano, 1976.

Bonomo, V., Ottaviano, P., Germanà, B., and Rapisarda, V.: *Trattamento dei soggetti in sovrappeso con il Mazindol*, Ed. Cedam, Padova, 1977.

Bruch, H.: *Patologia del comportamento alimentare*, Feltrinelli, Milano, 1977.

Binswanger, L.: *Per una antropologia fenomenologica*, Feltrinelli, Milano, 1970.

Cacciaguerra, F.: 'Schema corporeo e comportamenti relazionali nei gemelli', *Acta medica antologica*, Ed. Cappelli, Bologna, 1976.

Calandra, C., Colaciuri, V., and Rapisarda, V.: 'Anoressia mentale: un gioco adolescenziale mal riuscito', *Soc. Medico-chirurgica*, Catania, 1976.

Callieri, B., Castellani, A., and De Vincentis, G.: *Lineamenti di una psicopatelogia fenomenologica*, Ed. Il Densiero Scientifico, Roma, 1972.

Cargnello, D.: *Alterità ed alienità*, Ed. Feltrinelli, Milano, 1966.

Carp, E.: 'Troubles de l'image du corps', *Acta Neurol. Psychiat. Belgica* 52 (1952), 461.

Del Carlo Giannini, G.: 'Approccio antropofenomenologico alle dismorfofobie', *V Congresso Naz. Soc. Ital. di neuropsichiatria Infantile Rimini 23–27 sett. 1972*.

Di Stefano, G.: *Lo sviluppo cognitivo*, Ed. Giunti, Firenze, 1973.

Fischer, S. and Cliveland, S.: 'An Approach to Physiological Reactivity in Terms of Body-Image Schema', *Psychol. Rev.* 64 (1975), 26–37.

Gardner, G.: *Emotional Development in Adolescence*, Stuart, Harold C. (ed.), Harvard, Cambridge, Mass., in press.

Gardner, G.: *The Mental Health of Normal Adolescents*, Modern Library, New York, 1954.

Gesell, A.: *Il bambino nella società di oggi*, Bompiani, Milano, 1973.

Graumann, C. F. *Grundlagen einer Phänomenologie und Psychologie der Perspektivitat*, De Gruyter, Berlin, 1960.

Hoffer, W.: *Mouth, Hand and Ego-Integration*, Internation. Univ. Press, New York, 1949.

Husserl, E.: *Idee per una fenomenologia e per una filosofia fenomenologica*, Einaui, Torino, 1976.

Jasper, K.: *Psicopatologia generale*, Ed. Il Pensiero Scientifico, Roma, 1976.

Linn, L.: 'Some Developmental Aspects of the Body Image', *Intern. Journ. Psycho-Analysis* 36 (1955), 36–42.

Merleau-Ponty, M.: *La struttura del comportamento*, Bompiani, Milano, 1970.

Merleau-Ponty, M.: *Fenomenologia della percezione*, Ed. Il Saggiatore, Milano, 1965.

Meisser, O: *Psicologia cognitivista*, Ed. Giunti, Firenze, 1976.

Summerfield, H.: *La psicologia cognitiva*, Ed. F. Angeli, Milano, 1978.

Pick, A.: 'Zur Pathologie des Bewusstseins von eigenen Korper. Ein Beitrag aus Kreigs-medizin', *Neurol. Centralb.* **34** (1915), 257–265.

Riese, W. and Bruch, G.: 'Le membre fantôme chez l'enfant', *Rev. Neurol.* **83** (1950), 221–222.

PART V

THE PLACE OF THE SPIRIT
WITHIN THE SOUL–BODY ISSUE

RICHARD COBB-STEVENS

BODY, SPIRIT AND EGO IN HUSSERL'S *IDEAS II*

Husserl's analysis of the relationship between the various dimensions which comprise the unity of the person differs radically from traditional discussions of this topic. Of all of his works, *Ideas II* reveals most clearly the uniqueness of his approach.[1] This paper will first summarize Husserl's study in this text of the interplay between animate body, spirit, and ego, and then reflect upon the implications of his refusal to consider the unity of these strata from within a global ontological perspective.

I. DIMENSIONS OF THE PERSON

A. *The Animate Body*

Ideas II introduces the theme of the body in an oblique fashion. The body does not first appear as a topic of direct study, but rather as a necessary component in the determination of what we mean by a thing. We cannot fully appreciate what we mean by a material thing without allusion to the perceiving body. This is because things appear differently depending upon the normal or abnormal conditions of the body's sense organs, and upon its kinesthetic processes. Just as changing environmental circumstances such as lighting and atmospheric conditions alter the thing's appearance, so also corporeal transformations and states alter the angle and clarity of a thing's givenness. As a result, we think of a thing as what remains constant despite alterations in these two different types of circumstances. Such constancy is an essential component of what we mean by a thing.[2] This way of discussing the body should not be confused with an empirical study of kinesthesia or of the range of variations in organic performance. Husserl's purpose is not to contribute to empirical psychology, but rather simply to spell out the components of thinghood which emerge from ordinary experience. Of course, he recognizes that our ordinary experience has been affected by scientific interpretations. In fact, he explicitly situates his analysis of the thing's sense within the context of an understanding of nature influenced by the attitude of science. The contemporary sense of nature is the intentional correlate of an attitude which Husserl describes as both doxic and theoretical. It is

243

A-T. Tymieniecka (ed.), Analecta Husserliana, Vol. XVI, 243–258.
Copyright © 1983 by D. Reidel Publishing Company.

"doxic," because it is permeated by an unthematic belief in the existence of its object; it is "theoretical," because it prescinds from the practical, aesthetic, and axiological textures of its object.[3] While ordinary experience does not consistently maintain an exclusively theoretical stance, nonetheless the influence of science has generated the everyday conviction that the true sense of the thing is what remains when we bracket the useful, the beautiful, and the good. In later texts, Husserl strives to get beneath the sedimentations of sense imposed by scientific interpretations, in order to reveal a more original understanding of things.[4] Although full retrieval of a prescientific mode of experience is impossible, deconstruction of more recent scientific models, such as Galileo's mathematization of nature, helps us to recognize generally that something's sense is always a correlate of an interpretative attitude.

Further analysis of the sense of the thing reveals that its full objectivity requires an attitude which recognizes the intersubjective appreciation of nature. Part of the sense of a thing's objectivity is that it is a thing for everyone. The thing's spatiality is only incompletely constituted from my individual perspective which comprehends space as spread out around my body, as radiating out from that mobile center of orientation, my "here." The complete sense of nature's space requires the elaboration of a network of places that would permit demonstrative identification of my "here" with another's "there."[5] Things appear both as gathered around me and as located within a public system of coordinates. It should be noted that this analysis of the link between objectivity and intersubjectivity is not intended as a response to the problem of solipsism. There is no epistemological concern whatsoever in this passage. Husserl is engaging in an archeology of sense; his purpose is to exhibit the contribution of the sense of my body's perspective and that of others to the constitution of the sense of the thing. Perhaps the closest analogy to this method may be found in the typical procedure of semiotics, when it constructs the genealogical tree of some sememe (meaning content), by displaying the compositional structure of its denotative and connotative markers (meaning components). Semiotics legitimately brackets all metaphysical and epistemological considerations, when studying the semantic properties of various terms. This is true even of the properties of indexical terms. For example, in displaying the semantic structure of a term like "present" in that phrase dear to analytic philosophers, "The present king of France is bald," we may say that its markers are (1) time; (2) immediacy; (3) indexical locatability; (4) actual existence. This is what the word denotes and connotes, whether or not the sentence is applicable to the current circumstances of French politics.[6] In similar fashion, Husserl lists the various components of a thing's sense as (1) extension; (2) stability through changing circumstances; (3) correlate of my body's normal kinesthesia; (4) common to many actual or potential viewers; (5) situated within the coordinates of public space. He is surely justified in making such an analysis without getting involved in such issues as the Platonism of semantic markers or the status of meanings within the mind's "glassy essence." In an indirect fashion, however, Husserl's methodology yields something more than a breakdown of the semantic properties of the thing. The inclusion of reference to the perceiving body as one of the components of a thing's sense establishes that human consciousness is embodied. Moreover, only an embodied consciousness would require the roundabout route of imaginative construction of an intersubjective context, in order to grasp the thing's full objectivity. However, Husserl does not emphasize that thing-perception reveals the necessarily corporeal mode of our *existence*; he confines himself to explicating the sense of the body required by the sense of the thing. What differentiates Husserl's analysis from a scan of semantic markers, however, is his stress on the correlation between attitudes of consciousness and the various components of a thing's sense. An intentional analysis differs from a semantic explication in this: the intentional study takes the sense of the object as a transcendental guide to the intentive performances of consciousness which intersect in that sense.[7] The scientific sense of nature could not emerge without a corresponding attitude of objectification, i.e., a fusion of doxic commitment to existence and the bracketing of practical, aesthetic, and moral predicates. The sense of the thing's objectivity requires the assumption of an attitude which interprets the thing as given from other perspectives. To every nuance of a thing's sense, there corresponds an attitudinal modification of consciousness. In short, all seeing is interpretation. Forgetfulness of these attitudinal nuances leads to various forms of reductionism, for if we fail to take account of the interpretative performances which make various types of objectivity possible, it is easy to succumb to the conviction that some mode of objectivity is absolute, i.e., that it is what truly is apart from human interpretation. Whether we decide that the privileged zone of objectivity is nature or spirit is irrelevant; once we absolutize one or the other, we have initiated a process of reducing one to the other.

The second section of *Ideas II* takes a new sector of reality as its transcendental guide: the soul or the psyche. Just as an elaboration of the strata of the thing's senses required a description of that sense of the body implied in understanding of the thing, so now an elucidation of the semantic properties of the soul will require a further articulation of different senses of the body. By soul, Husserl means neither a formal principle of being, nor a *res cogitans*, nor a function of the body. All of these are explanatory or metaphysical concepts. For Husserl, soul refers simply to that objective dimension of reality which appears within the zone of nature as intertwined with, and yet other than, that privileged thing of nature, the body. Before unpacking the meaning components of the psychic order of reality, however, Husserl talks briefly about the status of the philosophical voice that conducts this investigation into the sense of the soul. He makes explicit what the reader familiar with phenomenology has already guessed, that this study is being conducted within the context of the phenomenological reduction. Psychic reality will be described as it appears for a pure ego, whose detached theoretical perspective has been made possible by a reflective gesture of distantiation. The exercise of the reduction is a discovery of our perfect freedom, an implementation of a pure interest in the achievement of a disinterested point of view.[8] We must be cautious, however, about construing the philosophical perspective as *a* perspective, especially as a detached perspective. It is true that Husserl's visual metaphors sometimes suggest that philosophical reflection generates a neutral, detached, nonhistorical perspective, a kind of theoretical gazebo from which even the philosopher's own psyche may be contemplated as an objective given. Husserl's analysis of the pure ego in *Ideas II*, however, does not invoke the image of an Archimedean point of view on every point of view. Although he briefly recalls his earlier description in *Ideas I* of the ego as a functional center from which radiating glances emanate, he quickly complements this analysis with a description of the pure ego as a perduring life, structured by "habitualities," a term which seems to suggest the cumulative momentum of a history of attitudinal stances.[9] How can the pure ego be both a detached spectator and a habit-laden life? Unfortunately, *Ideas II* only hints at how we might make sense out of these conflicting images. Husserl suggests, first, that the perspective of the pure ego is enjoyed or lived, rather than known in an objective fashion, and second, that its continuity derives from the dimension of original time.[10] A lived perspective does not reveal itself through spatial profiles, like an object. Moreover, in its operative state, it is not given as a moment within the constituted stream of consciousness, that flux of inner life which I describe as mine. It is true that, once objectified, the "focal" status of the pure ego can be contemplated from a fresh perspective, and thus constituted as *a* perspective. But as originally experienced, the "center" from which the pure ego acts cannot be located either in objective space and time, or within the subjective flux of psychic life. This is why, at every moment of my conscious life, I appreciate as much of the pure ego's status as I ever will. Husserl here describes a type of experience which is nonobjectifying, but nonetheless revelatory. How can philosophical reflection thematize this elusive mode of givenness without objectifying it, and thus treating it as *a* perspective, in the same way that one might speak of a medieval point of view, an American attitude, or an obstinate stance? The key to Husserl's understanding of the unique character of the philosophical perspective is his theory of how reflection can appreciate the difference between operative and thematized conscious life. We shall attempt to clarify this issue in the second section of this paper. Besides modifying the sense of the spatial metaphor of a point of view, Husserl also adds an important cautionary note about the temporal metaphor, which describes the ego's mode of living as "habitual." He makes it clear that the temporal structure of the ego's life is somehow different from the historicity of the person, by intimating that the ego's coherence derives from the self-constitution of a more original time than that of the constituted flux of subjective life. Later, we shall make some further comments on this notion of a proto-temporality which orders and differentiates in such a fashion that it makes subjective time, or the form of the present and of presence, possible.

Husserl next returns to the sense of the soul for the pure ego. Although the psychic stratum appears within the zone of nature, its objective status is unusual, for the pure ego understands the soul as its own objectification. Thus, Husserl's earlier intimation that the full objectivity of the thing requires the embodiment of consciousness is now confirmed by the unfolding of the sense of the soul. What we mean by the soul is that dimension where the ego is made mundane through the constitution of the body as animate body. However, we must be extremely careful about interpreting this incarnational motif. Husserl is not talking about some metaphysical fall of a separate soul into the prison house of the body. Indeed, his analysis makes no metaphysical claims about the relationship between ego, soul, and body. His sole intention is to accomplish a philosophical reconstruction of the layers of sense, interwoven in that complex domain of meaning, the animate body. The analysis next proceeds in two stages: (1) an examination of the sense of the combined corporeal and psychic strata, in comparison with the sense of the thing developed earlier; (2) a description of the relationship between the two strata of the animate body, the material dimension (*Körper*) and the psycho-organic dimension (*leiblich-seelichen*). Like the thing, the animate body presents itself as a stable reality that perdures through variations in circumstances. However, psychic properties, such as aptitudes, character traits, etc., differ from the properties of things, in that their sense includes the history of their development. It is because psychic properties have something in common with a thing's properties that empirical psychology can legitimately be modeled after the sciences of nature. Thus, phenomenology justifies the method of empirical psychology, on the grounds that consciousness can maintain the same mode of experience in investigating both physical and psychic properties. However, empirical psychology must respect its limits: it must not blur the difference in the sense of psychic properties as opposed to physical properties, a difference which manifests itself even within the naturalistic attitude. The psychic order has neither the same sort of perdurance, nor the same sort of localization, nor the same type of dependence upon circumstances. First, the stability of the psychic stratum is more like that of an historical continuity than like that of a permanent thing; second, the psychic dimension is experienced as localized in space, but not as extended in space or as spatially divisible; third, the soul's dependence on corporeal circumstances differs markedly from a thing's dependence on bodily transformations. The soul's circumstances are part of its appearance, whereas the thing's circumstances are conditions of its appearance.[11] When Husserl next considers the interplay of psycho-organic and material strata, he does not stress the opposition between the sense of the lived body and the sense given to the body by the natural sciences, in the manner of Sartre and Merleau-Ponty.[12] His purpose is rather to exhibit the sense of the body that emerges from the experience of organic functions involved in the appearance of things of nature. What is most intriguing about this analysis is that he takes tactile rather than visual experiences as paradigmatic: when I touch my body with my hand, my body appears twice, as an explored quasi-natural thing and as explorer. According to Husserl, this experience has no equivalent in the domain of vision: "Everything seen can be touched, and on this latter account refers back to an immediate relation to the body, but not, however, by means of its visibility. A purely ocular subject would not have a body that appears."[13] Husserl adds that a purely ocular body might understand kinesthesia indirectly, in that it could observe the correlation between movements of the body-thing and the appearance of other objects, but it could never grasp the owned character of the body. It would be as though a pure *res extensa* were to understand the body's freedom of movement as its pure freedom to move the *res extensa*. This brilliant exercise in imaginative variation demonstrates the danger of excessive dependence on the paradigm of vision.[14] Only tactile sensations reveal the spread-out character of the body's localization of the psyche.

The sense of the soul as localized within the space of nature, but not as extended or divisible, requires intersubjective levels of constitution. It is only by participating imaginatively in the manner in which others perceive things that I can fully objectify my body, and thus give a quasi-localization to my soul. By representing the mobility of my animate body as others perceive it, I understand my psyche as localized in that body which others experience. From my perspective, my animate body appears as an absolute "here," as a mobile center of a world whose coordinates are determined by my interested attention. However, I also interpret my "here" as an "over there" for others. Husserl does not affirm that either one of these attitudes has chronological priority over the other; rather, he stresses that they provide two different but complementary senses of the body. Indeed, feedback between the two attitudes enhances both the solipsistic and the intersubjective senses. I come to appreciate my own intimately experienced psycho-corporeal unity more fully by applying to it the understanding of the unity of psychic and corporeal strata that I acquire by observing another's facial expressions, gestures, speech, etc. Also, I transfer to the other the sense of lived body derived from my own experience.[15] In short, Husserl stresses that a perpetual oscillation between first-person and empathetically achieved second-person perspectives provides the full sense of the animate body.

Unfortunately, if we combine an excessive emphasis on the intersubjective objectification of the animate body with a scientific attitude toward nature, and if we interpret these senses not as correlates of conscious attitudes but as description of the true being of the body, then a philosophical behaviorism becomes possible. Although Husserl only hints at this danger, we might trace the steps involved in behaviorism's reification of the psyche as follows: (1) perceived qualities of thing or body are relegated to the realm of the "subjective," while scientific models are construed as the true underlying reality; (2) only the intersubjectively constituted sense of spatiality is afforded legitimacy; (3) the psychic stratum is dislodged from its experienced fusion with the body, on the grounds that the only true body is the one accessible to extrinsic observation; this makes of the perceived animate body an epi-phenomenon of the mathematicized body-thing; (4) after the eventual collapse of the animate body into the body-thing, for a while "mind" is given the status of epiphenomenon, on the grounds that operations like thinking and willing do not seem to depend on the body as fully as do seeing and feeling; (5) finally, mind seems to appear more and more as an oddity, or as a residue from a mythological or religious state of human understanding. As the behaviorists put it, mind is eventually understood as a "nonce locution" for behavioral reductions which the advance of the positive sciences will make evident. One of the great strengths of Husserl's methodology is that it makes it possible to display with extraordinary clarity the tacit attitudinal decisions involved in this particular self-understanding of man.

Husserl concludes his exposition of the stratified senses of the animate body, by reaffirming the polarity between this uniquely objectified reality and the pure ego for whom the regions of nature and psyche are constituted senses. Thus, the second section of the book ends with a clear distinction between the empirical and transcendental egos.

B. *The Region of Spirit*

The final section of *Ideas II* investigates the highest stratum of sense that belongs to human reality, i.e., the cultural dimension which overlays the whole of the natural world, and, in particular, that preferential segment of the natural world, the empirical ego. The early descriptions of the animate body had bracketed those senses that are contributed by the social and cultural dimensions of human life. Reference to intersubjectivity was confined to its role in the objectification of the animate body as a thing of nature. To emphasize the change in perspective, Husserl now describes the full concrete unity of soul and body as "person," and he refers to the objective region in which the dimension of person appears as "spirit." Like the earlier strata (*Körper, Leib, Seele*), *Geist* is also taken as a dimension of meaning, and not as an ontological category. The constitution of an individual's sense is dependent upon the sense of a community of persons and upon the whole realm of culture. We have seen that the individual's animate body appears as the referential center of a horizon of natural objects. In the region of spirit, the person as person (hence, as qualified by communal relationships) appears as the focal center of an *Umwelt*. This surrounding world is no longer the faceless world of nature, but rather an ambiance of humanized objects and social structures. The world that surrounds the person has been remade by technological transformations and cultural interpretations. The space of nature has been mapped and plotted in accordance with human needs, values, and aesthetic preferences. The appreciation of the sense of the person and the surrounding world of spirit requires an attitude of understanding, as opposed to explanation. In the attitude that makes empirical psychology possible, I am able to notice the psychic stratum as a modification of the corporeal, while still thinking of the corporeal as inserted within the texture of nature. In this perspective, I "observe" the soul in the body (sorrow in a face, meaning in a gesture, insight on a printed page). I recognize that sorrow, meaning, and insight are somehow other than the physical strata with which they are intertwined, but I nonetheless grasp these senses against the backdrop of nature, objective time, and the coordinates of mathematicized space. It is precisely because the psychic stratum does appear within the domain of objectivity opened up by the natural sciences that cognitive psychology finds it difficult to emancipate itself from the techniques of quantification appropriate to empirical psychology. This confusion is compounded by philosophers who have never reflected upon the different attitudes involved in the emergence of various strata of sense. Most contemporary philosophical reflection is carried out within the unquestioned framework of what Husserl called the natural attitude, i.e., the conviction that reality is what it is apart from human understanding. Reinforced by an uncritical acceptance of mathematicized nature as the true reality, this attitude tends to encourage, rather than to counter, the drift toward behavioral reductionism.

Husserl clarifies the concept of understanding, by contrasting motivation and causality. Naturalistic thinking grasps the relationship between psychic events as a causal chain occurring within the realm of nature; understanding grasps the same relationship as a history, shot through with teleological intention and progressively reconstituting its antecedents as a past. A cultural object is a locus wherein motivation and causality intersect. Understanding does not isolate the motivational dimension; on the contrary, it grasps the unity of nature and spirit in a cultural object. Ricoeur notes that this is a crucial moment in Husserl's study, for he finally points out the difference between the apperception of another soul in the gestures and facial expressions of a body taken as an object in nature, and the appreciation of another person by someone who *already* knows that people's faces are expressive of meaning, and for whom language is loaded with layers of cultural significance.[16] Husserl offers a striking example of the manner in which understanding grasps the fusion of nature and spirit. We may consider a book, with its pages made from wood pulp, its solid binding, its weight, as a thing. But when we read the book, we subordinate the thing-status of the book to the aura of signification that infuses this natural object, as the concatenation of words yields patterns of meaning and evokes affective responses.[17] Just as the printed page delivers meaning so immediately that its physical dimension seems to be absorbed by its communicative function, so also the body as cultural object is expressive in its very physicality. On the level of spirit, there is no question of priority of thing-perception over expressivity. The thing-body is no longer grasped as an index of the psyche;

rather the person is the primordial given, and the body is experienced only obliquely, as a kind of limit. Thus, the theme of motivation permits a re-reading of the senses of psyche, body, thing, and nature. From the perspective of spirit, the world of nature has a special sense because it is appreciated as the product of a cultural activity called science. Atoms, molecules, quarks, and DNA are part of the *Umwelt*, insofar as their discovery is a cultural accomplishment of human persons.

The human sciences ought to study not just the person in relation to the surrounding world of culture, but also and especially the mutual relationship among persons. The development of the stratum of personality cannot occur solipsistically. Husserl notes that one becomes a person only through inter-subjective struggle, dialogue, and recognition.[18] I begin to know my dignity as a person only through the demand of other persons that they be treated with respect. After this foray into the Hegelian province of spirit as the emergent product of dialectical recognition, however, Husserl repeats the refrain of transcendental phenomenology, that this priority of sociality over individuality is itself a constituted sense for the transcendental ego. The community of persons is still a sense for the ego, a region of reality made possible by an attitudinal stance of the ego. Transcendental philosophy refuses to be absorbed either by a sociology of personal reciprocity or a philosophy of objective spirit. On the contrary, Husserl sees his transcendental approach as justifying their methodologies on a level of discourse made possible, as Ricoeur puts it, "by the same act that destroys their absolute claim."[19]

Ideas II closes with a surprising twist. Up to this point, spirit has been described as the concrete reality of the person in relationship to other persons and to a surrounding world of culture. Just as the "human ego" (the objective unity of soul and body) is a reality having its place in the world of nature, so also the "person" is a reality having its place in the world of culture. But in the final passages of the book, individualized spirit is described almost as though it were identical with the pure ego itself. Just as the pure ego is not simply an empty pole, but a bearer of habitualities, so also each spirit has its style of motivation, its own individual history.[20] Indeed, the individuality of spirit derives from the unity of its temporal life. A thing does not possess individuality within itself; it becomes individualized only for a consciousness which synthesizes its unity in a flux of profile appearances. Thus, concludes Husserl, the absolute individuality of spirit, as opposed to the relative individuality of things, permits us to affirm that spirit is the absolute to which all of nature is relative. These texts seem to suggest that Husserl understands spirit as representing a concrete status of the pure ego itself. Could it be that the theme of motivation justifies an identification of transcendental ego and person? Has Husserl finally identified the proto-temporality of the pure ego with the historicity of the individual spirit, the person? Indeed, the above-mentioned analysis of how the priority of motivation in the realm of spirit subordinates the objective status of physical things to their expressivity suggests a Heideggerian appreciation of things as paths or obstacles to projects within a vast instrumental complex. It would seem that only a small step would be required for Husserl to begin to describe the person as a mode of being which contains within itself the possibility of transcendental constitution. But Husserl finally eliminates any possibility of interpreting his text as moving toward a fusion of transcendental ego and person. It is true that analysis of the stratum of the person required a break in the methodology of the book, in the sense that the realm of spirit cannot be described as founded on the realm the psyche, in the same way that the psyche is founded in the stratum of the body understood as a natural thing. The realm of spirit cannot be appropriately described as grafted upon the things of nature. Rather, as the analysis of the motivational world opened up by spirit suggests, the stratum of spirit can be linked with the realm of things only in terms of a perception of things that precedes scientific objectification. In other words, Husserl recognizes that the realm of spirit is grafted upon the life-world, not upon the world of science. But as all of the subsequent studies of the life-world clearly indicate, the recognition of the derived nature of scientific objectivity does not induce Husserl to abandon the subject-object paradigm. The structures of the life-world are still taken as transcendental clues for revealing attitudes of consciousness. Thus, even the stratum of the person and surrounding world are constituted as senses for the transcendental ego. In Husserl's view, just as empirical psychology cannot justify its validity without looking to the attitudinal stances which make possible the emergence of the kind of objectivity investigated by that science, so also a phenomenology of the person, or even a broader phenomenological ontology must analyze the attitudinal stances that make possible the sense of being-in-the-world. The person is too involved in the *Umwelt* by reason of pragmatic motivations to be completely identified with the philosophical voice that describes the stratum of spirit.

II. THE DIFFERENCE BETWEEN PERSON AND EGO

In the recently published *Geschichte des Zeitbegriffs*, the Marburg summer-semester lectures of 1925, Heidegger claims that Dilthey's project of a

personalistic psychology brought him close to formulating the question of the
ontological status of the person, and even of the meaning of being. Dilthey's
protest against the scientific naturalization of man, his recognition that the
person ought not to be described as a psyche linked to a natural thing, and
finally his stress on the opposition between explanation of nature and understanding of history — all testify to an unarticulated dissatisfaction with the
metaphysical prejudice that interprets being as *Vorhandenheit*. Knowing
that Husserl was favorably impressed by Dilthey's circumscription of the
limits to a natural science of the soul, Heidegger looked forward eagerly to
reading Husserl's long promised contrast of nature and spirit in *Ideas II*. In
February of 1925, Husserl sent him the manuscript along with a note, calling
Heidegger's attention to the radical changes in his treatment of the relationship between nature and spirit, as compared with his earlier Freiburg
lectures on the topic. Heidegger's reaction to the text oscillates between
admiration, "over against Husserl, I still consider myself a learner," and
manifest disappointment, "Husserl does not go beyond Dilthey . . . on the
contrary. . . . "[21] According to Heidegger, three unexamined factors prevent
Husserl from recognizing the unique ontological status of the person: (1) the
tendency to treat soul and spirit as superstructures built upon the foundation
of the natural thing; (2) an emphasis on immanent reflection as the method
of access to the nature of the person; and (3) a reliance upon the traditional
definition of man as *animal rationale*. Although Husserl brilliantly articulates
senses proper to natural things, nature itself, the psycho-organic strata of the
body, the stratum of the person, and finally of social unities, it never occurs
to him to ask about their modes of being. His purpose is always to describe
the essential correlation between these various regions of objectivity and the
attitudes of consciousness which make it possible for each type of object to
appear. According to Heidegger, this method of essential analysis presumes
that being is equivalent to objectivity. To be is always to be there for a
subject. The phenomenological reduction reveals a zone of absolute and pure
givenness, a kind of ideal scene where eidetic structures are exhibited for a
pure theoretical seeing. Thus, Husserl rejects the naturalistic interpretation
of being as the reality of things apart from our consciousness of them, but
still accepts a deeper prejudice of naturalism, the conviction that being
is presence-at-hand. Naturalism construes being as independent of our conscious attitudes, but also as the availability of things for objective disclosure.
Husserl's reduction criticizes the first presupposition, but his emphasis on
the zone of consciousness as a region where a more perfect seeing is possible,
tends to reinforce the latter presupposition that being means presence-at-hand.

Thus, according to Heidegger, the phenomenological reduction masks rather than reveals the being of the person. Husserl clearly distinguishes persons from things, and both from the ego, but he considers all of these as being in the same fashion. These remarks are, of course, reminiscent of the opening pages of *Being and Time*, where Heidegger claims that the entire metaphysical tradition has constantly dealt with the issue of being, while never making a careful inquiry into what is meant by being. The reason for this failure, he contends, is that all of the great philosophers have always taken it for granted that being is presence-at-hand. It matters little whether a philosopher claims that we are capable of knowing things in themselves, or that we only have access to things as they appear to us. In either case, things are construed as objects displayed before the gaze of a human or divine observer. It also matters little that most philosophers distinguish persons from things, appealing to soul, spirit, or subjectivity as the distinguishing trait of humanity. They do not really think of these unique traits as involving a different mode of being from the being of objects. Subjectivity may well differ from objectivity, in the sense that it is the point from which perspectival views on objects emanate, but its givenness too is construed according to the objective model. Thus, when Husserl attempts to describe subjectivity, his goal is always to show subjectivity for *what* it is. From Plato through Husserl, the meaning ascribed to being remains fixed: being is objectivity. Individual beings are mundane objects, essences are ideal objects, and being itself is construed as a display area wherein objective presence can occur.

Moreover, Heidegger objects to the description of the person's transcendental dimension as a disinterested theoretical point of view. He stresses that no freely assumed stance toward one's existence can emancipate itself from the facticity of that existence. The notion that the transcendental ego can take a distance from every structure of the person's involvement in the world (intersubjectivity, history, language, and culture) is the preposterous conclusion to the philosophical tradition which construes human beings as spectators. By deciding to drop the subject-object model entirely, and to shift the description into an ontological key, Heidegger interprets the transcendental dimension as a relationship with being that begins within being. The starting point is no longer Descartes' *cogito*, or the human being as onlooker, but the question of being. The human questioner who wonders about being is not designated as a consciousness, but by an ontological term that expresses the kind of being that makes the questioning of being possible. *Dasein* is the site or locus within being where the question of being emerges; *Dasein*'s involvement within being opens up a space within which beings can show themselves. All of the metaphors which guide this phenomenological description differ radically from those of Husserl. Instead of the image of an ego as a detached eye which stands apart and observes, we have the image of a clearing within being opened up by *Dasein*'s temporal structure of involvement. Thus, the theme of the "clearing" permits Heidegger to come to grips with the transcendental dimension of human thought, without having to appeal to the traditional theme of a detached spectator. The advantages of this interpretation are considerable. The ontological context permits Heidegger to express the unity of the person as both finitude and transcendence. The metaphor of human temporality as a clearing *within* being makes it possible to appreciate the complementarity of involvement and distantiation. Distantiation no longer implies a platform above or outside of the world; the human mode of being-in is no longer construed as a limitation, but rather as the condition of thinking.

In the light of this powerful critique, let us attempt to understand why Husserl consistently refused to describe the unity of person and ego from an ontological perspective. Before addressing this question directly, however, we must make some correctives to the above Heideggerian reading of Husserl's interpretation of the ego's status. Husserl's brief comments on habit and his lengthier remarks on motivation demonstrate that his reflections on the temporal structure of the ego gradually led him to recognize a dimension of depth to the ego's life that resists objective thematization. His stress in *Ideas II* on the fact that the ego's life is enjoyed or experienced, rather than known, suggests that there is an irreducible difference between lived experience and the subsequent reconstruction of experience in reflection. Meditation on this gap between reflecting and reflected yielded the thesis that the flux which makes it possible for the form of time to appear to immanent reflection cannot itself be within the form of time.[22] Although reflective consciousness can grasp the constituted temporal flux as an immanent object, it can only appreciate its own operative life in an oblique fashion. Thus, contrary to Heidegger's contention that Husserl implicitly reduces all modes of being to objectivity, or presence-at-hand, his study of the ego's temporal structure in fact culminates in the discovery of the impossibility of displaying the primal flux of the ego's life with total reflective clarity. Reflective consciousness can never focus directly on its own unthematic temporal life. This is why Husserl's understanding of philosophical reflection is far more complex than what is suggested by the model of a detached spectator consciousness. Philosophical reflection is a second-order reflection which directs its focus upon the interplay between reflecting and reflected dimensions, thus recognizing the structural necessity of an irreducible gap between the two. If we attempt to collapse this difference, we inevitably objectify the non-objective status of the ego's operative mode of being. We might say, therefore, that the difference between constituting and constituted temporality is the "space" within which the being of the transcendental dimension is disclosed. But we must not think of this space of disclosure as an objective region that precedes the reflection which reveals it, nor should we misconstrue the disclosure as a thematic display of an eidos. Rather, the difference between lived temporality and its objectification as the person's stream of consciousness is both the locus of disclosure and the essence of the disclosed. What is most significant about this analysis is that it shows how the "being-for-itself" of the transcendental ego cannot be objectified. The ego reveals itself as resistant to reflective exhibition, in the discovery by reflection of its own limit. We may conclude that Husserl's method yields an appreciation of the radical difference between transcendental and empirical modes of being, even though it does not explicitly formulate its insights as an ontology.

How does this analysis of Husserl's understanding of philosophical reflection advance our discussion of why Husserl always refused to collapse the transcendental dimension into the category of the person? The second-order character of philosophical reflection testifies to the fact that the uniqueness of the transcendental dimension can appear only at the intersection, or place of tension, between two attitudes of consciousness; the stance of reflective thematization, and the recognition of its limits by an attitude that appreciates the difference between lived experience and reflective reconstruction. If one were to remove the tension between transcendental and empirical dimensions of the person, one might be tempted to think that the person's unity could be thought from within a single attitude. Husserl could envisage no other means of preserving the tensional status of the transcendental dimension, than the resolute insistence on the impossibility of a global ontology that could do justice both to our being-in-the-world and our extraordinary ability to reflect upon that mode of being.

Boston College

NOTES

[1] Edmund Husserl, *Ideen zu einer reinen Phänomenologie und phänomenologischen Philosophie*, vol. 2 (*Husserliana*, vol. 4), ed. Marly Biemel (The Hague: Martinus Nijhoff, 1952).

258 RICHARD COBB-STEVENS

2 Ibid., p. 43.
3 Ibid., pp. 2–3.
4 Cf. Edmund Husserl, *Die Krisis der europäischen Wissenschaften und die transzendentale Phänomenologie* (*Husserliana*, vol. 6), ed. Walter Biemel (The Hague: Martinus Nijhoff, 1954).
5 *Ideen II*, pp. 79–84.
6 Cf. Umberto Eco, *Semiotics* (Bloomington: University of Indiana Press, 1979), pp. 84–86, 169–71.
7 Paul Ricoeur's commentary provides a lucid description of Husserl's transcendental methodology. Cf. Paul Ricoeur, *Husserl: An Analysis of His Phenomenology*, trans. Edward G. Ballard and Lester E. Embree (Evanston: Northwestern University Press, 1967), pp. 36–38.
8 Cf. Jürgen Habermas, *Knowledge and Human Interests*, trans. Jeremy Shapiro (Boston: Beacon Press, 1968). Habermas offers a comprehensive critique of the notion that the theoretical interest is a "pure" interest. Cf. pp. 301–17.
9 *Ideen II*, pp. 113–19.
10 Ibid., pp. 102–4.
11 Ibid., p. 139.
12 Cf. Ricoeur's commentary on this point, *Husserl: An Analysis of His Phenomenology*, pp. 60–61.
13 *Ideen II*, p. 150. "Jedes Ding, das wir sehen, ist ein tastbares, und als solches weist es auf eine unmittelbare Beziehung zum Leib hin, aber nicht vermöge seiner Sichtbarkeit. Ein blos augenhaftes Subjekt konnte gar keinen erscheinenden Leib haben . . . " (translation taken from Ricoeur, p. 61).
14 For an excellent commentary on this passage, cf. Tadashi Ogawa, "Seeing and Touching, or Overcoming the Soul–Body Dualism." *Phenomenology Information Bulletin* 4 (1980), 37–57.
15 *Ideen II*, p. 167.
16 Ricoeur, p. 75.
17 *Ideen II*, p. 238.
18 Ibid., p. 191.
19 Ricoeur, p. 72.
20 *Ideen II*, p. 299.
21 Martin Heidegger, *Geschichte des Zeitbegriffs*, ed. Petra Jaeger (Frankfurt am Main: Vittorio Klostermann, 1979), p. 173: "Husserl kommt über Dilthey nicht hinaus. . . . Im Gegenteil. . . . "
22 Cf. Edmund Husserl, *Cartesianische Meditationen und Pariser Vorträge* (*Husserliana*, vol. 1), ed. Stephen Strasser (The Hague: Martinus Nijhoff, 1959), sec. 18, and Edmund Husserl, *Zur Phänomenologie des inneren Zeitbewusstseins* (*Husserliana*, vol. 10), ed. Rudolf Boehm (The Hague: Martinus Nijhoff, 1966), no. 50, pp. 324–34.

HEIMO HOFMEISTER

DIE BEDEUTUNG DES GEWISSENS FÜR EINE LEIBHAFTE VERWIRKLICHUNG VON SITTLICHKEIT

Karl Ulmer zum Gedenken

I

Der archimedische Punkt einer modernen Theorie des Sittlichen ist fraglos die begriffliche Erfassung des Gewissens, denn an diesem entscheidet sich die individuell verpflichtende Kraft aller normativen Bestimmungen. Die Meinung, daß das Gewissen im Prozeß sittlicher Urteilsfindung keine Rolle spielt, ist Ausnahme, und kontrovers nur die Frage, welcher Stellenwert dem Gewissen in diesem Prozeß zuzusprechen ist. Setzen wir uns mit Husserls Überlegungen zu den Grundproblemen der Ethik auseinander, so fällt auf, daß er dem Begriff 'Gewissen' keinerlei Beachtung geschenkt hat, ja sogar, daß an den wenigen Stellen, wo auf dieses Bezug genommen wird, die Bezugnahme in einer abwertenden Weise erfolgt.[1] Wir werden uns daher fragen müssen, welche von Husserls ethischen Kategorien begrifflich jene Sphäre abdecken, die der Begriff 'Gewissen' z.B. bei Kant umschreibt, zählt doch gerade dieser Begriff zu jenen, die nicht nur in der philosophischen Reflexion einen festen Platz haben, sondern auch aus dem alltäglichen Sprachgebrauch und der Sprache des Rechtswesens nicht wegzudenken sind.

Die Bedeutung der Frage nach dem Stellenwert des Gewissens zu veranschaulichen, sei als Beispiel auf das Problem einer Wehrdienstverweigerung aus Gewissensgründen hingewiesen. Je nach Einschätzung der Bedeutung des Gewissens für die Verwirklichung von Sittlichkeit wird dieser Problemkreis eine andere Antwort erfahren.[2] Kommt dem Gewissen für die sittliche Urteilsfindung eine konstitutive Funktion zu, oder ist es nur der Ausdruck eines subjektiven Gefühls als der Relation zu einem objektiven Wert? Für diesen letzteren Fall könnte und dürfte der Staat — um bei dem genannten Beispiel zu bleiben — entgegen der Praxis der Mehrheit westlicher Demokratien das Recht individueller Meinungsbildung als Gewissensbildung nicht anerkennen.

In der Beantwortung sittlicher Fragen, für die die aufgeworfene letztlich nur stellvertretend die Schwierigkeit jeglicher sittlicher Entscheidung aufzeigt, scheint mir das Gewissen eine so eminente Rolle zu spielen, daß ich nicht umhin kann, in Auseinandersetzung mit Husserls ethischen Untersuchungen die Frage nach seiner philosophischen Erfassung der hier mit dem Begriff

259

A-T. Tymieniecka (ed.), Analecta Husserliana, Vol. XVI, 259–266.
Copyright © 1983 by D. Reidel Publishing Company.

Gewissen umschriebenen Aspekte für eine leibhafte Verwirklichung von Sittlichkeit aufzuwerfen. Die Vermittlung von individueller Einsicht und allgemeinem sittlichen Ansproch, die der Husserlschüler Scheler versucht, wird in ihrer Unzulänglichkeit gerade offenkundig, wo er seine theoretischen Überlegungen an dem auch hier erwähnten Beispielsfall veranschaulicht und schreibt, daß "der Staat (im äußersten Ausmaße) wohl das Opfer des *Lebens* der Person [zu] fordern (z.B. im Kriege)" berechtigt sei, daß der Staat jedoch "niemals das Opfer der Person überhaupt (ihres Heiles und Gewissens)" fordern könne.[3] Die Richtigkeit dieser Feststellung ist zwar nicht zu bestreiten, für sich genommen bleibt sie jedoch bedeutungslos. Sie basiert auf der von Scheler angesetzten Unterscheidung zwischen der ökonomischen Person, die unterstaatlich ist, d.h., den Forderungen des Staates zu gehorchen hat, und dem Kern der individuellen Geistperson, die ihrerseits als überstaatlich qualifiziert wird.[4] Doch so stellt sich die Frage: Läßt sich diese Unterscheidung durchhalten, wenn mit dem den ökonomischen Interessen eines Staates gehorchenden und aufzuopfernden Leib auch Gewissen und Heil der Person ihr leibhaftes Dasein zu verlieren drohen? Eine Antwort sieht voraus, daß nicht ungeklärt bleibt, wo und wie die überstaatliche Personssphäre in dem staatlichen Bereich zu ihrem Recht kommt. Wird das Gewissen als der Ort und die Weise dieses Bezuges angesehen, so lautet die Frage, welche Rechte und Möglichkeiten dem Gewissen zustehen, um sich auch als Gewissen leibhaft konkretisieren zu können.

Gerade hiezu scheint mir Husserl einige Erwägungen zur Ethik beigesteuert zu haben, die nicht in eine versöhnte Dichotomie führen und, obgleich unausgeführt und zum Teil nur andeutungsweise erkennbar, motivmäßig dem entsprechen, was Kant in seinem Gewissensbegriff in größerer Systematik ausgeführt hat.

II

Husserl unterscheidet zwischen dem Wert, dem Erfühlen eines Wertes und dem Einsehen des Wertes. So hebt er in seiner Kritik an dem Hedonismus hervor, daß zwischen dem "wertenden Erfühlen des Wertes und dem Werte selbst ein grundwesentlicher Unterschied ist", das Werten ist etwas "zeitlich Kommendes und Gehendes, der *Wert* aber als Wert einer ewigen, d.i. überzeitlichen Wesenheit selbst etwas Ewiges".[5] Dieser Unterschied umschreibt jedoch noch nicht die spezifische Differenz des sittlichen Urteils von anderen. So läßt sich eine solche Unterscheidung auch bezüglich des wertenden Hörens einer Symphonie treffen und dem Wert, der ihr als idealer Gegenstand eignet. Für das sittliche Urteil steht der wertende Bezug zu einem Wert nicht in der bloßen Relation meines Fühlens zum Wert einer Sache bzw. Handlung, sondern in der des Einsehens eines Wertes.

Meine These ist es nun, daß dem Begriff des *Einsehens*, oder, wie Husserl auch sagt, der "erkennenden Einsicht" jene Aufgabe zu erfüllen zugedacht ist, die in Kants ethischer Theorie das Gewissen einnimmt. "Wer gegen die mathematischen Gesetze verstößt, verfällt dem Irrtum. Wer aber gegen ein ethisches Gesetz verstößt, verfällt der Sünde",[6] schreibt Husserl und beleuchtet damit den Begriff des Einsehens näherhin, indem er die Differenz beider Urteilsarten analysiert. Sünde oder das Böse ist hier gefaßt als der Verstoß gegen ein sittliches Gesetz, dessen Richtigkeit erkannt wurde, wogegen die Übertretung eines solchen Gesetzes, wo es in seiner Wahrheit nicht erkannt ist, Irrtum heißt. D.h., daß jemand, dem die Forderung der Erhaltung des Staates und der verpflichtende Einsatz von Leben zu dessen Bestand in ihrer Richtigkeit nicht als sittliches Gesetz bekannt ist, im Falle der Verweigerung einem Irrtum verfällt, derjenige hingegen, der die Richtigkeit eines solchen Gesetzes eingesehen hat, unterliegt im Falle der Verweigerung der Schuld des Bösen, bzw. der Sünde.

Zum einen wird aus dieser Unterscheidung zwischen Sünde und Irrtum klar, daß Husserl dem Einsehen in die Richtigkeit oder Wahrheit eines *sittlichen Gesetzes* eine verpflichtende Forderung an das einsehende Subjekt zuspricht. Zum anderen schließt an diese Unterscheidung die Frage, inwiefern das "Einsehen" selbst konstitutiv in die Bildung eines sittlichen Gesetzes eingeht. Um diese Frage beantworten zu können, ist auf das Verhältnis von theoretischer und praktischer Vernunft bzw. auf die Verschränkung von Sachentscheidung und sittlich moralischer Entscheidung zu verweisen. Die Verschränkung, von der Husserl ausgeht, bedeutet für das sittliche Urteil, daß es für sich keine Anerkennung beanspruchen kann unabhängig von einer ihm zugrundeliegenden sachlichen Urteilsmotivation. Anders ausgedrückt, das praktische Urteil, soferne es als erkennende Einsicht eine bestimmte Verhaltensweise fordert, soll die Rechtfertigung und Begründung dieser Forderung zugleich in der Rechtfertigung jenes Urteiles finden, in dem die objektiven Inhalte und ihre Zusammenhänge zur Sprache kommen. In jenen Sätzen, in denen ich über das gegenseitige Verhältnis von Individuum und Staat spreche, über die Sinnhaftigkeit der Erhaltung von Freiheit mittels Unfreiheit, bringe ich demnach auch mich in meinem Einsehen dieser Inhalte zur Sprache.

Das "Einsehen" zeigt sich in dieser Hinsicht, obgleich es selbst, wie noch

zu erweisen sein wird, keine Leistung der Vernunft ist, rationaler Argumenta-
tion nicht unzugänglich, und Husserl beschreibt das sittliche Urteil näherhin
als eines, das nicht wie das Urteil einer Sachwissenschaft sich auf Sein oder
Nichtsein bezieht, sondern das eine Einstellung gegenüber sachlichen Werten
unter Einbezug sachlicher Aspekte bildet. Das sittliche Urteil formt sich
somit aus den in der gegenständlichen Sachlichkeit impliziten selbst "zu
schöpfende[n] Unterschieden des Gesollten und Nichtgesollten".[7] Sachgesetze
gehen, so heißt es ausdrücklich, auf Sachliches, das "eventuell auch als
Substrat von ethischen Akten den ethischen Charakter von Gesolltem und
Nichtgesolltem annimmt, während es selbst . . . nicht selbst Gesolltheit . . .
ist".[8] Dieser Einbezug, wie Husserl es nennt, des sachlichen Urteils in das
ethische kann nur durchgehalten werden, wenn sich eine Norm sittlicher
Urteilsfindung aufzeigen läßt, die "neben den Gedankenformen, die sie rein
der Vernunft entnimmt, auch eine Materie enthält, die dem *Gefühls-* und
*Begehrungs*vermögen entstammt".[9]

Gesucht werden also Sollensgesetze, die als "Gesetze des möglichen Wertes
. . . zum Sinn oder Wesen des Wertes als solche[m] gehören",[10] jedoch im
"Gemüt" Entsprechungen finden, "die jedes Wesen überhaupt angingen,
das überhaupt fühlt, begehrt, will, möge es diese oder jene materiellen
Besonderungen von Gemütsakten besitzen".[11] Ein formelles Wollen des
Guten zu denken erscheint Husserl hiebei "genauso widersinnig, wie ein
Ton ohne jede Tonqualität oder eine Farbe ohne Ausbreitung oder eine
Vorstellung ohne jedes Vorgestellte".[12]

Gesucht wird eine vermittelnde Instanz, die sowohl die Gedankenformen
der Vernunft als auch die Materie des Gefühls enthält, und die damit dem
"Verstandesurteile eine Kraft" gibt, "daß es seine Triebfeder werde, den Willen
zu bewegen." Diese Instanz zu finden, heißt Kant, den "Stein der Weisen"
finden.[13] Husserl bietet als diesen Stein das "Einsehen" an, das, nach seiner
Auffassung im Kant'schen Sinne durch einen intellektuellen Grund gewirkt
wird, zum anderen aber im Gefühl situiert ist. Hiebei rekurriert Husserl hier
in Anlehnung an Kants Begriff der Achtung auf einen Begriff von Gefühl,
der "Einsehen" nicht in einer sensualistischen Gefühlsmoral sich erschöpfen
läßt, sondern aus dem sich Wesensgesetze herauslesen lassen sollen, die
aussagen, "wie gefühlt werden soll, welches Fühlen ein richtiges, ein gültiges
sei".[14]

In Anlehnung an Hume nannte Husserl das Gefühl der *Sympathie* das
Grundprinzip jedes moralischen Urteils, doch versteht er nicht, wie dieser,
unter Sympathie Gleichsinnigkeit und Gefühlsresonanz, sondern ein *Einsehen*
und Anerkennen, das dem Anerkannten und Bejahten Geltung verschafft.

Ein Gegenstand der Sympathie ist in diesem Sinn "Grund eines positiven Gefühls", "das nicht empirischen Ursprungs ist und a priori erkannt wird",[15] also von der Vernunft durchdrungen ist und andererseits durch Umorientierung der Gefühlsantriebe deren Kraft in sich aufsaugt und diese in den Dienst der praktischen Vernunft stellt. Das "Einsehen" ist so gewissermaßen das im Gemüt beheimatete Korrelat zur praktischen Wahrheit. Wie es bei Kant die Funktion des Gewissens ist, die Einstimmung hinsichtlich des von ihm unterschiedenen moralischen und logischen Urteils herzustellen und zu bezeugen, ist Husserls Begriff der vernünftigen Sympathie konzipiert, der Einstimmung von wertendem Erfühlen und Wert Ausdruck zu geben. Hiebei ist das Gefühl der Sympathie als Einsehen nicht sowohl als Resultat als auch als Prinzip des Handelns zu sehen, und widerspricht in diesem Punkte dem Kant'schen Gewissensbegriff, der sich seinerseits ebenso von sittlichem Bewußtsein als dem Gegenstand seiner Beurteilung unterscheidet, wie das Einsehen vom Wert als seinem Gegenstand unterschieden ist. Die Analogie zum Kantischen Gewissensbegriff hält also nur bis zu diesem Punkt.

III

Die Möglichkeit, das Gewissen als Prinzip und Resultat des Handelns zu denken, ergibt sich für Kant aus seiner Bestimmung des Gewissens als sich selbst richtende moralische Urteilskraft. Das Gewissen kann so einerseits "als das Gesetz in uns"[16] angesprochen werden, und andererseits ist ihm ebenso zuzumuten, die Applikation unserer Handlungen auf dieses Gesetz zu leisten. Die Prüfung und Legitimierung einer Handlung in ihrem sittlichen Charakter vollzieht sich solchermaßen durch ein doppeltes Beurteilungsverfahren und unter zwei verschiedenen Aspekten. Entsprechend der auch bei Husserl sich findenden Unterscheidung zwischen dem Sachurteil und dem moralischen Urteil richtet das Gewissen nicht die Handlung als Casus in ihrer Sachhaltigkeit, sondern fragt (1.), ob das Sachurteil in seiner sittlichen Zuordnung im Horizont des Allgemeinen erfolgt, und (2.), ob sowohl in der sachlichen Beurteilung als auch in der sittlichen Zuordnung die nötige "Behutsamkeit" aufgewendet worden ist. Die Aufgabe des Gewissens ist es folglich zu fragen, ob die Übereinstimmung mit dem Maß sittlichen Handelns nicht nur eine vorgegebene und gespielte ist, und ob das sittliche Bewußtsein als konkretes Urteilsmaß, das den Reichtum erfahrener Weltbezüge in sich aufgenommen hat, die Anforderung der Allgemeinheit antizipiert.

Wenn Kant dem Gewissen die Fähigkeit abspricht, das sittliche Urteil zu fällen, und diese ausschließlich der praktischen Vernunft zuspricht, so handelt

es sich hiebei nicht nur um eine Kompetenzbeschneidung des Gewissens,
sondern zugleich um eine Erweiterung von dessen Aufgabenbereich. Die
Entlastung des Gewissens von der Verpflichtung, den sittlichen Urteilsspruch
zu fällen, setzt dieses frei, sich ganz der Überprüfung der Übereinstimmung
des Urteils der praktischen Vernunft mit den allgemeinen Werten zu widmen
und gleichzeitig eine Prüfung dieser Werte vorzunehmen. Mit anderen Worten:
für Kant existieren keine Werte, die nicht auf das Gewissen hin relativ sind,
sodaß keine Gewissensentscheidung umhin kann, diese Relativität, aber auch
das aus ihr entspringende Risiko jeden sittlichen Handelns reflektiv in ihre
Entscheidungen eingehen zu lassen. Husserl vollzieht gewissermaßen die
Kompetenzbeschneidung des Gewissens mit Kant, sieht sich aber ob seiner
Lokalisierung des "Einsehens" im Gemüt außerstande, auch die Kompetenzer-
weiterung nachzuvollziehen.

Die Folgen dieser einseitigen Beschneidung sind, daß Relativität und
Risikohaftigkeit jeden sittlichen Entscheidens im Begriff des Einsehens nicht
reflektiv erfaßt und verarbeitet werden können, sondern dem Einsehen
gegenüber äußerlich bleiben. Zwar gilt für Husserl unbestreitbar der Satz,
daß "ein richtig motivierter, begründeter und zuhöchst vollbegründeter Wille
. . . besser [ist] als ein bloß richtiger",[17] aber das Gefühl des Einsehens, an
dem als dem Probierstein der Angemessenheit eines Wertes diese Differenz
erfahren wird, gibt für sich keine Basis ab zur Überwindung oder zum
Einbezug dieser Erfahrung in das Handeln. Denn das Gefühl des Einsehens
ist auch als intellektuell konzipiertes nur das Gefühl der Empfänglichkeit für
Sympathie aus dem Bewußtsein der Übereinstimmung einer Handlung mit der
zutreffenden Norm. D.h. das "Einsehen" ist zwar gegenüber jeder argumenta-
tiven Vermittlung offen und findet in diesem Sinne seine Rechtfertigung in
der Tauglichkeit des Sachurteiles und seiner Übereinstimmung mit den
allgemeinen Normen, aber, und dies scheint mir das Entscheidende, es selbst
wird nicht imstande gesehen, in die Bildung sittlicher Werte konstitutiv
einzugehen. Das "Einsehen" ist gleichsam nicht das Gesetz in uns.

Es wäre verfehlt, aus diesen Überlegungen den Schluß ziehen zu wollen,
Husserls Ethikansatz könne ob dieses Mangels, sei es im Vergleich zu Kant
oder einem anderen Denker der Neuzeit, zu den Akten gelegt werden.
Nicht dies zu sagen, war die Absicht der vorgetragenen Ausführungen. Diese
wollten vielmehr den Hinweis liefern, (1.) daß die landläufig mit dem Begriff
Gewissen umschriebene Problematik ihren systematischen Ort auch bei
Husserl findet, selbst wenn wir der Auffassung zuneigen, daß die Art und
Weise ihrer Lokalisierung im Gefühl des Einsehens defizitär ist, (2.) daß
Husserls Beitrag zur Ethik vor allem im Versuch liegt, den Begriff des

Gewissens als nichtsensualistisches Gefühl der Sympathie, des Einsehens zu bestimmen. Die Weiterführung eines Weges, der bei Kant trotz seiner Hervorhebung des Gefühls der Achtung im Ganzen seiner Ethik zu kurz gekommen ist, öffnet nicht nur neue Perspektiven für eine Weiterentwicklung der im Achtungsbegriff wie im Begriff des moralischen Gefühls angesprochenen Problembereiche, sondern enthält ebenso für die konkrete sittliche Urteilsfindung als leibhafte Verwirklichung von Sittlichkeit Aspekte, die so wenig vernachlässigt werden dürfen wie die Frage eines Einbezugs von Risiko und Wertrelativität in die sittliche Entscheidung selbst. Daß für Husserl letztlich beide Aspekte die jeweils ihnen adäquate Berücksichtigung finden müssen, geht aus jener Feststellung hervor, die zentral für seine "Gefühlslogik" ist: Sittliches Handeln begründet sich nur aus einem Wollen, das "vernünftig im Bewußtsein seiner Normhaftigkeit gewollt und durch die Normhaftigkeit auch motiviert ist".[18]

Universität Wien

ANMERKUNGEN

[1] Siehe hierzu die Arbeit von Roth, Alois, *Edmund Husserls ethische Untersuchungen*, Den Haag, 1960. Es sei nicht verschwiegen, daß diese Arbeit vor allem bei der Auswahl der Textstellen Pate stand.
[2] Vgl. hierzu Hofmeister, Heimo, 'Das Gewissen als Ort sittlicher Urteilsfindung', *Zeitschrift für Evangelische Ethik* 25, 1 (1981).
[3] Scheler, Max, 'Der Formalismus in der Ethik und die materiale Wertethik', *Jahrbuch für Philosophie und phänomenologische Forschung II*, 1930, S. 533.
[4] ebd.
[5] Husserl, Edmund, 'Freiburger Vorlesung zur Einleitung – Ethik', 1920, 1924, Ms F I 28, Husserl Archiv, S. 93.
[6] a.a.O. S. 181.
[7] a.a.O. S. 184.
[8] a.a.O. S. 187.
[9] Husserl, Edmund, 'Kritisches aus Ethik-Vorlesung', Sommersemester 1902, Ms F I 20, Husserl Archiv, S. 260.
[10] a.a.O. S. 233.
[11] ebd.
[12] Husserl, *Einleitung Ethik*, S. 283.
[13] Menzer, Paul, *Eine Vorlesung Kants über die Ethik*, Berlin, 1924, S. 54.
[14] Husserl, a.a.O. S. 300.
[15] Kant, Immanuel, Werke, Akademieausgabe, Berlin, 1968, V, S. 73.

[16] Kant, a.a.O. IX, S. 495.
[17] Husserl, Edmund, 'Beilagen zu den Grundproblemen der Ethik (insbesondere der formalen Ethik)', Wintersemester 1908/9 Ms F I 21, Husserl Archiv, S. 17.
[18] Husserl, 'Freiburger Vorlesung', S. 327.

FILIPPO LIVERZIANI

VALUE ETHICS AND EXPERIENCE

When we want to speak of the foundations of an ethics, it seems to me that one of the first problems that we have to face up to is to decide what we mean by *founding*. In the philosophical sense, surely, founding one or more judgments means to *justify them*, that is to say, it means to *explain them* by tracing them back to or deriving them from more evident judgments, deriving them from judgments that express more originary evidence. The ideal case, indeed, would be to derive such judgments from absolute evidence that can be expressed in a perfectly rigorous manner even in its selfsame formulation. But I do not think that this is quite as easy to achieve; or, rather, I have the distinct impression that in present conditions it is humanly impossible. What seems absolutely evident to us today may no longer appear quite as obvious tomorrow. We are not participating in a quiz program where the emcee will eventually tell us whether "our answer is correct" or otherwise. We are embarked on a boat that we are obliged to repair and perfect little by little while we ply the seven seas, without ever getting the chance of pulling it up on dry land and entrusting it to a shipyard. In such circumstances one simply does the best one can.

Let us now see what truly originary evidence we could use as the basic anchorage of an ethics or, indeed, of any other organic complex of propositions. Husserl would surely consider this anchorage to consist of the *Erlebnisse* (the "experiences", that is) of the phenomenologically reduced consciousness. Indeed, Husserl defines this consciousness as absolute. It is this (pure, reduced) consciousness, this transcendental "I" that highlights the significance and confers the sense of being upon all things. But here one has to observe that the consciousness, no matter how reduced it may be, will always and inevitably present the *Erlebnisse* of a human consciousness. In reality this reduced consciousness appears as the consciousness of a man who, for the most part, suffers his own *Erlebnisse*, inasmuch as these derive — as far as he is concerned — from specific conditionings, that is to say, the very conditionings inherent in his living in the world. This consciousness, even though Husserl proclaims it as absolute, does not by any means succeed in explaining itself and its own *Erlebnisse*.

Our consciousness, then, does not succeed in explaining itself and yet

267

A-T. Tymieniecka (ed.), Analecta Husserliana, Vol. XVI, 267–282.

wants to understand itself and its *Erlebnisse*, it longs to attribute to itself, as indeed to all things, a significance that will no longer be a purely subjective one, but rather objective and real, that is to say, it wants to give them a significance that will be profound, true and exhaustive or, if you prefer, the sense of being that they have *in se* and not just for me, for you, or for us.

The phenomenologist realizes a great interior discovery: everything that *is*, is only in relation to some consciousness. Nothing can *be* unless it is related to a consciousness. It is the consciousness that gives significance and sense of being (and therefore being) to all things. A consciousness is indispensable, but a human consciousness (no matter how much one wants to reduce it) is not sufficient to confer upon the various realities their significance and their sense of being and, consequently, the being that they have *in se*. A human consciousness that becomes aware of its own insufficiency (an irremediable insufficiency in spite of every reduction) cannot but refer matters to a universal Consciousness that, by thinking, calls all things into being just as they really are. This will be the only consciousness that really deserves to be defined as absolute, a consciousness that, as far as its essential aspects and attributes are concerned, may well reveal itself, little by little, to be similar to the essential features associated with our traditional image of God.

I am convinced that, if we really want to understand anything whatsoever, if we really want to *found* any proposition or doctrine whatsoever, the ultimate term of reference that we must seek as an anchorage cannot be the phenomena of a consciousness that — no matter how reduced — remains always relative, but only phenomena inasmuch as they are phenomena of a truly absolute Consciousness, inasmuch as they are phenomena of God.

This is not a case of limiting ourselves to *inferring* God. We may infer him, we may argue his existence, always using the phenomena of our human consciousness as our starting point, but then we can also seek some experimental confirmation in the experience that — in some way or other — we have of God himself; in other words, in metaphysico-religious experience.

Even a metaphysico-religious experience can be relevant in a phenomenological sense. It is true that, as far as Husserlian terminology is concerned, the data of metaphysico-religious experience cannot be defined as originary on a par with the *Erlebnisse*, the phenomena of the human consciousness, but they nevertheless merit phenomenological consideration and can constitute phenomenological foundations because, albeit not in the idea of "being as experience" (*Sein als Erlebnis*), they can be subsumed — in the wider sense — under the concept of what Husserl calls the knowledge of "being as thing" (*Sein als Ding*):[1] if we want to use this language, which in the present context is certainly not the happiest one, even God can be defined, in the widest possible sense, as a transcendental "thing" that yet manifests itself in some way through "adumbrations" (*Abschattungen*), albeit in a manner that we can grasp only very imperfectly (and one can never insist sufficiently on the abyssal imperfection of the knowledge that we may gain of God).[2] It is quite clear that this application is my own and not that of Husserl; but I do believe that it can be developed from premises that he postulated — and in critical continuity of his thought — by means of a more personal research, a more theoretical research, but one that to all intents and purposes avails itself of his method, that is to say, the phenomenological method.

As regards the problem of how one can attribute phenomenological and philosophical citizenship to the experience of God, to the experience of the Absolute or the Sacred, to metaphysical, religious and mystical experience, etc., this represents a matter that I cannot discuss here at length, although I have dealt with it extensively in other writings.[3] I must therefore limit myself to saying that only in an experience of God can we find the true foundation of a system of ethics, that is to say, in something that will truly *provide a reason for it*, that will truly justify the system and clarify its significance at the most profound level, something that will explain its *whys* and *wherefores* in a satisfactory manner, in the most exhaustive manner possible.

In the ambit of ethics, in the ambit of moral action, indeed, one should note first of all that we human beings feel something like a voice from deep within ourselves that tells us to do certain things and not to do other things. We feel that this imperative that manifests itself in our intimacy is a categorical imperative: it is not *hypothetical* and it is not *conditional* like the technical rules that tell us that we have to behave in a certain manner if (that is to say, on the *hypothesis* that) we want to obtain a certain result that we are morally free not to desire. In the case of the ethical imperatives, however, we are not morally free to circumvent them, we are morally bound to obey them: these imperatives manifest themselves as *unconditional, absolute*. This is the famous contestation that Kant makes with such great clarity. From the overall context of Kant's ethics we can here abstract (or, with a little good will, explicit) a phenomenological moment: this is the moment in which Kant contests the moral fact, the moment in which he notes the absoluteness, the categoricity of this command that we hear coming from deep within ourselves and which tells us that we must (or must not) do this or that. But why must we do it, in view of what must we do it? Because we have to. We have to do it because it is our duty. We must not do it because certain things are not done. Even if we admit that one's duty may change as the situations vary and become modified, it always remains possible to say that in this particular situation it is our specific duty to do this or that. We are in war and have to kill the enemy, the same people who in times of peace would invade us as tourists, thus obliging us to tell them the way or to show them how to use a telephone booth. In any case, we always come face to face with an unconditional, categorical, absolute *you must do* this.

This, then, is the moral fact, the indubitable fact that all can testify and which can be the object of phenomenologic investigation. At this point, indeed, we have to explain the fact, we have to give the reason for it, to found it, to derive it from other phenomena that stand a little less in need of being founded and explained or, better still, don't have this need at all because they explain themselves in an exhaustive manner by virtue of their own (inherent) evidence.

Now, for as long as it was desired to found the absoluteness of the ethical imperative on the absoluteness of God, that is to say, on Him who alone can be (properly) defined as absolute, we were concerned with a solution that was debatable if you like, but undoubtedly coherent and not contradictory. For in that case God would be conceivable as the originary absolute, the absolute from which the moral norm would derive an absoluteness at least as regards the formal aspect of its imperativeness; as regards its contents, on the other hand, one can always hypothesise that it is capable of being perfected and even that for the time being it remains rather far removed from perfection: indeed, as far as the mere content of the moral imperative is concerned, one can hold it to be the result of an authentic divine inspiration, but one that has been filtered through the imperfection of human vehicles, of men whose mentality, sensitivity and receptivity are all extremely limited by a whole series of conditionings, be they bio-psychological, historico-geographical, socio-cultural, or of any other kind whatsoever.

We know, however, that modern thought, at least in its most characteristic trends and expressions, has tried as far as possible to do without God. This observation should really be further qualified, articulated, shaded and exemplified, but I cannot do this here, since I have to develop my reasoning within given limits of space. For the moment I shall therefore limit myself to noting that in modern philosophy God is invoked less and less as the foundation of ethics, this not least because people believe less and less in Him and also live Him less and less. As the foundation of ethics is becoming detached from God, attempts are being made to anchor it to natural, mundane and human realities. But, seeing that these are relative realities, how can one possibly found the ethical imperative in all its absoluteness on any one of them?

It is true that the attempt to absolutize realities that are anything but absolute (i.e. Reason, Nature, Science, Art, Society, the Nation and the Race, Freedom, Mankind, Production and Consumption, Love and Sex, Sport, and so on) is a rather characteristic attitude of modern man; but here we are always concerned with pseudo-absolutes that sooner or later will reveal their character of "idols" (even though this attitude may have proved to be functional, at least in certain respects, for the autonomous constitution of many sciences and arts and forms of the human spirit and reality and, consequently, a more articulated and richer *regnum hominis*). Anyway, once certain claimed (or pretended) absolutes have revealed their falsity, once we have highlighted the decidedly non-absolute character of certain natural and human factors from which people would derive a system of ethics, on which they would found such a system, the insufficiency of these foundations always ends up by becoming rather obvious.

Indeed, what really ends up by becoming rather clear is the contradictory nature of trying to derive the ethical imperative, which all of us feel to be absolute, from relative and contingent factors. Paolo Valori, in his book entitled *L'esperienza morale* and bearing the subtitle *Saggio di una fondazione fenomenologica dell'etica* (*The Moral Experience: An Essay about the Phenomenological Foundation of Ethics*), also examines the principal ones among those that in our day and age appear to be "insufficient hermeneutics of the moral fact". He points out that for neo-positivism in general a moral judgment does nothing other than express a feeling, a subjective state of mind, and that it cannot therefore be verified, just as one cannot verify a scream of pain or a word of command; in a conception of this kind, therefore, ethics represent nothing more than a sector of psychology and of sociology. As far as sociologism is concerned, the ethical norms derive from social pressure, which provides the ultimate explanation of what underlies every moral phenomenon. Structuralism, likewise, derives these phenomena from the anthropologico-social structures. For certain psychoanalytic interpretations, again, the moral consciousness (with all its prohibitions) is determined by purely psychologic (and often psychopathologic) factors. Valori, quite rightly, stresses the useful services that have been rendered by these new interpretations in throwing better light on many of the empirical aspects of the moral fact; but he goes on to point out that these interpretations, each in its own way, contradict the phenomenological fact that expresses the essential and specific aspect of the moral experience and makes this experience appear as the reception of an absolute imperative (which, among others, may also command the subject to maintain a nonconformist attitude vis-à-vis society and its structures, or an attitude of contestation and rebellion). Insufficient hermeneutics of the moral fact are also certain interpretations that, while trying to explain it, are in contradiction with what specifically emerges from it; while endeavouring to explain it, they thus deny it in the very thing that is peculiar of it.[4]

I have already said that in the ethics of Kant there is (or, at least, one can extract from them) a moment that can be defined as phenomenological, i.e. the moment in which Kant highlights the moral fact with the particular type of imperative that distinguishes it. For Kant, indeed, the moral experience is the experience of a voice inside ourselves that prescribes us something like a duty, like an unconditional and absolute duty to be discharged in a certain action or in refraining from such an action. Kant throws proper light on certain essential characteristics of this duty, of this moral law. And up to this point his reasoning seems to me to be substantially good phenomenology. But Kant then wants to *found* all this, he wants to explain it and justify it and to discover its more profound reason. He considers that he cannot found his own ethics on God, because God can neither be *experimented* nor *argued* (as is indicated by the confutation, be it true or claimed to be such, of the principal types of arguments in favour of the existence of God in his *Critique of Pure Reason*). Kant holds that God can only be *postulated* by a system of ethics that has already been founded by other means. For him, therefore, ethics are founded on themselves, on a moral law that reveals itself directly through the voice of duty and sets itself up as an absolute. The human subject must act only in respect of this moral law. This is the only motive for which man, a rational being, can act rationally and therefore morally. Any other motive is associated with the senses and would therefore compromise the moral purity of the intention. Just as conforming to norms that are not the expression of human rationality, but derive from the divinity for example, would compromise ethical action in its autonomy; the result would be a system of ethics that is heteronomous and no longer autonomous.

In this conception it seems to me that ethics become absolutized, an end in itself. Thus, as Gianfranco Morra points out, Kantian morals become moralism, hypermoralism and even moral fanaticism that subordinates religion to morals, replaces religion by morals,

attributes the 'sacral' characters of religion to morals: the *tremenda majestas* of duty, the *sanctity* of good will, the absoluteness of the pure intention, the solemn and awe-inspiring *majesty* of the moral law, the *sacrality* of the reign of the ends (the moralized 'communion of the saints').[5]

At this point one cannot but ask oneself whether one can really speak of duty, or the moral law, as something that is absolute in itself, to be pursued solely for its own sake and for the love of it (or, better and as Kant himself prefers, for *respect* of it: for we are here concerned with something that is far easier to respect than to love). One may well ask whether Kant is not running the risk of turning duty, the moral law or, if you prefer, a certain rationality into an idol. In fact, a tendency of this kind would seem to be present in Kant, even though it must be doubted that this corresponds to his true and fundamental intention. But one may observe that duty and the moral law do in fact reveal a certain absoluteness: they appear absolutes by participation. Something absolute or, better, an absolute dimension expresses itself in them, even though one cannot say that they are absolutes in an originary manner since only God can be defined as such. We now have to see to what extent the absolutization to which Kant subjects duty can be considered as a mere testimony of what is really (and albeit even by participation) the absoluteness with which the ethical imperative is endowed, an absoluteness that cannot be reduced to any relative or contingent factors, even though this is precisely what the aforementioned insufficient hermeneutics of the moral fact endeavour to do. For it is to this extent I believe, that the conclusions of Kant can be accepted and utilized, always provided however that they are to be integrated into a theistic metaphysical vision. Because, considering Kantian morals from another and different point of view, one might note that to a certain extent they do not by any means limit themselves to testifying to a derived, participated absoluteness of duty and the moral law but, quite the contrary, attribute to these latter a type of absoluteness that can be properly attributed only to the originary absolute, to God that is.

In this connection it is interesting to note the criticism that Scheler levels against Kantian formalism and the undue absolutization to which Kant subjects duty and the moral law. Scheler points out that "to speak, as Kant does, of a duty that is floating-in-nothingness as it were, that is not a duty vis-à-vis any person and has not been commanded by any authority, is simply saying something that has *no* meaning ('*keinen* Sinn')".[6] Kant substantially says that "*good is* what thou canst want everyone (in thy situation) to want";[7] he does not say "want the good!".[8] Now, as far as Scheler is concerned, what is morally "good" is not "the inclination" but rather the "act-of-wanting"; but "our wanting is 'good' only to the extent to which it chooses the highest among the values that form part of its inclinations".[9] One cannot speak of duty other than in relation to something that is good in itself, and good is everything (and, indeed, only that) in which a value is present; this value has to be grasped, it has to be perceived in an experience *sui generis*, an experience that we could define as metaphysical and not sensorial. Quite rightly, Kant notes that human action cannot be considered moral to the extent to which it obeys empirical motives, that is to say, contingent, relative, natural and purely human factors. Accepting a limitation of the concept of experience that for a long time past has been traditional in Western thought, Kant conceives only these contingent and finite realities as properly capable of being experienced and therefore as empirical. In fact, an absolute reality (like God, for example) will not be capable of being experienced, or it will be capable of being experienced (as when one perceives the voice of duty, for example) but only in a rather wide and improper sense of the expression. So much so that — in keeping with the whole of this tradition — Kant is inhibited from speaking not only of *experience* but also of *empirical* and *a posteriori*; and therefore Kant speaks of the voice of duty and of his categorical and unconditional imperative as something that is grasped *a priori*.

Scheler, in turn, speaks of "phenomenological experience",[10] of "original moral experience";[11] and he speaks of values as "de-facto realities that form part of the domain of a certain sort of experience".[12] Of values he even says that they are the object of an "affective perception" (of a *fühlen* as he puts it).[13] More readily, however, he uses the term "intuition" in relation to values. Husserl had already begun to speak of an "intuition of essences"; and the *values* of Scheler are precisely *essences* that a phenomenology can grasp intuitively (even though Scheler tends to accentuate their ontological character well beyond the limits of Husserl's concept of intentionality). And thus Scheler speaks of an "intuitive knowledge" of values that endeavours to be "adequate",[14] and also of "ethical intuitions" that are "material"[15] and at the same time *a priori*.[16]

Scheler feels that knowledge of values is an experience, and he cannot avoid calling it an experience, a form of perception. But we are all well aware that for a very long time past Western philosophy has been manifesting a very strong trend to identify all experience with sensorial experience: all this is defined as *empirical* and *a posteriori* and is tendentially depreciated by every philosophy oriented in the opposite direction, i.e. towards the rational and the metaphysical. Even Scheler, although he speaks of moral experience and perception of values, seems concerned to protect his own ethics against the accusation that they are bound up with empirical factors and therefore, notwithstanding the fact that he defines them to be material, wants them to be *a priori*.

But is it not a contradiction of terms to speak of an *a priori experience*? Would it not be better to conclude quite frankly that even a moral, spiritual, metaphysical, religious or mystical experience, each in its own peculiar manner, is something that comes to us *a posteriori*? We could do this only once we succeed in ridding ourselves of the prejudice (an inveterate one, unfortunately) that one can only have an experience of the physical, natural, mundane realities. If even values, as Scheler defines them, are de facto realities and possible objects of experience, and are so precisely because they are values, why should it not be possible to qualify the Absolute in an analogous manner? Why should we not speak of an absolute reality, object of an experience, object of an experience of the Absolute?

However this may be, Scheler — in critical continuity with Kant — sets out to found an ethics that will not be a formal one (like the one that prescribes duty for the sake of duty, absolutizing it) but rather material (that is to say, having a content, value, with respect to which duty is finalized and relativized), a system of ethics that will be both material and *a priori*. Values are essences: essences that are no longer merely intentional, as they were for Husserl, but real, objective, ontologic, metaphysical. Scheler sees them structured in accordance with a precise "objective hierarchy".[17] This "aprioristic hierarchy"[18] ("values of the pleasant and the unpleasant", "vital values", "spiritual values", "values of the sacred and the profane")[19] suggests a spontaneous comparison with the hierarchy of the a priori forms of knowledge in which Kant finds his delight and which, to be truthful, seems a little too precise and specific; and this would give rise to a rather unpleasant impression if one were not to note that there is far less rigidity in Scheler's classification and if one were not comforted by finding some further qualifying remarks like the following:

... As a philosophical discipline, ethics are by their own essence incapable of drawing up an exhaustive list of moral values; they are concerned only with values and with the universally valid preferential correlations.[20]

In spite of correctives of this kind, and in spite of a constant reference of the values to the human person and (as foundation and ultimate term) to God, it nevertheless seems to me that in Scheler one can glimpse at least a tendency to hypostatize values, to absolutize them in some manner, to conceive them a little in the manner of something that exists and subsists *in se* and continues as such even as time passes and the contingencies change, in a manner similar to that of the Platonic ideas and the Aristotelian forms. It is true that Scheler levels the reprimand against formalistic ethics that they ignore the essential historicity of the *ethos* as a form of lived experience of the constant values and their immutable hierarchy;[21] but Scheler's historical vision nevertheless remains that of a humanity that, notwithstanding the variation of the subjective valuations, progresses towards the knowledge and the realization of a universe of objective values, a universe of their objective hierarchy.[22]

In this vision, then, we have on the one hand the subjective valuations of men immersed in the world of historical relativity, while on the other we have "the values", immutable and transcendental as in a kind of hyperuranian world. But to conceive a hierarchy of values as something that stands by itself, fixed and immutable and above the historical contingencies, is this not going a little too far in absolutizing values? Is this not a way of putting them, if not in the place of God, at least by the side of God, and quite unduly so? Does this not mean that, side by side with the absolute one, there would thus be a multiple absolute, a kind of coeternal hypostasis?

Personally I would prefer to conceive values as something absolute, certainly, but absolute by participation; in other words, as something that in certain respects is absolute and in other respects is incarnate in things multifarious and continuously becoming, incarnate in historical humanity. But are we really sure that, every time we have grasped or identified a value, we have in hand (as it were) a *slice of the absolute*? This would undoubtedly please many people with a clearly logical mind, who tend to see all things in an intellectualistic light.[23] Unfortunately for them, however, in this human condition of ours the absolute never presents itself in forms that are very clear and distinct, but rather appears incarnate in the relative in a manner that is not very readily discernible at first sight and it reveals itself only after being filtered through the diaphragm of relativity and the extreme imperfection of our means of knowledge. Such a situation can be symbolically expressed by the image of sunlight, which is unique and always the same, but which can nevertheless penetrate in different ways into many different rooms, passing through a variety of windows with glasses of different colours and, here and there, even rather dirty.

Values are that which has value for man, and in them there undoubtedly is a human aspect, even though — through them — man pursues something that is decidedly above him. A careful consideration of human values and of everything that is of importance for man, as also a careful consideration of how these values come to be historically delineated, should make it possible for us to realize that these values, even though they may express and harbour within them something absolute, they nevertheless always appear — in another respect — as relative to the mentality of each individual and each group that live their lives in their own particular situation.

The ideal may well be all the time that of realizing oneself, living better as it were. But what exactly do we mean by realizing ourselves, living better, developing ourselves to the full of our possibilities? Once he has guaranteed a minimum of vital space to his own person, his family and his community, and also the means that ensure their survival (health, children, essentials of life), man pursues a given value rather than others and commits the greater part of his energies to the attainment of a goal: a king wants to become more powerful, subjecting other countries to his dominion; a politician wants to increase his power, enlarging his following (and also his clientele) by means of initiatives that, even though they are not always perfectly moral, are at least legal and possible in a democratic country; a gangster aspires to controlling all the gambling houses, the prostitution and the drug trade in his particular zone; a bully wants to be the strongest and the more feared in his neighbourhood or a gunman in his Western village (and even in a type of this kind there expresses itself, albeit in a manner very different from that of the author of the *Oratio de hominis dignitate*,[24] the aspiration to prove himself "a true man"); a bank clerk with wife and two children wants to buy (on the never-never, of course) a car that is bigger than his neighbour's, and then a second house by the seaside (what kind of a "true man" is a person who does not earn, does not show that he earns lots of money?); a Don Giovanni desires, all for himself, a collection of the most beautiful women; a woman who has chosen to live for her own beauty offers on this altar sacrifices worthy of a heroine or a protomartyr; a small farmer kills himself with work to enable him to buy another field with his savings; and the keeper of a delicatessen store does the same in the hope of having a larger and more modern store with a neon light and three entrances; and then, at a somewhat higher idealistic level, there are the goals set themselves by the scientist and the philosopher, the painter, the musician, the poet, the yogi, the saint; indeed, this listing could continue almost ad infinitum.

Even the egoist who has turned his own egoism into an idol that requires the sacrifice of his entire existence is no longer, certainly, an *egoist* in the ordinary acceptance of the word, and even less so could one classify him as a *hedonist*; obviously, he will continue to be an *egocentric*, but his egocentricity will be in the service of an *ego* that has come to transcend him, that transcends his empirical "I". By virtue of his abnegation, by virtue of an ever more heroic fidelity to what for him has become the voice of duty, our immoral egoist has almost become a moralist: a *moralist of immorality*, as one might define him by means of an expression that may seem a little far-fetched, and which is only apparently contradictory. Undoubtedly, his "morality" will be as impure and suspect as you like; but then, where can one find a morality that is a hundred percent pure, that is not polluted by gratification or by fear of the consequences (even if only the judgment of others), or by fanatic drunkenness, or by masochism or infantilism or complexes of every kind? Indeed, here we have at our disposal the entire baggage of the psychoanalysts and of the great "masters of suspicion", and also of the not so great, of every type and school.

Each one of us realizes himself in what he does, and each one of us feels a kind of vocation. It is not only the religious who feels himself "called", who feels that deep within him there is an intimate voice that enjoins him in a categorical manner to maintain himself faithful to the response he has given and the commitment he has assumed by overcoming every difficulty and resisting every temptation, every impulse to betray his duty and to harken to the inclinations of his senses that counsel him to seek an easier and more comfortable life, but at the price of renouncing his goal or doing so at least temporarily, putting it off as it were. When the religious does not develop a superior sensitivity, or when this sensitivity becomes blunted and, consequently, his originally lofty aims become clouded and even fall into oblivion, it may happen (and, indeed, it generally does happen) that a different vocation begins to take shape. Often the religious motive remains as if it were confined in the background, and so evanescent and weak as no longer to be able to inspire a vocation; and so a different vocation begins to take shape more in the foreground, a vocation that is very different from the other one that has collapsed, that has been missed, and very often even in blatant contrast with it. Thus the king pursues his dominion, the captain of industry pursues the expansion of his undertaking, the gangster pursues the control of the local rackets (and so on) with an energy and a tenacity and a spirit of sacrifice that, as it were, no longer have anything egoistic in the hedonistic sense of the word (as I have suggested above) and which one would not succeed in explaining except by pointing out that they, too, derive from a vocation (however strange or funny it may seem to speak of the "vocation of a gangster"). And each of these professions may have its own peculiar "ethics" and its own code of honour, so much so that the very Mafia self-defines itself as the "Honoured Society". Even in some of the less noble "vocations" there may express itself something very similar to a voice that speaks with the absoluteness of an unconditional and categorical imperative that seems to come from an absolute, even though it will eventually reveal itself to be a false absolute, an idol; but an idol that exacts many sacrifices, an idol that can induce its faithful to sacrifice the whole of themselves and every instant of their lives.

In many cases, moreover, what at the beginning is nothing but a weak voice may gradually come to manifest itself in an ever more vehement, possessive and exclusive manner, as if it were an invisible force that takes possession of the entire personality of a man and eventually drags him where he never wanted to be dragged. Certain men seem to be possessed by their vocation to such an extent that they give the impression of being hallucinated, obsessed, individuals driven on by a demon. On the other hand, even some of the most genuine among the saints may give an analogous impression: the strength that possesses them and drags them to where they could never arrive by their own initiative could be defined as a divine energy or as the selfsame Spirit of God; in any case, we are clearly concerned with a force that is invisible, active, that has an initiative of its own and manifests itself as an interior voice, but one that does not limit itself to being a mere voice and gradually comes to manifest itself as an energy that becomes ever more powerful, real, visible, and physical. A simple reading of the Acts of the Apostles can give us a very vivid idea of this, but then there are also numerous other testimonies regarding the phenomena of charismatic sanctity in both the Judeo-Christian ambit and in other spiritual traditions.

In each of these vocations (be it a vocation to sanctity, to politics, to the arts, to commerce, or to brigandage) the subject is always stimulated and driven on by an inner voice that speaks to him with unconditional and categorical imperativeness (even when it does not take complete hold of him with the impetus of an all-pervading and almost physical force as we have seen). Such a voice appears to him as the manifestation of an absolute, and face to face with this absolute he may assume a religious attitude, that is to say, an attitude of adoration (more or less), of unconditional obedience and full availability. The voice, of course, may be a false absolute, an idol. In that case even the Christian religion — which proclaims itself to be the true religion and at times does so with accents of exclusivity that many people find unbearable — this selfsame Christian religion and theology (and especially the Catholic variety) recognize that the good faith of the person fallen into error and his unbound availability for a false god have the same value, from the subjective point of view of the intention, as the assumption of an analogous religious attitude before the true God, the God of Christianity. This makes it clear that, even from the Christian point of view, morality consists of following the voice of one's conscience or, which amounts to the same thing, following one's vocation.

To do this, however, means that one has to dig ever more profoundly in one's own soul, enclosing in parentheses and thus suspending the effectiveness of all the motivations that, as the examination proceeds, do not appear to be wholly absolute and coincident with the voice of duty, but rather suspect in one way or another: one therefore has to subject oneself with great severity to a kind of asceticism and to be extremely diffident as regards the seeming absoluteness of the motivations that underlie one's actions; there must be complete and effective readiness, not just by way of lip service, to listen to the inner voice of our best inspiration and to detach ourselves, fully so, from our habits and our customary mental schemes. This means that we must maintain ourselves in a constant state of listening to the inner voice and do this without ever enclosing ourselves in our own points of view as if these represented absolute truth.

Whoever adopts an attitude of this type and succeeds in maintaining it in spite of all the urgings to the contrary, will have a better chance of discovering the true God beyond all the innumerable idols, of discovering his own genuine vocation beyond everything that adulterated and deformed it. He will then come to realize that, even when he followed an idol and a mistaken ideal, his attitude, even though erroneous from an objective point of view, was subjectively right: in its own peculiar way it was an ethical attitude, certainly to be held in greater respect than that of a person who follows the "true religion" and yet follows it only halfheartedly, offering himself only partially and with everything subordinated to his own wellbeing. He will realize that the true and supreme value can be pursued through each and every human value, because God participates himself to all beings and to all values and is present in the whole of the universe, albeit (as Dante says) "more in one part, and less in another"; the important thing, therefore, is to arrive "in the sky that take more of His light".[25] Nor can it be excluded (and, indeed, I would say that it is very probable) that a truly religious life (to be discovered and lived to the very plenitude of everything it implies) includes also a commitment to the pursuit of humanist values, technical, artistic, scientific, political, economic or social values, and so on, each one of which is false only when it is falsely absolutized, but is true when it is integrated, together with the others, in the pursuit of a full life; a full life that seeks to be an imitation of the fullness of God and a help offered to God by his creature to bring the whole of creation to its ultimate completion.

Many sacrifices are required of those who place themselves at the service of their vocation. And perhaps the greatest of these sacrifices is that of maintaining a constant phenomenological attitude vis-à-vis this selfsame vocation. Every now and again they have to ask themselves: "But is this really my vocation, and is it all of it? Could it be that, for a long time past, without really wanting to do it and almost without realizing it, I have made myself insensitive to other aspects of my vocation and to some higher instance, to some voice that came from an even more intimate and profound part of my being and that made itself heard in as yet too weak a manner when, inadvertently or almost, I allowed it to be smothered, as it were, by more superficial but more urgent instances?". The problem therefore is that of maintaining a continuous attitude of suspicion vis-à-vis the vocation we believe to have and the things that are most striking in that vocation, but not for that reason necessarily also the truest. Not always the voice that shouts loudest is the right one. And the continuous readiness to place these strongest urgings into parenthesis, into *epoché* if you prefer, is the true phenomenological attitude. We can assume it also in our ethical life, for it will make us more apt to discover what truly founds it or, alternatively, to wait for these foundations to reveal themselves.

Centro italiano di ricerche
fenomenologiche, Rome, Italy

NOTES

[1] See E. Husserl, *Ideen I*, chap. 2 of the second section, and especially § 42.

[2] See my article 'Teleology and Inter-subjectivity in Religious Knowledge', *Analecta Husserliana*, vol. IX, pp. 235–47.

[3] See especially ibid., as well as my *Esperienza del sacro e filosofia* (Roma: Liber, 1970).

[4] See P. Valori, *L'esperienza morale* (Brescia: Morcelliana, 1971), pp. 55–108.

[5] G. Morra, *La crisi dell'autonomismo etico*, 'Ethica', V, p. 98.

[6] M. Scheler, *Der Formalismus der Ethic und die materiale Wertethik; Neuer Versuch eines ethischen Personalismus*, Vol. II of *M. S. – Gesammelte Werke* (Bern: Francke, 1954, pp. 225–226) (IV, 2, b). On Kantian formalism and Schelerian phenomenological foundations of value ethics see A. Lambertino, *Il rigorismo etico in Kant* (Parma: Maccari, 1970) and *Max Scheler; Fondazione fenomenologica dell'etica dei valori* (Firenze: La Nuova Italia, 1977).

[7] Ibid., p. 497 (VI, B, b).

8   Ibid.
9   Ibid., p. 63 (I, 3, b).
10 Ibid., p. 71 (II, A).
11 Ibid., p. 196 (IV, 1).
12 Ibid., p. 202 (IV, 1).
13 Ibid., p. 80 (II, A), p. 182 (IV, 1) e p. 257 (V, 1).
14 Ibid., p. 190 (IV, 1).
15 Ibid., p. 68 (II).
16 Ibid., p. 68 (II) e p. 83 (II, A).
17 Ibid., p. 310 (V, 6).
18 Ibid., p. 130 (II, B, 5, 4).
19 Ibid., pp. 125–130 (II, B, 5).
20 Ibid., p. 499 (VI, B, b).
21 Ibid., p. 318 (V, 6, a).
22 Ibid., p. 310 (V, 6).
23 Remember, for example, that Aristotle was convinced that we can adequately grasp
(θιγεῖν) the (perfectly simple and indivisible) forms or essences of things. One either
grasps them or one doesn't, *tertium non datur*. Every time one grasps an essence, one
simply cannot deceive oneself (see *Metaph.*, IX, X). Closely analogous concepts are
expressed by Scheler with reference to the intuition of the values in *Form.*, II, A, at
the beginning of n. 1. This is an aspiration (and also a temptation) that often recurs in
Western philosophy, albeit in different forms. Even Scheler does not by any means
seem to be immune.
24 Giovanni Pico della Mirandola (1463–1494).
25 Dante, *Paradiso*, I, 1–6.

PETER KAMPITS

THE SIGNIFICANCE OF DEATH FOR THE EXPERIENCE OF BODY AND WORLD IN HUMAN EXISTENCE[1]

An ancient definition of philosophy understands it as learning to die, indeed, as a striving toward death. Death is also understood as the separation of body and soul. This interpretation of death, soul and life, powerful ever since, shows that particular views of death, man, his body and his soul are inseparably joined: still further, that the issue of death, with all its implications and consequences, decides not merely incidentally, but intrinsically, what it means to exist as body. For whether it is the body, or the entire human being which dies, whether or not I or something of me is immortal, is important not only for dying itself, for the moment of death, but affects the course of our existence as a whole, our relation to ourselves, to our fellow human beings, and to the world. Whether or not death in the final analysis matters, or affects our existence crucially, whether it be conceived as "natural" or as fundamentally contradictory to our entire existence, is also of incalculable significance for a philosophically directed inquiry into body and soul.

We are mortal, and know that we are. We finally terminate like all living creatures, but lead our lives in knowledge and understanding to this end. We pass, and we know about our passage. Thus death remains at once that which is most familiar and most hidden, not so much a phenomenon as something which, to a degree, suffuses all the phenomena of human life. It disturbs our understanding of being, truth, and world. It constantly comes up, not only in change, loss and separation, but also in happiness, joy, and fulfillment. Out of and within this dedication of our being to death, philosophical questioning insists on the attempt to understand death, to answer the challenge, the scandal that it represents for thinking, for reason itself.

For even if it is obvious at first glance that we are bodily constituted as mortal like all living creatures, whose material body lingers a little while as a lifeless corpse, there is still the question whether we have thereby grasped the meaning of our death, our being mortal, adequately. Is it the body which in its limitations, its vulnerability and frailty, makes us mortal? Or could it be the other way around, that our mortality decides how we experience our body, that our constitution as finite beings qualifies the human spirit itself as mortal? Succinctly, it may be stated, with regard to the history of philosophy, that the equation of body with mortality, of soul with immortality, has likely

283

A-T. Tymieniecka (ed.), Analecta Husserliana, Vol. XVI, 283–294.
Copyright © 1983 by D. Reidel Publishing Company.

underlain that devaluation of the body relative to spirit which philosophy as
metaphysics features in all its forms, and which Nietzsche, the defender of the
"grand reason of the body", so vigorously opposed. Nietzsche's dictum, that
"the soul is merely a word for something about the body" deeply questions
the tradition in which animal and divine are confused, dividing the perishable
animal side of man from the imperishable and eternal.

Phenomenology, as we know, assigns an important place to the problem
of body and the problem of soul. From Husserl's painstaking and richly
varied analyses through Merleau-Ponty's investigations of perception and
comportment, from Scheler's thematization of the body in the horizon of
personality through Sartre's ontological theory of the body and Marcel's
idea of incarnation, there is a growing emphasis on the value of body, in
which at the same time the metaphysical dualism of body and soul, especially
in its modern form as initiated by Descartes, is shattered and overthrown. For
some of those named this also takes the form of a changed understanding of
death. From Heidegger through Marcel this begins to assume central importance within each overall philosophical perspective.

In a nutshell, the changed understanding of death, principally through its
having become related to body, might be put like this: the Platonic notion of
death as a separation of body and soul is replaced both by Gabriel Marcel and
after him by Fridolin Wiplinger, with another, namely, death as separation
from the beloved. That the body, its co-activity, its oneness with and difference from the soul, should play a pervasive part, accords with the thrust of
phenomenologically oriented thought. The basic maxims of phenomenology,
since become shibboleths, viz. "to the things themselves", "the self-givenness
of phenomena", and "life-world", point in their original ontological reference
toward our constitution as bodily, despite the difference between sensory
perception and perception (*Anschauung*) in Husserl's sense. That here soul
and spirit are manifoldly thematized and investigated, without ever being
given a univocal or unilevel status, follows the fundamental project of
phenomenology, particularly that of Husserl.

Simply put, let us say that the problem of embodiment inaugurated by
Husserl, and basically modified by his students and successors, is increasingly
viewed together with that of death, even though in Husserl death is scarcely
thematized.

In this connection, I shall only emphasize here in general that Husserl
characteristically never tired of stressing the role of the body within the
problem of transcendental constitution, and pursued in detail its link with
ego and world, without ever at the same time abandoning the privileged

status of the phenomena of consciousness. Especially in recent literature, from Strasser and Boehm to Theunissen and Waldenfels, there have been repeated references to the conflictful position of Husserl, from the *Ideas* through the *Crisis*, even when Husserl's admission of transcendental-phenomenological idealism has been accepted as merely a blanket label for the course of his thought taken as a whole.

In this framework I shall only outline Husserl's thematization of the body as well as of the soul in rough strokes. That Husserl gives a crucial role to the body, both in reference to the constitution of world, as well as in its relationship to soul, to ego, to consciousness and mind, may be followed from the *Ideas* to the *Crisis*. This also holds for his project of a phenomenological psychology, as well as for the paramount importance he assigns to the life-world in his later work. The doxic-experiential attitude remains the determining foundation for Husserl with respect to the issue of originary founding moment, even though the guiding thread of the extramundane and transcendental subjectivity also remains decisive for the ontological reality of the world. With Husserl's distinction between the originary, processual (*fungierendum*) body, and body as an objective material body, appears a theme which, from Merleau-Ponty's distinction between *corps phenomenal* and *corps objectif* or "physical", to Marcel's *avoir un corps* and *être mon corps*, is not to abate. Even if one might properly object that Husserl here only opens the way to insights that he will later abandon (Waldenfels) or that it is doubtful whether Husserl has in fact preserved the unity of the concretely existing human being by tracing it back to the transcendental ego (Strasser), it is undeniable that the body for Husserl is more than a mere corporeal thing, of which each of us is more or less accidentally the partner.

Already within the constitutions of the so-called ontic regions with their constantly shifting perspectives, principally the basic naturalistic and personalistic perspectives, the body appears as more than merely corporeal. Beginning with the "lowest level" of animal constitution, of "man as natural reality", it is clear that the body is not merely a "physical thing, matter", but also points over and beyond. To be sure, the body is also definable as a "real thing", inserted into the causal nexus of material nature (*Ideen* II, p. 159), at the same time that specific body events standing out in relief from merely physical ones are apparent in the "sentiments" (*Empfindnisse*).

From perception to movement, from the body as center of orientation and thereby as null-point for all orientations (*Ideen* II, p. 158), the body, both in relation to the world and to the psychic ego subject as Husserl puts it, occupies a special position. Here the soul appears not in contrast to the body as material nature, but as "the concrete unity of body and soul, the human (or animal) subject" (*Ideen* II, p. 139).

The body on one side is the organ that opens the world to me simply and purely; it is, as Husserl repeatedly reminds us, necessarily present in all perception (*Ideen* II, pp. 55f., 128ff.). It is my corporeal organ, through which and in which I am an active and suffering "entity in an utterly unique and own manner" (B III, 2/21).

As such the body is so closely interwoven with the soul that Husserl, considering the problem of the real psychic ego, acknowledges that this ego embraces the entire human being, body and soul. And yet, references to the thing-character of the body run through all his investigations of constitution. The privileged status Husserl gives from the beginning to the physical leads, among other things, even to calling the body a "peculiarly imperfect constituted thing" (*Ideen* II, p. 159). The body slips once more onto the side of nature, thereby becoming the bearer of a substrate for the ego subject in the personalistic attitude, in which the field of a subjectivity no longer part of nature constitutes itself as mental ego.

Husserl here defines the basic rule of motivation for the mental realm in contrast to causality which is attributed to nature; he inflects the problem of body and soul in a way which shows its basic order to the oriented to transcendental subjectivity. The body appears first for the personal ego as a possession, simply as given for me like anything else ego-alien. The principal effect of the personalistic attitude is to direct inquiry concerning the unity of the ego toward that transcendental sphere in which body and soul are both finally encompassed. The pure ego therein revealed, to whose constituting activity both psychophysical and physical reality owe their ontological validity, reveals itself as constitutive for all other ego qualities. To be sure, the body remains for Husserl the "place where mental causality reverses into that of nature", spatially exterior and subjectively interior, at once a spatial thing, a live corporeal body and an inwardly vital body, organ and habitual system of subjective functions, always about to pass over into actual subjective functioning, as Husserl puts it (cf. *Ideen* II, p. 286). The importance Husserl gives to the body for the constitution of the person is certainly considerable; yet, after all, the soul is finally approximated to mind (*Geist*) so that embodiment becomes the objective constituent of the constituting consciousness. The different levels of being — world, nature, space, time, animal being, man, soul, body — derive ontological validity in relation to transcendental subjectivity. It is also just in the problem of individuation that the priority of the mind as pure ego appears, within itself and removed from itself in the mode of the "disinterested spectator".

Much the same could be shown within the framework of Husserl's theory of intersubjectivity. Thus the often remarked transcendental-idealistic direction of Husserl's phenomenology, despite its incisive insights and suggestions concerning embodiment, finally remains committed to a modified dualism of the Cartesian or transcendental-idealistic variety. Whether the unity of the concretely existing human being is not thereby forfeited, through being ultimately guaranteed only by the unity of the transcendental ego, is the question, as well as that of whether Husserl did justice to the autonomous reality of the body.

This impression is amplified when one considers various remarks by Husserl about the connection of death and life. Here too we see that Husserl does not simply link death with the body, putting it together with what he called the natural substratum, the "underground" or "support" for mind. Still a statement from his unpublished work remains definitive: "Man must die, but transcendental pristine life, the ultimately world-productive life and its ultimate ego cannot arise out of nothing and pass away into nothing; it is immortal, because dying has no meaning for it" (K III 399).

This crystal clear announcement of immortality in the realm of the transcendental ego cannot of course be taken as an argument for the devaluation and deprecation of human — which also means for Husserl bodily and embodied — life. But its parallelism with a traditional metaphysical interpretation of the mortality of the body and immortality of the soul is unmistakeable. Here body has been exchanged for mundane subjectivity or animal nature, psychophysical and social reality, while the soul has been replaced with the transcendental ego, the ultimately absolute transcendental consciousness.

To be sure experience tells us, as Husserl expressly notes, that the body dies, ceases to be a body, no longer a material body ruled by an ego (K III 18 32). But such experience remains external, inasmuch as we experience death, like birth, as world events, occurrences in the world. Human beings are born and human beings die. But birth and death as world events are originally constituted in my intrinsic life (cf. B I 14); they are "limiting cases" of it. As such they must be viewed "from within", as "bounding instances", in Husserl's words. Only within the horizon of this life can they then be anticipated, be it only as the most extreme possibilities of existence. That in this the constitution of temporal consciousness with its specific forms of retention and protention, as also the historicality attributed to soul, play a significant part, is no less evident than the debt which Heidegger's conception of "being unto death" owes to these considerations. To be sure, Husserl acknowledges the difficulties confronted by the constitution of death, also defined as "cessation of the process of consciousness", a constitution which, unlike that of birth as a beginning, cannot be directly broached. Now, from the sphere of one's own experience the phenomenon of sleep is invoked, so that death, through the phenomena of sleeping and waking in incessantly streaming life, can be considered within the perspective of an overarching unity. Here there are frequent references to "letting go of the world", to an "end of being worldly", so that in fulfillment of Husserl's methodic requirement of leaving behind a "natural world life", the detachment from the life-world can also be understood as a dissolution of life itself in the sense of a universal epoché. This would be nothing else but an anticipatory prolepsis of death, one inseparable from the natural bodily life of human beings within the life-world. The liberation of the ego from this natural bodily human life would then be not only a victory over death, but also the beginning of true life, of life in the spirit.

It is certainly no accident that Scheler finally took this route, or in any event a similar one. In his impressively detailed analyses of the phenomenon of the body, it is particularly in relation to death that he stresses the superiority of the person to the body, and in a certain sense the independence of the human being as person. Scheler's basically theonomic Christian orientation leads him to a factual conquest of the body and world through the person who survives the disintegration of the body, or, as Scheler once put it in his *Death and Survival*, "soars beyond death as the limit of life and lives on."

The second path taken by Husserl relates to the death of another, one opened to me through empathy, in which it is first of all evident that a live body ceases to be a live body. In Husserl's theory of intersubjectivity the body also plays no slight part, since it is via the transcendental constitution that the body enables the other to be primarily given and opened to me in original presence. Within the constitution of the "objective world" it guarantees the humanness of another and ultimately my own being as human. That in Husserl's doctrine of intersubjectivity transcendental subjectivity has priority, one which finally determines the ontological value of everything, and that even further, the function of the stranger is to finally mediate my own constitution, only strengthens the suspicion that even the death of another is something I emphatically experience, and is thus incapable of placing once more into question the very relation of death, body and soul.

Husserl's investigations of body and soul, or their unity, definitely leave no central place for the phenomenon of death. All the same, Husserl's moves are subsequently decisive for a sequence of philosophical positions, even when a particular position deviates considerably from his own conception of phenomenology, as was the case with Heidegger's fundamental-ontological project.

Heidegger's interpretation of death as the extreme, absolute, and ineluctable possibility of existence, as *existentiale*, with all its familiar implications, of existence as "being unto death", certainly incorporates important suggestions from Husserl's analyses of time experience. It could also be seen as an effort to institute the proleptic anticipation as the uniquely adequate approach to death.

The central significance of death in Heidegger for the whole of human existence is accompanied by an abandonment of metaphysical-anthropological categories, demonstrating in this way the inadequacy of the body/soul schema. Death is now no longer a separation of body and soul, but on the contrary the uttermost possibility of existence, one encompassing all other possibilities. It is seen not in terms of a "natural" event, but as one occurring within our understanding itself, distinguishing man as being unto death for all other transient modes of life. Without going into Heidegger's invaluable analyses regarding death in greater detail here, I wish only to point out how human finitude is critically thought through here even while it remains detached from any anchorage in bodily constitution. Death as the end of an existence for which being is intelligible, is neither decay nor a passing away of the human body, no mere event of nature.

Whatever one may think of Heidegger's fundamental-ontology as a whole, it is undeniable that death here is illumined in its full significance for human existence. Husserl's interpretations of the live body are modified and radicalized in a still different way by Merleau-Ponty. Merleau-Ponty's thought as a whole revolves about our bodily constitution. The body, according to Merleau-Ponty, not only mediates the world for us, is not only the medium for our appropriation of the world, but our very anchorage within one (*notre ancrage dans un monde*). The body opens to us at the same time the sense of everything there is; it determines the genesis of sense, which for Merleau-Ponty takes place in the world not through thetic acts of consciousness, but through our interwovenness within it as bodily beings.

Merleau-Ponty also refers to the body in its relationship with the ego as our *moi naturel* which, unlike Husserl's empirical or mundane ego as contrasted with his transcendental world-alien ego, denotes the "visible expression of a concrete ego", expressive of a processual egoity of the body which Merleau-Ponty also calls *l'incarnation perpetuelle*.

With his bodily structured *être-au-monde* Merleau-Ponty sees every recourse to transcendental consciousness as derivative in view of the bodily concretization of subjectivity. We cannot go any further into his thought here; what is important for him, however, is that the significance of the body for our existence is not simply equated with our finitude which is nevertheless also very important for Merleau-Ponty. Indeed to be sure our body also makes itself known by its resistances, its frailty and vulnerability; it collaborates with our finitude and facticity. At the same time the latter stems from an intentionality and transcendence also grounded in the body, accounting for our openness to the world and for others.

Merleau-Ponty's phenomenology thus turns out, not least by virtue of his interpretation of the body, to be one of finitude, even though he does not indeed include death in any decisive fashion. Death is rather discovered in the interplay of integration and disintegration of the body and soul. The now more factually present body as merely material is the corpse; the decomposition, the disintegration into material body and soul happens only in death. But this at once also raises the suspicion that Merleau-Ponty has not firmly enough integrated the tie of death and finitude into his conception of human existence.

Gabriel Marcel pursues an altogether different path. His interrogation of death continues the direction of Husserl's inquiry via the death of another. Only his concern is not about another in general, or a constitutive theory of intersubjectivity, but a deeper penetration into the very experiencing of death. Fridolin Wiplinger, after Marcel, and unquestionably influenced by phenomenology, has developed this experience in its originary, pristine givenness, with all its indisposability, its irreducibility, and inevitability. Such an experience of death occurs only with the death of a person whom we love.

For Marcel, and also Wiplinger, the experience of the other's death is not merely the way in which the phenomenon of death becomes at all accessible to us, but the way through which the question of the meaning of everything that exists is raised. The death of the other generates a crisis in our understanding of being, one also affecting the mode of our own bodily being.

The radicality of this engagement with death is already to be found in Augustine, whose lament in his *Confessions* for the death of a beloved friend is not to be taken as merely metaphorical when in describing his relationship with him, he speaks of one soul in two bodies (*animam meam et animam illius unam fuisse animam in duobus corporibus*). Augustine also discovers that in this experience he has himself become one great question. Gabriel

Marcel, sensitive to the ensuing conflict between love and death, grasps the problem in a perspective which is of supreme importance for the interpretation of embodiment. Death, which in metaphysics, quite simply put, separates a natural biological mortality of the body from an immortality of the soul, becomes, through the bodily death of the beloved other, a central focus. Marcel's distinction between *être* and *avoir* thus involves precisely the bodily character of my own being, thereby also controlling my interpretation of subjectivity and consciousness. "Having", as finally equivalent with the structure of intentionality, is rendered problematic by its connection with our bodilyness, which may be seen as the hinge between being and having. Marcel's notion of "incarnation" seeks to express the merely experienceable, non-objectifiable relationship between ego and material body, in which his *être incarné* also means "being in the world."

The body, also viewed as *corps instrument*, the very prototype of instrumental possession, is also discovered as that which, neither through a discrimination of ego and live body, nor by their identification, is capable of being adequately determined.

Throughout Marcel acknowledges that the body (*Körper*) as my body (*Leib*) is capable of being "felt", but that this does not permit their identification. Thus I am neither able to simply say either that I *have* my body, or that I *am* my body, even while within the fundamental givenness of incarnation a possible change of attitude in regard to the own body between having and being always remains open.

It is just this *être incarné* in the counterpoint of being and having, the body now as corporeal object, now as my body, which becomes central in my relationship with the other. The other can appear for me as an object within an intentional horizon of having or as the personally addressed "you". Simply put, in the first instance we come upon another object, say an alien material bodily subject, an "it", any bearer or owner of a body whatsoever. But wherever such another is a "you" for myself, he or she does not face me as merely one who bears a body by means of which a strange ego, a soul or a spirit, expresses itself. Marcel's key expression of the "presence" of "you" in a personal-dialogical relationship stands for the integral presence of "you" in concrete bodilyness, immediately apparent in bodily gestures, a glance, a smile, a handshake, as well as instantaneity, that mode of temporality which remains decisive for encounter.

Here, the bodilyness and personal presence of the other are contrapuntal within an indivisible whole. Therefore such "presence" cannot be identified with the complications, implications, and explications, of the spatial-temporal relationship, as in the sense of now-ness, nor be reduced to any "inwardness" independent of the body, in the sense of my realization, my recollection. This becomes critical for how death is experienced since it now can no longer be simply conceived as destruction of body, but only as betokening, above all, the utter annilihation of the other's living presence. Because of this the death of the beloved, to whom I am not merely related in constitutive intentionality, also stands for a lesion of my own being, a transformation of my self-being, in a certain sense, my own annihilation.

Wiplinger heightens this, even beyond Marcel, into a paradoxical extreme: it is just where the bodilyness of such being together is taken seriously, no longer misinterpreted as physical bodyness or materiality, that death necessarily plunges this personal and dialogical being-together into crisis, this because of its finality which has nothing to do with an end in terms of the biological phenomena of decomposition and decay. Death as "end" stands not only for the finality of separation in the sense of "never again", but also for its finality in the sense of fulfillment. This double meaning can initiate a changed understanding of the temporality and historicality of the personal-dialogical relationship.

For the bodilyness of being together as personally understood, death necessarily appears meaningful, while for the relationship being realized in personal love it must seem absurd; death stands for a consummation of both relationships, and also separation from the beloved "you".

That this does not entail any kind of decision concerning a possible life after death, or about its finality as dissolution, must, in light of various possible misunderstandings, be emphasized. For death shows itself here to be equivocal: it is annihilation and breaking off, consummation and entirety at one and the same time. What matters is that this ambiguity be upheld to countervail both a derealization of our bodily make-up through a banalization of death, and a relativization of the claims of love, which would simply mean surrender to death.

In the face of this paradox it should be clear that death at the extreme of life, as its end, crucially supports and determines the entire bodily character of our lives. We are not mortal by virtue of the superaddition of a body to our spirit, our soul, or ego, but we are mortal in our bodily constitution: mortality is intrinsic to our bodilyness. So that Marcel's careful words concerning a transcendence of death are by no means to be understood as a return to a dualism of body and soul, even when his gestures toward the presence of the beloved "you", triumphant over death in fidelity, often appear to suggest this.

The fidelity of which Marcel here reminds us, a fidelity capable of ensuring the presence of "you" in despite of death, even while it attributes to the post-mortal form of the vitality of "you" a more pure and direct mode of "presence", does not however allow any sort of reduction to a soul as its equivalent. That it is important to proceed with caution precisely where death is not interpreted as a natural event in regard to the body, is conveyed in Wiplinger's suggestions of a real transcendence of death which, within the ontological experience of "being together" deriving from a changed understanding of temporality, is not to be confused with immortality understood from the standpoint of eternity. Here however we reach a limit of philosophical speculation, one which can only be crossed in our own dying.

We are who we are, bodily. We exist bodily. And this existing is at the same time disclosure of world. Consequently our mortality, if it is not merely external, can neither be dissociated from the way in which we comprehend ourselves, the world, our ties with others, and their meanings for us. And it is only where death is taken seriously, no longer simply as an event in nature nor as a chance punctate occurrence, that our bodyness can become opened to us in its very transience and finitude.

Death alone holds such power to determine the significance of our body and therewith the significance of the world. A philosophy which takes bodilyness seriously must also try to illuminate it from the position of death. Where death engages us in our very being, where it is taken as more than a merely natural biological event, our body is differently considered and our bodily make-up is no longer judged from the position of a remote spirit.

Death as departure from the one we love, and not as a division of body and soul, indicates in any event a level of experience for which "body" and "soul" as basic categories of human life are no longer adequate.

Husserl could have meant this when he wrote:

The doctrine of immortality, if it is not to contradict the sense of the world, as this is realized through universal objective experience, must then assume a completely different meaning . . . (IX, 109).

University of Vienna

Translated by Erling Eng

NOTE

[1] Throughout, "body", "bodily", "bodilyness" will be used to translate *Leib, leibhaftig*, and *Leibhaftigkeit* respectively, all of which refer to the *live* body. Wherever *Körper* or related words are employed in the original, reference will be made to the "material" or "physical" body.

BIBLIOGRAPHY

Heidegger, M.: 1927, *Sein und Zeit*, Tübingen.
Marcel, G.: 1953, *Présence et immortalité*, Paris.
Merleau-Ponty, M.: 1959, *Phénoménologie de la perception*, Paris.
Scheler, M.: *Tod und Fortleben*, in *Werke*, VI.
Wiplinger, F.: 1971, *Der personal verstandene Tod*, Freiburg.

MARIO SANCIPRIANO

LA TRANSFIGURATION DU CORPS DANS LA
PHÉNOMÉNOLOGIE DE LA RELIGION

CONSIDÉRATIONS PRÉLIMINAIRES

La phénoménologie de la religion s'est appliquée, jusqu'ici, par d'excellentes
études, à déceler sourtout les "figures" propres de l'objet religieux dans
l'histoire. Maintenant l'analyse phénoménologique doit être reconduite à sa
tâche primordiale, à partir de l'intentionnalité de la conscience religieuse qui
opère aussi, par les voies d'une synthèse transcendentale, la transformation
du corps. Il s'agit donc d'analyser l'origine du rapport réciproque entre
l'âme et le corps dans l'unité psycho-physique fondamentale, qui est plus
mystérieuse que les origines préhistoriques de l'humanité. On arrivera à
comprendre tout ce que le corps offre et tout ce par quoi le corps s'offre
lui-même afin que s'accomplisse l'acte de religion dans sa plénitude essentielle,
avec la profession de la foi, la prière et le culte.

Le but de cette relation c'est de montrer que le mysticisme nous donne
une expérience de la vie spirituelle, qui n'oublie pas le corps, mais simplement
le transfigure, tandis que le corps, par une sorte de *mémoire* qui lui est propre
et particulière, c'est-à-dire par la "rétention" de ses perceptions et de ses
tendances, agit sur le psychisme et sur l'être spirituel de l'homme.[1] Dans
la tradition la plus orthodoxe du catholicisme, s'épanouit une *mystique de
la vie*, qui a d'abord son ressort dans une communion avec la nature: bien
entendu avec une nature sublimée par ses points de repère à la transcendance.
"Laudatu si, mi Signore, cum tucte le tue creature", s'écrie St. François
d'Assise, dans ses *Laudes creaturarum*, où sa louange a un objet sensible dans
la splendeur du soleil, la jouissance du feu et la pureté de l'eau; et elle a
aussi un objet intelligible dans la prière. On aperçoit ici le point d'arrivée de
la mystique de la vie. L'inhérence du mysticisme à la vie sensible n'est pas
seulement une expression poétique: on y trouve la relation du sujet mystique
avec la nature, par le sentiment des créatures. C'est bien vrai que le sujet
mystique, pour se disposer à la Grâce, devrait arriver à l'oubli de la nature,
dans la "nuit des sens" (St. Jean de la Croix); mais, pour lui, il est toujours
possible une perception poétique et aussi religieuse, qui est conduite à son
but par une certaine sensibilité, selon les lois de la connaissance et en accord
avec la morale. Au contraire, un mauvais commencement au niveau sensible

295

A-T. Tymieniecka (ed.), Analecta Husserliana, Vol. XVI, 295–297.
Copyright © 1983 *by D. Reidel Publishing Company.*

ne pourrait pas aboutir à une bonne destination religieuse. En effet, l'oubli de la nature n'arrive pas sans laisser une trace dans la *mémoire*. Celle ci est, dans sa signification la plus étroite, une propriété essentielle de l'âme; pourtant, dans le cas que nous considérons, la "mémoire" appartient au *corps propre* et lui est particulière.

C'est le souvenir, même sensible, des bons résultats, qui encourage à de nouveaux exploits moraux et religieux, pour l'avenir. La sensibilité peut donner la matière imaginative à l'intuition mystique de deux façons: (1) en conservant les *images physiques* convenables dans les symboles de la foi, par exemple le signe sensible de l'eau dans le baptême; (2) par l'exercice effectif de l'*ascétisme*, qui agit sur l'association des idées et conditionne, *d'en haut*, les réflexes psycho-physiques.

L'homme parvient à cela par cette sorte de "mémoire" qui s'imprime dans le corps et le rend plus docile et malléable. Cette *mémoire du corps* est plus forte et puissante que le corps naïf lui-même et agit pour favoriser les réalisations, les plus essentielles, de la vie humaine. La douleur, la souffrance et le renoncement peuvent fournir de conditions préalables pour cette victoire sur les sens, qui s'accomplit par les sens mêmes. La mémoire du corps est donc la "rétention" des tendances et des images qui ont agi déjà de façon positive et qui ont prédisposé certaines formes de la vie spirituelle, en donnant le souhait et l'espoir de rendre toujours, par leur "souvenir", notre vie a nous mêmes. Au niveau de la perception, dans cette genèse passive, l'homme reçoit l'impression des sens; mais enfin il s'élève, par son activité créatrice, à ses buts les plus propres et particuliers. Par le contrôle de soi même l'homme *compos sui* dispose d'un "autre" *corps propre*, qui satisfait mieux que le corps naïf aux exigences de sa nature.

Nous sommes passés ainsi de la *perception* à l'*imagination* et aux *fonctions intellectuelles et symboliques*, qui sont les plus étroitement connexes avec la vie psycho-physique et avec les fonctions organiques et nerveuses. La phénoménologie de la religion devra analyser l'imagination, qui peut être reproductive ou productive et créatrice. Les images religieuses se réalisent en métaphores et en allégories, selon un système qui utilise les images physiques dans la plus pure expérience du mysticisme. Il y a donc: (1) des *symboles élémentaires*, par exemple le sang en tant qu'instrument de la vie et le sang de Christ en tant que moyen de la redemption humaine: (2) des *symboles complexes* dans les allégories, par exemple la vision dantesque du char, qui est attaché à l'arbre mystique et qui réprésente l'Église.[2]

Une enquête phénoménologique doit donc rendre compte de la synthèse active qui donne une image physique du corps humain compatible avec l'image religieuse et même sublimée en celle-ci. On a dit justement que l'homme est un esprit incarné: cela signifie qu'il reçoit la matière de son corps et qu'il agit sur celle-ci en tant qu'il est *plastes sui*, capable de configurer sa nature. Cela ne signifie pas que tout soit possible à la volonté et à l'intelligence de l'homme, mais seulement qu'il peut, avec l'aide de la Grâce, atteindre beaucoup de ce qu'il est nécessaire à une vie digne d'être vecue. L'activité de l'homme ne s'arrête pas à l'augustinien *"agere in passionibus corporis"*, qui est déjà un *agere* orienté à la perfection humaine mais elle veut parvenir enfin à la vision de Dieu et à l'*amor Dei intellectualis*, par la discipline des passions de l'âme (Spinoza). L'homme, en tant qu'homme, est tout dans son unité psycho-physique; mais il se dépasse lui-même lorsqu'il s'assimile à l'unité transcendante, dans laquelle s'accomplissent toutes ses intentions religieuses.

En conclusion, la phénoménologie de la religion ne nous donne pas seulement les "figures" des mythes et des anciennes traditions: bien plus, elle utilise le rayon intentionnel de la conscience pour explorer ce qu'il reste de purifié et de sentimental, après l'*epoché*, dans l'expérience religieuse et ce qui s'accomplit de "matériel", par les symboles et les allégories, même dans la prière et dans l'extase la plus pure. Le mystique n'est pas un sujet insensible: son *epoché* n'anéantit pas son corps, ce qui signifierait la mort. Au contraire, le mystique réalise une union plus étroite entre l'âme et le corps, par une compréhension sentimentale, qui dure encore, toujours actuelle, dans l'expérience religieuse. Il confirme enfin cette unité en se référant à l'unité transcendente et divine. L'image du corps, avec ses perceptions, ses tendances et ses symboles sensibles, n'est pas effacée dans le mysticisme, mais elle se transfigure dans la "mémoire" du corps même, en tant que celui-ci conserve les empreintes d'un passé qui est encore désirable, comme préambule des formes de la vie religieuse.

University of Siena

NOTES

[1] On peut voir, à propos de cette *Wiedererinnerung*, les analyses de Husserl: " . . . Aber gegeben bin ich mir auch durch reproduktive Akte, z.B. durch Wiedererinnerung als vergangenes Ich und nach meinem vergangenen Wahrnehmen, Wiedererinnern, Fühlen, Hoffen, etc." (E. Husserl, 'Analysen zur passiven Synthesis', *Husserliana*, XI, p. 366).
[2] Dante, *Purg*, XXXII, 38.

PART VI

THE HORIZON OF NATURE AND BEING

THÉODORE F. GERAETS

MERLEAU-PONTY'S CONCEPTION OF NATURE

The purpose of this paper is to provide a "sounding" of Merleau-Ponty's philosophy, sounding effectuated on a point that seems to me of strategical importance: Merleau-Ponty's conception of nature. We shall see that while trying to understand this conception, we are lead to question, and possibly understand better, the very nature and foundation of "conception" itself.

Although referring to many other passages, I want to concentrate initially on a paradoxical statement in the *Phenomenology of Perception*: "l'unité de l'expérience ... *ne me libère* de chaque milieu particulier *que parce qu'elle m'attache au monde de la nature* ou de l'en-soi qui les enveloppe tous."[1] This idea of an attachment that liberates us is repeated later in a slightly different form:

La vie humaine se définit par *ce pouvoir qu'elle a de se nier* dans la pensée objective, et ce pouvoir, *elle le tient de son attachement primordial au monde lui-même*. La vie humaine "comprend" non seulement tel milieu défini, mais une infinité de milieux possibles, et elle se comprend elle-même, *parce qu'elle est jetée à un monde naturel*.

C'est donc cette compréhension originaire du monde qu'il faut éclaircir.[2]

Two interrelated theses are here clearly expressed. First, the freedom and distanciation with regard to our vital milieu, which are at the basis of objective thought, are seen by Merleau-Ponty as grounded in our primordial attachment to the world itself, to the natural world or the world of nature or of the in-itself. We have to remind ourselves of the passages consacrated by Merleau-Ponty in his first book to the properly *human* perception of *things* as revealed in man's symbolic behavior.[3] Man transcends the purely functional milieu (*Umwelt*): he recognizes the same thing as such and the same function ("sens") as such and is able to make the same thing perform various functions and to use various things to perform the same function. This bi-polar variability makes possible a behavior which is at the same time cognitive and free (*SC*, 133; *PP*, 103).

The second idea that emerges from the two texts we are reflecting upon is that all our comprehension, or understanding, of a particular milieu, and of ourselves as capable of understanding an infinity of possible or virtual milieus, is grounded in the primordial attachment (to the natural world)

A-T. Tymieniecka (ed.), Analecta Husserliana, Vol. XVI, 301–312.
Copyright © 1983 by D. Reidel Publishing Company.

which is *itself* the original comprehension, or understanding, of the world. These texts, therefore, not only intend to reveal the foundation of comprehension, but also that this foundation is *already comprehension*: it is the original comprehension of the world (*Welt*) as opposed to our functional milieu (*Umwelt*), i.e., of nature itself.[4]

Il faudra comprendre comment d'un seul mouvement l'existence projette autour d'elle des mondes qui me masquent l'objectivité, et l'assigne comme but à la téléologie de la conscience, en détachant ces 'mondes' sur le fond d'un unique monde naturel. (*PP*, 340)

It is true that Merleau-Ponty speaks, in the texts quoted above, of the natural world, rather than of nature — although he uses this term in the first sentence of the passage of which the second text is the conclusion: "Ce qui est donné, ce n'est pas la chose seule, mais l'expérience de la chose, une transcendance dans un sillage de subjectivité, une nature qui transparaît à travers une histoire" (*PP*, 376). This nature that thus reveals — and conceals — itself ("transparaît") is more than the collection of objects, or even things, that can be qualified as "natural" in opposition to what is taken to be "cultural." *Any thing*, even the most cultural of all, is still part of nature. Merleau-Ponty has described this all-pervading transparency of naturalness, e.g., in our perception of a painting (cf. *PP*, 339).

This naturalness is expressed elsewhere as the inhumanness hidden in everyday life, but revealed to us when we assume "une attention métaphysique et désintéressée" (*PP*, 378). All things are, as such, rooted in the soil of inhuman nature (cf. *PP*, 374) which, itself indestructible (cf. *VI*, 321), seems to confer on them their reality.[5]

Now, this nature, in its inhumanity, or absolute otherness (cf. *PP*, 376) — is it perceived by us, or is it essentially a concept? And what light can this consideration of nature shed on the relation of perception and "conception"?

Merleau-Ponty, who speaks so often of the perceived world, only rarely speaks of perceived nature or of nature being perceived (e.g., *SN*, 224; *RC*, 176). Such statements are related to his criticism of a certain concept of nature, the one mentioned at the beginning of *The Structure of Behavior*: "Une multiplicité d'événements extérieurs les uns aux autres et liés par des rapports de causalité" (*SC*, 1). Thus Merleau-Ponty opposes to the nature of the sciences, or rather of a certain ontology they carry with them, "celle que la perception me montre" (*PP*, 494). But this does not mean that in the deconstruction, or reduction, of what is often called the Cartesian idea of nature (cf. e.g., *RC*, 97ff.) all conception is eliminated. There *is* only the nature that perception shows us, but perception could never show *nature*, and the naturalness of things, unless it is already somehow *conception*. As soon as we realize the naturalness of what we perceive, conception has started and reveals itself as immanent in, and essential to, human perception (cf. *PPCP*, 150). Before trying to establish this thesis, something has to be said about Merleau-Ponty's "return to perception."

It is true that, just like the idea of nature, our idea of perception and, to a certain extent, our perception itself have to be deconstructed or reduced. How is it possible, asks Merleau-Ponty, to return from perception as molded by culture to what he calls "la perception brute ou sauvage" (*VI*, 265)? What is this act of deconstruction? And is its result the elimination of all conception from perception?

We cannot hope to answer all these questions, raised but never in extenso dealt with by Merleau-Ponty himself, within the limits of this article. Some brief indications, requiring a follow-up, must suffice.

I have tried elsewhere[6] to describe the return to perceptual experience in view of establishing the meaning, or meanings, of the "primacy of perception." I think that there are two meanings to this expression. First, the primacy of perceptual faith ("foi perceptive"), i.e., of the whole realm of original experience ("expérience-source"), with regard to the interpretation or expression of this experience − it being the case that this interpretation and expression grafts itself onto experience ("s'inscrit dans l'expérience"). Second, perception as encounter with the things of nature, i.e., with "les choses naturelles," "les choses sensibles," is a privileged, archetypal part of original experience. This latter point, and its relatedness to the first, require elaboration.

"Nature," in the first and original sense of the world, is for Husserl the domain of original presence (*S*, 216; cf. *Ideen*, II, 163), "la sphère de l'*Upräsentierbar*" (*HNN*, 267). Nature comprehended in this sense is the truth of nature in the sense of the natural sciences (*S*, 222). This sphere is "la plénitude insurpassable du perçu en chair et en os par moi, plus par les autres" (*HNN*, 267).

All this is summed up by Merleau-Ponty toward the end of the historical part of his course on the concept of nature (1956−57):

A la source et dans la profondeur de la Nature cartésienne, il y a une autre Nature, domaine de la "présence originaire" (*Urpräsenz*) qui, du fait qu'elle appelle la réponse totale d'un seul sujet charnel, est présente aussi par principe à tout autre. (*RC*, 116)

What does "total response" here mean? It does seem to rhyme with "plénitude insurpassable" and to refer back to an earlier passage where

Merleau-Ponty indicates how Husserl, in his efforts to understand "le thème de la pensèe objective et savante," is lead to uncover "la vie intentionnelle qui le porte, le constitue et en mesure la vérité" (RC, 112).

It is in virtue of this intentional life that the world is *Lebenswelt*. The mode of being of the *Lebenswelt* is all-encompassing (*Ideen*, II, 134). This mode of being carries in its flux all perceived things, and nature itself, but also all our historical constructions of theory and practice (cf. *RC*, 151; *Ideen*, II, 134).

Could it be true that the latter (these constructions) are grounded on the former (perception and nature), *more* than the former on the latter, and that this is why nature calls for our total response?

We reach here the problem of what Husserl calls *Fundierung*.[7] It is the paradox of a relationship that is *as foundational* at the same time mutual or reciprocal, and asymmetrical. Such a relationship reveals itself to exist between reflection and the unreflected, between reason and fact, between thought and perception (*PP*, 451), between "l'objectivité logique" and "l'intersubjectivité charnelle" (cf. *S*, 218).

Husserl has insisted on the reciprocal relationship (*Wechselbezogenheit*) between nature and the body (or the soul) within the context of constitution (*S*, 223; *Ideen*, III, 124; cf. *HNN*, 268). But the circular character of this relationship owes something to the natural science conception of nature (cf. *Ideen*, II, 210; quoted by Merleau-Ponty, *S*, 222–23), that is, at the origin of the often discussed dilemma between the realist and idealist perspectives (cf. *PP*, 489) which Merleau-Ponty has, throughout his career, endeavored to overcome.[8]

It is true that the circularity reappears in Schelling's critique of a totally external nature: "Les esprits sont les parents de la nature, comme ils en sont les enfants ... production de l'esprit par la nature et de la nature par l'esprit."[9] However, this circularity opens up into a spiral as soon as true awareness of naturalness is reached.

In spite of other tendencies in Bergson's thought, Merleau-Ponty feels that his precept of a return to the evidence of the actual is to be understood as "une allusion à la pré-existence de l'être naturel, toujours déjà-là qui est le problème même de la philosophie de la Nature" (*RC*, 111).

We are born from within nature. Even when nature becomes an object the above truth reasserts itself: "c'est un objet dont nous avons sùrgi" (*RC*, 94). A fortiori, when nature, in its full and original sense, is retrieved — which retrieval requires us to disengage it from categories like substance, accident, cause, end, potency, act, object, subject, "en-soi," and "pour-soi" (cf. *RC*, 141) – the circle between nature and spirit or consciousness is opened up.

Intentional analysis, and phenomenology insofar as it identifies itself with such analysis, leads us in two diametrically opposed directions: "D'un côté elle descend vers la Nature, vers la sphère de l'*Urpräsentierbare*, pendant que de l'autre elle est entraînée vers le monde des personnes et des esprits." [10] But this means that such analysis, and phenomenology as limited to it, are not the ultimate method; and we are reminded of the primacy of original experience:

> C'est quelquefois de l'*expérience* que Husserl se réclame, comme du fondement de droit dernier. L'idée serait alors celle-ci: Puisque nous *sommes* à la jonction de la Nature, du corps, de l'âme et de la conscience philosophique, puisque nous la vivons, on ne peut concevoir de problème dont la solution ne soit esquissée en nous et dans le spectacle du monde, il doit y avoir moyen de composer dans notre pensée ce qui va d'une pièce dans notre vie. Si Husserl se tient ferme aux évidences de la constitution, ... c'est que le champ transcendantal a cessé d'être seulement celui de nos pensées, pour devenir celui de l'expérience entière, c'est que Husserl fait confiance à la vérité *dans* laquelle nous sommes de naissance et qui doit pouvoir contenir les vérités de la conscience et celle de la Nature. (*S*, 223–24)

This truth we inhabit from our birth onward – is it not *more* the truth of nature, of *true* nature, than the truth of consciousness?

Phenomenology, understood and practiced as a philosophy of consciousness, has to transform itself by giving to "l'être naturel, le principe 'barbare' dont parlait Schelling" its due place (cf. *S*, 225). This transformation reveals itself most clearly in the change of meaning Merleau-Ponty attributes to constitution itself: "Projet de possession intellectuelle du monde, la constitution devient toujours davantage, à mesure que mûrit la pensée de Husserl, le moyen de dévoiler un envers des choses que nous n'avons pas constitué" (*S*, 227).

This "envers" is exactly their naturalness. "Nature" expresses and epitomises the awareness that "there is" ("*il y a*"): a medium "où il peut y avoir l'*être* sans qu'il ait à être posé" (*VI*, 267). This is why Merleau-Ponty interprets Husserl as rediscovering "le sensible comme forme universelle de l'être brut" (*S*, 217), and this is why perception of natural things is a privileged part of original experience.

The true foundation is not the act of perception but the perceived itself: "Le propre du perçu: être déjà là, n'être pas *par* l'acte de perception, être la raison de cet acte et non l'inverse" (*VI*, 272). Man is nature's own way of freeing its own "magic" knowledge: "La Nature ne sait pas par science, elle

sait par son être-même. Elle a un savoir magique. Le sens n'est pas sans
l'homme, mais il n'est pas rien sans lui";[11] "L'homme est le devenir conscient
de la productivité naturelle" (*RC*, 107). We ourselves are, with all our theoretical and practical activities, part of Nature's self-expression. It is true that
nature's productivity is both oriented and blind; blind as opposed to what
Merleau-Ponty calls "la téléologie proprement dite, comme conformité de
l'événement à un concept" (*RC*, 117). But this does not exclude teleology
in another sense, harder to define precisely because it expresses the very
mystery of Being:

L'irrélatif, désormais, ce n'est pas la nature en soi, ni le système des saisies de la conscience absolue, et pas davantage l'homme, mais cette "téléologie" dont parle Husserl,
qui s'écrit et se pense entre parenthèses – jointure et membrure de l'Etre qui s'accomplit
à travers l'homme. (*S*, 228)

Being is not "nature en soi," it is at the same time "monde sauvage" and "esprit
sauvage" ou "esprit brut": "Ce renouveau du monde est aussi renouveau de
l'esprit, redécouverte de l'esprit brut qui n'est apprivoisée par aucune des
cultures, auquel il est demandé de créer à nouveau la culture" (*S*, 228).
    What is the relationship between this "esprit brut" and Nature? I think
that it is necessary to read here, in extenso, a working-note leading up to the
idea of our *Urstiftung*:

La philosophie n'a jamais parlé – je ne dis pas de la *passivité*: nous ne sommes pas des
effets – mais je dirais de la passivité de notre activité, comme Valéry parlait d'un *corps
de l'esprit*: si neuves que soient nos initiatives, elles naissent au coeur de l'être, elles sont
embrayées sur le temps qui fuse en nous, appuyées sur les pivots ou charnières de notre
vie, leur *sens* est une "direction" – L'âme pense toujours: c'est en elle une propriété
d'état, elle ne peut pas ne pas penser parce qu'un *champ* a été ouvert où s'inscrit toujours
quelque chose ou l'*absence* de quelque chose. Ce n'est pas là une activité de l'âme, ni
une production de pensées au pluriel, et je ne suis pas même l'auteur de ce creux qui se
fait en moi par le passage du présent à la rétention ce n'est pas moi qui me fais penser
pas plus que ce n'est moi qui fais battre mon coeur. Sortir par là de la philosophie des
Erlebnisse et passer à la philosophie de notre *Urstiftung*. (*VI*, 274–75)

    The critique of the phenomenology of "Erlebnisse" (*VI*, 235, 275, 293,
296, 307) and of the psychological "bric-à-brac" of concepts, judgments, etc.
(*VI*, 289, 307, 324; *HNN*, 266), of those spiritual things (*VI*, 289), of those
"prétendues 'réalités' psychiques positives" (*VI*, 307), consists in recognizing
these, as well as the objects, the things represented – i.e., *all idealisations of
the Psyché and of Nature* – as "en réalité découpage s'abstrait dans l'étoffe
ontologique, dans le 'corps de l'esprit'" (*VI*, 307).

We recognize the expression, borrowed from Valéry in the text quoted above at length (cf. *RC*, 177; *S*, 21). If "esprit" is defined as "l'autre côté du corps" (*VI*, 312), this means, at the same time, that the body is de-objectified and that spirit needs the body, is really *of* the body: "Il y a un corps de l'esprit, et un esprit du corps et un chiasme entre eux" (*VI*, 313). Analyzing the notion of "chair" which expresses this "chiasme," this "*Ineinander*," Merleau-Ponty finds at its very heart the "il y a": "En précisant l'analyse, on verrait que l'essentiel est le *réfléchi en bougé*, où le touchant *sur le point* de se saisir comme tangible manque sa saisie, et ne l'achève que dans un *il y a*" (*VI*, 313).

This experience of "il y a" (*Es gibt*) is essentially a gift — and, let us say it, a gift of nature, a natural light:

Ce qui est vrai au total, c'est donc qu'il y a une nature, non pas celle des sciences, mais celle que la perception me montre, et que même la lumière de la conscience est, comme dit Heidegger *lumen naturale*, donnée à elle-même. (*PP*, 494)

Reflection itself is a gift of nature:

La réflexion n'est pas absolument transparente pour elle-même, elle est toujours donnée à elle-même dans une expérience, au sens du mot qui sera le sens kantien, elle jaillit toujours sans savoir elle-même d'où elle jaillit et s'offre toujours à moi comme un don de nature. (*PP*, 53)

Reason itself is experience elucidating itself:

Ce qui est donné, c'est un chemin, une expérience qui s'éclaircit elle-même, se rectifie et poursuit le dialogue avec soi-même et avec autrui. Donc ce qui nous arrache à la dispersion des instants, ce n'est pas une raison toute faite, c'est — comme on l'a toujours dit — une lumière naturelle, notre ouverture à *quelque chose*. (*PPCP*, 129)

We therefore do not only reconquer the consciousness of rationality through seeing its contrast with a background of inhuman nature (cf. *PPCP*, 133), we have to realize that all "lumière instituée" finds its origin in the "lumière naturelle" which is "la sourde réflexion du corps sui lui-même" (*VI*, 202), "l'adhérence charnelle du sentant au senti et du senti au sentant" (*VI*, 187; cf. 157).

This identity and difference (identity of identity and nonidentity?) "fait naître un rayon de lumière naturelle qui éclaire toute chair et non pas seulement la mienne" (*VI*, 187). Communication and language itself are founded in, i.e., *within*, this "natural light":

En un sens, si l'on explicitait complètement l'architectonique du corps humain, son bâti ontologique, et comment il se voit et s'entend, on verrait que la structure de son monde muet est telle que toutes les possibilités du langage y sont déjà données. (*VI*, 203; cf. *PPCP*, 150)

I want to draw attention here to a little known report by J.-B. Pontalis of remarks made by Merleau-Ponty only a few months before his death:

La perception, à condition de ne pas le concevoir comme une opération, comme un mode de représentation mais comme doublée d'une imperception, peut servir de modèle, et même le simple fait de voir: "voir, c'est n'avoir pas besoin de former une pensée." M. Merleau-Ponty rappelle qu'à ses yeux, l'ouverture à l'être n'est pas linguistique: C'est dans la perception qu'il voit le lieu natal de la parole.[12]

It is in harmony with this perspective that we have to read passages like the analysis of the cube (*VI*, 255–56). I quote the conclusion, which is most relevant here:

C'est pour ma chair, mon corps de vision, qu'il peut y avoir le cube même qui ferme le circuit et achève mon être-vu. C'est donc finalement l'unité massive de l'Etre comme englobant de moi et du cube, c'est l'Etre sauvage, non épuré, "vertical," qui fait qu'il y a un cube.

Saisir sur cet exemple le jaillissement de la pure "signification" – la "signification" cube (telle que la définit le géomètre), l'essence, l'idée platonicienne, l'objet sont la concrétion du *il y a* .... (*VI*, 256)

Merleau-Ponty does not only convey his interpretation of Husserl but his own views and intentions when he says: "Le logos articulé dérive d'un sens immanent au perçu du monde primordial, de l'*Erfahrungsboden*" (*HNN*, 266); "Husserl cherche un fondement "esthétique" et réhabilite une philosophie de la Nature, une membrure du monde perçu" (*HNN*, 267). We reach here what Merleau-Ponty calls the most difficult point: the bond between "chair" and "idée," between "le visible" and "l'armature intérieure qu'il manifeste et qu'il cache" (*VI*, 195):

Comme la nervure porte la feuille du dedans, du fond de sa chair, les idées sont la texture de l'expérience; son style, muet d'abord, proféré ensuite. Comme tout style, elles s'élaborent dans l'épaisseur de l'être et, non seulement en fait, mais en droit, n'en sauraient être détachées pour être étalées sous le regard. (*VI*, 159–60)

Are language and thought then to be understood as *emerging* (*from*) *within* experience, "chair," nature? Merleau-Ponty certainly has not fully developed his thought on this relationship. However, he clearly raises the question and continues to give some indications of the direction his thinking is taking:

Certes, c'est une question de savoir ... par quel miracle notamment à la généralité naturelle de mon corps et du monde vient s'ajouter une généralité créée, une culture, une connaissance qui reprend et rectifie la première. Mais, de quelle façon que nous ayons finalement à la comprendre, elle fuse déjà aux articulations du corps esthésiologique, aux contours des choses sensibles, et, si neuve qu'elle soit, elle se glisse par des voies qu'elle n'a pas frayées, transfigure des horizons qu'elle n'a pas ouverts, elle emprunte au mystère fondamental de ces notions "sans équivalent," comme dit Proust, qui ne mènent dans la nuit de l'esprit leur vie ténébreuse que parce qu'elles ont été devinées aux jointures du monde visible. Il est trop tôt maintenant pour éclairer ce dépassement sur place .... (*VI*, 200)

Let us remind ourselves also of what Merleau-Ponty says about the true role of reflection:

Si donc la réflexion ne doit pas présumer de ce qu'elle trouve et se condamner à mettre dans les choses ce qu'elle feindra ensuite d'y trouver, il faut qu'elle ne suspende la foi au monde que pour *le voir*, que pour lire en lui le chemin qu'il a suivi en devenant monde pour nous, qu'elle cherche en lui-même le secret de notre lien perceptif avec lui, qu'elle emploie les mots pour dire ce lien prélogique, et non pas conformément à leur signification préétablie, qu'elle s'enfonce dans le monde au lieu de le dominer, qu'elle descende vers lui tel qu'il est au lieu de remonter vers une possibilité préalable de le penser — qui lui imposerait par avance les conditions de notre contrôle sur lui — qu'elle l'interroge, qu'elle entre dans la forêt des références que notre interrogation fait lever en lui, qu'elle lui fasse dire, enfin, ce que dans son silence *il veut dire* .... (*VI*, 61)

This is what philosophy as interrogation is all about: "aménagement, autour du ceci et du monde qui *est là*, d'un creux, d'un questionnement, ou ceci et monde doivent *eux-mêmes* dire ce qu'ils sont ... " (*VI*, 314); "Il faut une connaissance philosophique qui se place à l'intérieur de l'intuition: qui soit la chose même se mettant à se penser." [13]

The legitimate function of the fixation of "invariants eidétiques" is the following: "mettre en évidence l'écart entre eux et le fonctionnement effectif, et nous inviter à faire sortir l'expérience même de son obstiné silence ... " (*VI*, 71).

We are here reminded of the constant leitmotiv borrowed from Husserl: "c'est l'expérience ... muette encore qu'il s'agit d'amener à l'expression pure de son propre sens." [14]

All modes of expression, including philosophy itself, are called forth by the perceived world, i.e., by nature:

Le monde perceptif "amorphe" dont je parlais à propos de la peinture — ressource perpétuelle pour refaire la peinture — qui ne contient aucun mode d'expression et qui pourtant les appelle et les exige tous et re-suscite avec chaque peintre un nouvel effort d'expression — ce monde perceptif est au fond l'Etre au sens de Heidegger qui est plus que toute peinture, que toute parole, que toute "attitude", et qui, saisi par la philosophie dans son universalité, apparaît comme contenant tout ce qui sera jamais dit, et nous laissant pourtant à le créer (Proust) (*VI*, 223–24; cf. 251).

Our philosophy of the *Lebenswelt* which certainly is our construction and our creation, the thematizing of experience, is at the same time "produit extrême du monde" and "plus que simple produit partiel du *Lebenswelt*." But although it is true that it is this philosophy which reveals the *Lebenswelt* "comme Etre universel," the latter comprehends the first rather than the other way round:

En un sens il (*Lebenswelt*) est encore impliqué comme non thématisé par les énoncés mêmes qui le décrivent: car les énoncés comme tels vont être à leur tour sédimentés, "repris" par le Lebenswelt, seront compris en lui plutôt qu'ils ne le comprennent — sont déjà compris en lui en tant qu'ils sous-entendent toute une *Selbstverständlichkeit*. (*VI*, 224)

This is why philosophy is, for Merleau-Ponty, de-possession rather than possession:

l'idée du *chiasme*, c'est-à-dire: tout rapport à l'être est simultanément prendre et être pris, la prise est prise, elle est inscrite et inscrite au même être qu'elle prend. A partir de là, élaborer une idée de la philosophie: elle ne peut être prise totale et active, possession intellectuelle, puisque ce qu'il y a à saisir est une dépossession. (*VI*, 319)

This de-possession is inaugurated once and for all in the recognition of the naturalness, the belonging to Nature, not only of everything perceived, but of perception itself, not only of everything expressed, but of expression itself:

Ce n'est pas nous qui percevons, c'est la chose qui se perçoit là-bas — ce n'est pas nous qui parlons, c'est la vérité qui se parle au fond de la parole — Devenir nature de l'homme qui est le devenir homme de la nature. (*VI*, 239)

Do Husserl and Valéry express the two sides of what Merleau-Ponty calls "la réversibilité qui est vérité ultime"? Or do they express the same movement, and is the apparent opposition only due to the two different meanings given to the word language?

En un sens, comme dit Husserl, toute la philosophie consiste à restituer une puissance de signifier, une naissance du sens ou un sens sauvage, une expression de l'expérience par l'expérience qui éclaire notamment le domaine spécial du langage. Et en un sens, comme dit Valéry, le langage est tout, puisqu'il n'est la voix de personne, qu'il est la voix même des choses, des ondes et des bois. (*VI*, 203–4)

It is true that in projecting the work he prepared, Merleau-Ponty indicated two distinct parts, one dealing with Nature, the other with Logos (cf. *VI*, 328, 10, 11): "Redécouverte de la φύσις, puis du λόγος . . . . " These parts would have been "total parts," "parties totales" (cf. *VI*, 177, 271), recognizing in Nature and Logos just such total parts or "éléments" (cf. *VI*, 184, 320; *S*, 239) of Being. However, a definite ontological primacy is given to Nature and our speaking, as questioning, is listening to Nature's silent self-expression: "la philosophie est la reconversion du silence et de la parole l'un dans l'autre" (*VI*, 171).

The silent awareness of the *naturalness* of everything perceived is itself the very root, foundation, and true beginning of all conception. Although, of course, all somewhat developed conception is construction of meanings expressed by us, no exclusively constructivist comprehension of knowledge is possible. We are able to construct, conceptually and otherwise, only because we perceive things *as natural* — which means that our perception *is always already conception*.

Nature is thus at the same time perceived and conceived, not as a thing, but in all things as what makes them things and really present. To realize this is to return to "la perception brute et sauvage," and the deconstruction (of our culturally overdetermined perception) which is involved in this realization is the recognition of *the original identity of perception and conception*, of "l'unité de l'expérience (perceptive)" which is "la compréhension originaire du monde (naturel)."

The originality of Merleau-Ponty's concept of nature consists in the retrieval of this identity, and it is because of this identity that Merleau-Ponty can say that nature calls forth our *total response* and that within the calling forth of this response — perception and conception — of one "sujet charnel" *all others* are called upon: "La Nature est cette chance offerte à la corporéité et à l'intersubjectivité" (*HNN*, 267).

University of Ottawa

NOTES

[1] *Phénoménologie de la perception*, p. 340, my italics; (cited as *PP*). Other abbreviations used are: *La structure du comportement* (*SC*); *Sens et non-sense* (*SN*); *Résumés de cours*

(*RC*); *Le primat de la perception et ses conséquences philosophiques* (*PPCP*); *Le visible et l'invisible* (*VI*); *Signes* (*S*); 'Husserl et la notion de Nature,' *Revue de Métaphysique et de Morale*, 1965, pp. 257–69 (*HNN*). The extracts from Maurice Merleau-Ponty, *Le visible et l'invisible* are © 1964 by Éditions Gallimard.

2 *PP*, 377, my italics.

transcendantale, Phaenomenologica 36 (The Hague: M. Nijhoff, 1971), pp. 43ff.

4 This comprehension is to be distinguished from the comprehension by which our body "incorporates," through the acquisition of "habit," a new hat, a car, a blind man's stick, a typewriter, a musical instrument, etc. (cf. *PP*, 167ff.). What takes place *there* is the integration of a new functional milieu (*Umwelt*).

5 Cf. 'Sens perçu, profondeur et réalité dans la Phénoménologie de la perception,' *Studi Filosofici*, 1982.

6 'Le retour à l'expérience perceptive et le sens du primat de la perception,' *Dialogue*, December 1976, pp. 595–607.

7 Cf. *PP*, 451; *S*, 218. For Husserl, see e.g., *Ideen*, vol. 1, pars. 37, 93, 95, 117, 119.

8 Cf. Th. Geraets, pp. 31ff.

9 Lecture notes, taken in 1956–57 by Madame Michèle Jalley-Crampe.

10 *S*, 224. Cf. P. Ricoeur, *Esprit*, December 1953, p. 837, cited by M. Dufrenne, Maurice Merleau-Ponty, in *Les Etudes philosophiques*, 1962, pp. 91–92.

11 Lecture notes (Jalley-Crampe). Cf. *RC*, 107.

12 *L'inconscient*, ed. Henry Ey (Paris: Desclée de Brouwer, 1966), p. 143.

13 Lecture notes (Jalley-Crampe, 1956–57).

14 *PP*, x; cf. *Husserl*, Cahiers de Royaument (Paris: Minuit, 1959), p. 157, and O. E. 225. I intend to question, in another article, the interpretation of the use of this quote by Merleau-Ponty as proposed by J. Taminiaux in 'L'expérience, l'expression et la forme dans l'itinéraire de Merleau-Ponty,' *Le regard et l'excédent* (The Hague, 1977), pp. 90–115.

YNHUI PARK

MERLEAU-PONTY'S ONTOLOGY OF THE WILD BEING*

I

In *The Visible and the Invisible*, an incomplete but powerful book, written, as one of his commentators pointed out, "in a poetico-ontological incantatory style,"[1] Merleau-Ponty intended to fulfill, in his own words, "the necessity of a return to ontology,"[2] that is, "the necessity of bringing them [the results of *The Phenomenology of Perception*] to ontological explicitation."[3] According to him, "*The Phenomenology of Perception*, which can be considered as psychology, is in reality an ontology."[4] Given these words, we must consider Merleau-Ponty's entire philosophical project as essentially ontological in orientation. Continuity between the above two works must be found.

The object of *The Phenomenology of Perception* was to establish an epistemological foundation for our ontological beliefs, whereas the intention of *The Visible and the Invisible* is to show the nature of the reality we believe in on a phenomenological basis. Indeed, there is a continuity between these two projects. The epistemological phenomenology in the former work already implies its ontological thesis, and the ontological thesis in the latter is inevitably delayed by lengthy and repeted phenomenological descriptions. Merleau-Ponty's ontology thus becomes a phenomenological ontology or a phenomenology of Being. Insofar as the knower and the known are insepa-rably connected to one another in the act of knowing, it is indeed logically impossible to separate phenomenology as an epistemology from ontology as theory of Being. They are two sides of the same coin. The difference between them is a matter of focus.

The purpose of this paper is to elucidate Merleau-Ponty's ontology. But what is ontology? At least some preliminary clarification of the very concept of ontology is in order. Ontology literally means "study of Being," and it is usually not distinguished from metaphysics defined as an inquiry into the ultimate reality. However, it is impossible to accept this identification since the Heideggerian deconstruction of Western metaphysics. Heidegger's ultimate concern is the Ultimate Being, whereas according to Heidegger, metaphysics has been concerned only with beings or entities within the Being,

313

A-T. Tymieniecka (ed.), Analecta Husserliana, Vol. XVI, 313—326.
Copyright © 1983 by D. Reidel Publishing Company.

not with the Being as such.[5] Although nowhere does Merleau-Ponty make clear the concept of ontology, let alone the distinction of ontology and metaphysics, as will become apparent, Merleau-Ponty's ontology is intended to inquire not into beings but into Being as such, the "wild Being." If we follow another Heideggerian distinction between the ontic which is the subject matter of metaphysics dealing with entities or beings, on the one hand, and the ontological which is the inquiry into the Being as such, on the other, the Merleau-Pontian ontology is neither metaphysics nor the ontic, but the ontological. Within analytic philosophy one often talks about "ontology of art," "ontology of drama," for instance. Ontology in this sense amounts to a definition of certain objects. It is obvious that Merleau-Pontian ontology is entirely different from analytic ontology.

Since *The Visible and the Invisible* many studies on Merleau-Ponty's philosophy focusing particularly on his ontology have been published.[6] Although they are illuminating and enriching, they are not sufficiently systematic and clear to provide us with a coherent and transparent picture of Merleau-Ponty's ontology. The burden of this paper is to fill this gap in the studies of Merleau-Ponty's philosophy, to complement, rather than reject, existing scholarship.

Traditionally there are three ontological controversies: between realism and nominalism; between monism and pluralism; and between idealism and materialism. In order to relate Merleau-Ponty's ontology to these different traditional ontologies and thus to make it more intelligible, I shall place Merleau-Ponty's ontology in the context of these traditional controversies, which will put in relief the originality and the value of Merleau-Ponty's phenomenological ontology.

II

A. *Aesthetic Realism or Perceptual Faith*

One of the fundamental questions in ontology lies in the controversy between realism and nominalism. Traditionally this controversy centers on deciding the nature of universals as opposed to particulars. The entity denoted by a general term "man," for example, is said to be a universal, and concrete different men are said to be the particulars of this particular universal. Realism believes in the reality of this universal, which nominalism denies. Nominalists hold that this universal is merely a name, not a reality. Typical nominalists include Quine and Goodman.

The nominalist's argument is epistemological. In order for something to be asserted as existing, that is, as real, it must be known; in order for it to be said to be known, it must be stated; in order for it to be so stated, it must be expressed in a language. Reality cannot be separable and independent from a language which represents it. It is a language which determines reality. Thus, in the famous Quinean words, "to be is to be the value of a [linguistic] variable." [7] It is along this line of thinking that Whorf has argued that different linguistic communities have different worlds, that more recently Rorty proclaims that "the world [is] well lost," and that recently Goodman had talked about "ways of worldmaking." [8] Nominalism thus understood is a Berkeleyian idealism under a linguistic disguise. It is absurd to accept that Hopi Indians and Yankees have different worlds, and that we make many different worlds. In referring to the above Quinian nominalist formula, Butler convincingly argues:

Since we do not wish to take the Quinian formula to mean that all existents necessarily depend on the existence of linguistic entities, we must take it to mean that to be is to be in some sense *capable* of being a value of a variable. A value of a variable is what would be named by a constant truly or falsely substitutable for the variable. But if the predicate is truly predicated, the entity of which it is predicated must have properties, in particular the property connoted by the predicate. And if the predicate is falsely predicated, the entity again must have properties, namely properties other than what is connoted by the predicate. Thus to be a potential value of a variable is to have properties. [9]

Contrary to the nominalist assumption, what there is cannot be identified with what is said or described.

The difficulty in nominalism stems from the confusion between what is known and what there is, between epistemology and ontology. Although what there is cannot be asserted without being said and described, that is, known, they are not identical. Berkeleyian ontological idealism is one of the best examples of such a confusion. It is due to the fact that the relation between language and reality, between epistemology and ontology is seen only from the linguistic, epistemological point of view.

For Merleau-Ponty the ultimate starting point is "perceptual faith." There are things and the world related to our linguistic representation of them, but separable and distinguished from it. Reality does not coincide with what is represented or described. It is ontologically prior to language even though it is epistemologically ulterior to its linguistic representation. "The words, vowels and phonemes are so many ways of 'singing' the world." [10] Language does not construct the world, or make things, but it is rather grounded in them. Merleau-Ponty unequivocally argues for the restoration of "the world

as a meaning of Being absolutely different from the 'represented,' that is, as
the vertical Being which none of the 'representations' exhaust and which all
'reach,' the wild Being."[11] The "perceptual faith" affirms this: "We see the
things themselves, the world is what we see."[12] What we see and distinguish
in the world are not what we construct by language. On the contrary "it is
the things themselves, from the depth of their silence, that it [reflection]
wishes to bring to expression."[13] The intelligibility of the world does not
originate purely from our epistemological consciousness superimposed over
the world. In reality "the only pre-existent Logos is the world itself."[14] For
Merleau-Ponty there are things and the world prior to their linguistic representations. What there is is thus really real, not nominally "real." Therefore
the task of philosophy is not merely to analyze concepts or linguistic practice,
but to ask "of our experience of the world what the world is before it is
a thing one speaks of and which is taken for granted, before its has been
reduced to a set of manageable, disposable significations."[15] Clearly Merleau-
Ponty is an ontological realist.

Merleau-Ponty's reality is, however, neither Platonic nor scientific. It is
neither ideal nor physical in character. What is real and exists is neither
Platonic-Husserlian idea-eidos nor the scientists' molecules. For Merleau-
Ponty both of them are the unreal products abstracted from reality which
cannot be reduced either to ideality or to physicality. The Platonic ontology
recognizes the impossibility of identifying as real particular objects independent of our perception of them, and emphasizes the intelligibility,
that is, the meaningfulness of them. As a result it forgets its perceptual
root. Scientific ontology goes in an opposite direction. It is oblivious of
the constituted dimension of molecules. Against the Platonic realism Merleau-
Ponty calls for "the ontological rehabilitation of the sensible,"[16] and in
opposition to scientific realism he writes: "A thing is . . . not actually given
in perception, it is internally taken up by us, reconstituted and experienced
by us insofar as it is bound up with the world, the basic structure of which
we carry with us, and of which it is merely one of many possible concrete
forms."[17] What is necessary for the Platonic realism is to recognize that "that
which is called an idea is necessarily linked to the act of expression,"[18] and
that since all acts of expression involve particular concrete media, "there is no
one rational truth which does not retain its coefficient of facticity."[19] What
is crucial for scientific realism is to admit that "the natural world . . . is the
scheme of intersensory type-relations."[20] Ultimately the only real things
are the perceptual world, the realities as perceived, whose meanings can
be elucidated by phenomenological descriptions. The perceptual world is

neither mental nor physical; perceived realities are neither transcendental nor empirical. It is the world which is irreducible to either ideality or materiality. And the only certain thing is the world. Thus, "to ask oneself whether the world is real is to fail to understand what one is asking, since the world is not a sum of things which might always be called into question, but an inexhaustible reservoir from which things are drawn."[21] Merleau-Ponty's ontological realism, which is neither transcendental-idealistic nor scientific-physicalistic, can be characterized as "aesthetic" realism, which at once transcends and integrate two traditional realisms into a single whole.

B. *Dynamic Monism or Chiasm*

The aesthetic realism of Merleau-Ponty by itself does not settle one of the classical questions of ontology concerning the controversy between monism and pluralism. To be sure, we see many things. There are Peter and Jane, trees and mountains, humans and animals, chairs and pencils, artworks and pizzas, hands and legs, women and men. Reality seems to be legion, and Leibniz's monadology is a philosophical theory of our commonsensical view of the plurality of reality. But it is not easy to swallow the preestablished harmony of the infinite number of windowless monads. It is also a commonsensical view that Peter and Jane, men and women are the same human being, and that trees and mountains, artworks and pizzas, Peter's body and Jane's body are things, and finally that there is our consciousness to which we seem to have immediate access, without which even the very denial of its own existence seems to be contradictory. It seems that if ontological pluralism is easily refuted, ontological dualism is inevitable. If Leibniz is wrong, Descartes must be right.

One of the central features of Merleau-Ponty's philosophy is the rejection of any sort of reductionism. He fights against the reduction of mind to body or body to mind, against the reduction of physical things to living things or living to physical things, against the reduction of color to its component molecules or molecules to its color, language to its semantical content or semantical content to language. Contrary to Grene's thesis,[22] Merleau-Ponty's rejection of reductionism is intended as a refutation of pluralism rather than as a defense. His rejection consists in denying the existence of any clear-cut ontological boundaries among all the things we categorize as *T1, T2, T3,* and so on, which he believes is presupposed in reductionism. It is his philosophical intention "to mix up all our categories"[23] in order to liberate reality artificially divided and thus reduced to certain linguistic categorial prisons.

If Merleau-Ponty can successfully reject ontological pluralism, is he also able to erase the seemingly radical distinction between consciousness and thing, between mind and body, between the knower and the known? In short can he overcome ontological dualism? The hardest and most central ontological task in Merleau-Ponty is to combat this dualism.

The burden of *The Phenomenology of Perception* was to fight against the epistemological split between what he calls empiricism and intellectualism, which he believed to be derived from ontological dualism. Merleau-Ponty's goal was thus to fight against the purity of the Cartesian ego and against the irremediable split of the Sartrean *pour-soi* and *en-soi*.[24] We witness once again the same fight in the major portion of *The Visible and the Invisible*, particularly with reference to Husserl and Sartre. The transcendental field of the subject is not separate from and independent of the empirical field. "It is by the combination of words," Merleau-Ponty writes, "that I *form* the transcendental attitude, that I *constitute* the constitutive consciousness."[25] Against the Sartrean conception of consciousness as nothingness, he writes: "The visible can . . . fill me and occupy me only because I who see it do not see it from the depth of nothingness, but from the midst of itself; I the seer am also visible."[26] The ego is intrinsically opaque. The transcendence of consciousness is intertwined with the concrete world. It is necessary to admit the wordliness of subject and the invisibility, that is, the intelligibility of the visible, that is, the empirical. There is no ultimate ontological, that is, real separation between consciousness and object, ego and the world, which are absolutely distinctive only from the epistemological point of view. Reality is then ultimately *One*. The difference among the things we observe is only the difference of their respective dimensions of the same indivisible Being. Merleau-Ponty's ontological problem was to ask, as Grene put it, "what it means to be *before* all the neatness of conceptual order has been contrived."[27] Merleau-Ponty is definitively an ontological monist.

This brings Merleau-Ponty's ontology into an inevitable comparison with Spinozist monism. The one and single Spinozistic Being, whether called God or Nature, is substance. It is something transparent, abstract, rational, serene, placid, civilized, constant, and stable for ever. For Merleau-Ponty, Being is not a substance in the way in which consciousness and things are not a permanent stuff which would subsist beneath the turmoil of phenomena. In Merleau-Ponty, as Heidieck puts it, "the ontology of substance is rejected."[28] Being is, metaphorically speaking, alive. It is not something stable as opposed to something changing, something essential as opposed to existential. For this such distinction has no meaning. Being is everything and everything is Being with its turmoil and chaos as well as infinite complex of others, with its "system of equivalence" as well as relations among themselves. The seer and the seen, the invisible and the visible, consciousness and object, the physical order and the vital order, the touching hand and the touched are altogether in constant chiasm. There is a movement of criss-crossing taking place. "Through this criss-crossing of the touching and the tangible, its own movements incorporate themselves into the universe they integrate."[29] Every relation with Being is *simultaneously* a taking and a being taken."[30] Chiasm is nothing other than this relation with the Being. The Merleau-Pontian Being is thus more Hobbesian than Spinozistic, more like a musical performance than a intact Greek sculpture. It is more Heraclitean than Parmenidean.

Insofar as it is indivisible and undifferentiated, this Being, as the Taoist, Lao-Tzu, told us, ceases to be what it is as soon as it is named. Therefore, as Heidegger writes, "if man is to find his way once again into the nearness of Being he must first learn to exist, in the nameless,"[31] and yet, the Merleau-Pontian Being is not exactly like either the Taoist's Being, that is, *Tao* or the Heideggerian Being. With the notion of "call" of Being, and with the idea that the Being uses *Dasein* as its tool, the Heideggerian Being appears to be contaminated by the Western teleological attitude or by a certain anthropomorphism. Being seems to be seen somehow as analogous to a kind of ur-person. Although the Heideggerian notion of *Offenheit* of the Being in and through human beings makes one think that the Merleau-Pontian Being is very Heideggerian, the latter seems to be more naturalistic, and nonteleological in character. For this reason the Merleau-Pontian Being seems closer to the Taoist's Being than to any other Western conception of Being. In spite of the absence of either theocentrism or anthropomorphism in both ontologies, the Taoist's Being, that is, *Tao*, is, in the Taoist's own expression, smooth like water or like woman, whereas Merleau-Ponty's Being is in the state of a whirlwind, or a vortex in depth on its layers, levels, articulations. If the Taoist *Tao* is comparable to a serence and simple Japanese melody, the Merleau-Pontian Being is comparable to a Wagnarian opera; if the former calls for an analogy with a Chinese monochrome Zen painting, the latter calls for an analogy with a tortously articulated Cezannian landscape. The latter is stormy while the former is transparent. With its depth and criss-crossing, Merleau-Ponty's monism can be characterized as "dynamic."

C. *Carnal Idealism or Flesh*

To say that Being is real rather than not nominal, and that it is not many but one is to say nothing about its nature. But our characterization of Merleau-Ponty's realistic monism as both "aesthetic" and "dynamic" already suggests the nature of Being. Our ontological question here is, what *kind* of reality is the aesthetic and dynamic Being?

There are many things in the world. Ultimately these things are reduced to either matter or mind. The concepts of "matter" and "mind" are the most universal, the most general categories of reality we can have, and think of. What can anything or stuff be if it is neither material nor mental? How could anything be ultimately described if it is not characterized either as "material" or "mental." It seems to follow that Being, whether aesthetic or dynamic, must be either material or mental, insofar as it is something, and as soon as it has to be described. Therefore, even if an ontology has overcome nominalism and pluralism, and even if it has overcome certain difficulties in classical realism and monism, it has to decide between ontological materialism and ontological mentalism, which is more generally called "idealism."

Typical materialism is phenomenologically not true. Insofar as it is a thesis, a belief, conceptually it faces a logical impossibility. To assert that materialism is true is to presuppose at least one certain mind which is aware of it, and asserts its truth. In the mind-body issue, physicalist reductionism is an impossible dream. If there are only two alternatives to materialism and idealism in ontology, and if materialist ontology is unacceptable, it appears that the only conclusion we can draw is that ontological idealism is true.[32]

The typical monistic idealists are the Buddhist, Berkeley, and Hegel. For both the Buddhists and Berkeley there is only mind. The former argues that the so-called nonmental phenomena, the appearance in the Platonic sense, are mere illusions. The really real is nothing but the undifferentiated stream of consciousness or mind. For Berkeley material things *are* in fact *ideas*. They seem here to be asking us to regard physical things *as* mind. They are proposing a dissolution of the problem instead of the solution of it, in transforming everything into ideas just like Midas transformed whatever he touched into gold. Yet philosophical things are more stubborn than Midas' things. Material things must be dealt with more seriously. Indeed, Hegel was more cautious. Hegelian idealism does not treat material things as illusions or linguistic fabrics. They are real, although they are conceived of as the manifestions of Spirit. They are not different from Spirit in the way in which the mentalist conception of person tends to think that body is secondary to soul. Thus Hegel's idealism does not do justice to the phenomenal world, and Hegel remains an idealist not fundamentally different from other idealists. And, as we have seen, idealism is unsatisfactory.

Neither materialism nor idealism is acceptable. However, if there are no other possible ontologies, we are in a total impasse. But, are materialism and idealism really only two possible alternatives? The whole point of Merleau-Ponty's ontology is to show that that is not true, that there is an alternative other than the traditionally available two. For Merleau-Ponty, Being is neither material nor ideal in kind, but "fleshy." The concept of "flesh," which is essential in Merleau-Ponty's ontology, and which is alien to traditional ontology and philosophy in general, becomes intelligible and illuminating when we realize that the ontological issue between materialism and idealism crystallizes into the mind-body problem, for it is primarily in the relation between mind and body that the ontological problematic of the dichotomy between materialism and idealism is found, and it is through the most immediate lived experience of body that the problematic relation between mind and matter arises. The concept of "flesh" in Merleau-Ponty's ontology is intended to show precisely the un-split reality of our bodily experience. It is something unitary which at once transcends and blends both mind and body, both the subjective knower and the objective known, both the transcendental ideality and the empirical materiality. Flesh is ontologically prior to body as a physical thing as well as to mind construed as ideality. Thus, flesh is rather "a general thing midway between the spatio-temporal individual and the idea, a sort of incarnate principle that bridges a style of being wherever there is a fragment of being."[33] Merleau-Ponty writes: "It [flesh] is the calling over of the visible upon the seeing body, of the tangible upon the touching, which is attested in particular when the body sees itself, touches itself seeing and touching the things."[34] But here we must not be misled by the term "flesh," which is, in ordinary usage, related to living beings, which are only part of Being. The concept in question in Merleau-Ponty's ontology does not refer only to the biological domain. It is an ontological concept in the Heideggerian sense. It denotes the nature of Being as a whole. Thus Merleau-Ponty talks about "the flesh of the world,"[35] and also writes: "Between the alleged colors and visibles, we would find anew the tissue that lives them, sustains them, nourishes them, and which for its part is not a thing, but a latency, a *flesh* of things."[36]

From this Merleau-Pontian ontological perspective matter and mind, consciousness and thing, subject and object are merely abstractions of the flesh of Being. The distinction between mind and body, idea and thing is merely a product of epistemological necessity, for without such a distinction the very notion of knowledge becomes unintelligible. But this epistemological split, though necessary, does not entail any ontological division. Traditionally, as we have said earlier, matter and mind, which, transferred to a region of Being, that is, to a living human body, correspond to body and mind, are the most universal and general ontological categories both in philosophy and ordinary thinking. In place of such categories, which for Merleau-Ponty are only linguistic and epistemological, he introduces the concept of "flesh" as a concept which ontologically underlies and unifies the existing concepts. To be sure, the characterization of the ultimate nature of Being as "flesh" is metaphorical, but often a metaphor is more illuminating than a nonmetaphorical expression. In the case of Merleau-Ponty this is more than true. To say that in our vocabularies "matter" and "mind" are the most general ontological categories, and to say, nonetheless, with Merleau-Ponty that the ultimate nature of Being is "fleshy" seems to imply that Being as flesh cannot be adequately described and thus made intelligible in our traditional available philosophico-ordinary language. In simpler words this means that Being is not represented, that it is, as Lao-Tzu and Heidegger insist, nameless. Thus, for Merleau-Ponty, Being could not be given to us *as ideas* except in a carnal experience."[37] It is for this reason that Merleau-Ponty thinks that art, painting in particular, better represents the way Being is than scientific and linguistic representation does.

The ontological concept of "flesh" is intended precisely to overcome the conflicting classical categories of matter and mind. It is intended to free ontology from the traditional perspective altogether. Yet, it seems to be useful to locate and relate this "fleshy" ontology in and to the classical debate. Recently Margolis proposed what he calls "emergent materialism"[38] in order to account for the ontology of the person against both physicalistic reductionism and the dualistic-idealistic account of person. It is an ontological thesis, which is monistic and which is intended to overcome dualism, idealistic monism, and materialistic monism, all at once. Margolis' ontology looks, then, similar to, if not identical with, Merleau-Ponty's ontology. Yet, as his term "emergent materialism" suggests, his ontology seems to be closer to materialism than to idealism. If his Being is neither materialistic nor idealistic, why not call his ontology "emergent idealism"? The reason for his choice of "emergent materialism" rather than "emergent idealism" is his desire to square his ontology with scientific achievements, which he believes overwhelming and unshakable. It is to justify the scientific description of reality without accepting the physicalistic reductionism.

By contrast Merleau-Ponty's ontology can better be called "carnal idealism" instead of "carnal materialism." It is carnal because it is neither material nor mental, but rather "fleshy": it is closer to idealism than to materialism. One of the focal points of Merleau-Ponty's philosophy is to combat a materialistic ontology which he believes implicit in science. However, his rejection of materialism does not entail an idealism in the classical sense of the term. In order to avoid an idealistic interpretation, it seems appropriate to characterize his sort of idealism as "carnal," precisely to the extent that he sees the nature of Being as "flesh" before its conceptual split into matter and mind.

III

In characterizing Merleau-Ponty's ontology at once as "aesthetic realism," as "dynamic monism," and as "carnal idealism," I have tried to relate Merleau-Ponty's ontology to different classical ontologies. One of the possible objections to this approach to Merleau-Ponty's ontology is that I have misunderstood the very point of Merleau-Ponty's philosophy, which consists in going beyond the classical framework of ontology, and thus introducing a radically new way of thinking. In putting Merleau-Ponty's ontology into the classical categories of ontology, and thus in trying to interpret his ontology in the light of and according to the classical categories of ontology, I would have misunderstood and distorted Merleau-Ponty's ontology, even if I have qualified it as being "aesthetic," "dynamic," and "carnal." Such an objection would be misplaced. If Merleau-Ponty's ontology has been examined in relation to the classical ontologies, it is not to judge and evaluate its adequacy, but only to highlight the originality of Merleau-Ponty's phenomenological ontology and at the same time the inadequacy of the classical ontologies.

The entire project of Merleau-Ponty's ontology, which is in the last analysis identical with his entire philosophical project, is to show that Being in general is one single indivisible whole, and that, as a consequence, Being can never be describable in language. This ontological and linguistic view is exactly analogous to the Taoist conception of reality and language, according to which *Tao*, that is, the ultimate reality, is distorted as soon as it is represented by language. We find more or less similar ideas in Plotinus, Bergson, and Heidegger. On this view everything in the world is ultimately blended and encompassed in linguistic abstractions. Such words as "pre-objective," "prepersonal," "preconceptual," "conceptualness," and "precategorial," which abound in the writings of Merleau-Ponty, are intended to show the gap between reality and its linguistic representation, and the necessity of seeing

reality as it is, that is, prior to its linguistic representation. Hence Merleau-
Ponty's ontology can be seen as a philosophy of language, as a radical critique
of language.

Whether we are aware of it or not, ordinarily and philosophically we also
tend to confuse reality with its linguistic representation, and thus to believe
in the reality of linguistic representations such as "mind," "matter," "tree,"
"man," "ego," "object," "subject," "red," "hot," "atom," "energy," and so
on. We are Quinian nominalists without being philosophers to the extent that
we surreptiously accept the Quinian view that "to be is to be the value of
a variable." The importance of Merleau-Ponty's ontology, which is also a
radical critique of language, lies in the fact that it helps us realize the logical
difference between reality and language, thus corrects our misconception
about the nature of the reality prior to its linguistic distortion. From now on
we can see matter, mind, ego, tree, and so on, truely as they are.

Rorty argues that philosophers traditionally have assumed the possibility
of absolute knowledge, that is, absolute representation of reality or things,
and that the greatness of Wittgenstein, Dewey, and Heidegger's thought
consists in the fact that they have shown the historic-cultural origin of the
traditional problems themselves and their philosophical solutions, in other
words, in having deconstructed our philosophical ideas and beliefs, which
are in fact determined by our culture and language.[39] It is equally from
this perspective that the importance of Merleau-Ponty's ontology, and his
philosophy in general can be evaluated. Merleau-Ponty's ontology is a radical
critique of Western philosophy and the Western conceptual framework in
general.

Merleau-Ponty's ontology boils down to erasing the radical distinction
between consciousness and its object, between the transcendental and the
empirical in blending them into something one and undefinable. Many critics
of Merleau-Ponty point out the internal incoherence in such thinking. For
to say that consciousness and its object are one continuum is to argue that
such a view or fact is true. But as soon as it is presented as true, this claim
implies the distinction between consciousness and its object which it is
conscious of. Thus, Olafson, among many others, mentions "the difficulty
of dropping the concept of the *en-soi*"[40] in opposition to the concept of the
pour-soi. But this criticism seems entirely misplaced. It is based upon the
confusion between what I call ontological continuity and epistemological
discontinuity.[41] The distinction between consciousness and its object must
be radically maintained from the epistemological perspective. But this epistemological necessity does not entail the ontological distinction between

them. Merleau-Ponty's discussion of consciousness, its objects, and their relationship is an ontological discussion. He is arguing that although epistemologically they are distinctive from one another, ontologically they are continuous with one another. The concept of "wild Being" or "savage" in the later writings of Merleau-Ponty is intended to show precisely this ontological continuity among all things we epistemologically, that is, conceptually, divide, although it is necessary to distinguish everything from one another in order to talk not only about these particular beings but also about wild Being itself.

Simmons College, Boston

NOTES

* This paper was written at the NEH Summer Seminar for College Teachers under the direction of Professor Calvin Schrag in 1980 at Purdue University.

[1] Xavier Tilliette, *Merleau-Ponty* (Paris, 1970), p. 88.

[2] Merleau-Ponty, *The Visible and the Invisible*, trans. Alphonso Lingis (Evanston, 1968), p. 105; hereafter cited as *VI*.

[3] *VI*, p. 183.

[4] *VI*, p. 175.

[5] Martin Heidegger, *Basic Writings*, ed. David Farrell Krell (New York, 1976); see in particular "What is Metaphysics?" "Letter on Humanism."

[6] For instance, X. Tilliette, *Merleau-Ponty*; Michel Lefeuvre, *Merleau-Ponty: au delà de la phénoménologie* (1976), Gary Grant Madison, *La phénoménologie de Merleau-Ponty* (1973); François Heidsieck, *L'ontologie de Merleau-Ponty* (1971); Samuel B. Mallin, *Merleau-Ponty's Philosophy* (1979); John Sallis, *Phenomenology and the Return to Beginnings* (1973); Thomas Langan, *Merleau-Ponty's Critique of Reason* (1966), Marjorie Grene, "The Renewal of Ontology," *Review of Metaphysics*, June 1976; Fredrick A. Olafson, "Merleau-Ponty's Ontology of the Visible," *Pacific Philosophical*, January–April 1980.

[7] Willard Van Orman Quine, *From a Logical Point of View* (New York, 1955), p. 13.

[8] Richard Rorty, "The World Well Lost," *Journal of Philosophy*, and see Nelson Goodman, *Ways of Worldmaking* (Indianapolis, 1977).

[9] Clark Butler, "On the Impossibility of Metaphysics without Ontology," *Metaphilosophy*, April 1976, p. 128.

[10] Merleau-Ponty, *The Phenomenology of Perception*, trans. Colin Smith (London, 1962), p. 181; hereafter cited as *PP*.

[11] *VI*, p. 253.

[12] *VI*, p. 3.

[13] *VI*, p. 4.

[14] *PP*, p. xx.

[15] *VI*, p. 102.

[16] Merleau-Ponty, *Signs*, trans. R. C. MaCleary (Evanston, 1964), p. 166.

[17] *PP*, p. 326.

[18] *PP*, p. 390.   .

[19] *PP*, p. 394.

[20] *PP*, p. 327.

[21] *PP*, p. 344.

[22] Grene, p. 618.

[23] Merleau-Ponty, *The Primacy of Perception*, trans. James E. Edie (1964), p. 168.

[24] See *The Phenomenology of Perception*, pp. 383, 377, 403–4, 400.

[25] *VI*, p. 17.

[26] *VI*, p. 113.

[27] Grene, p. 618.

[28] François Heidieck, p. 107.

[29] *VI*, p. 133.

[30] *VI*, p. 266.

[31] Heidegger, p. 199.

[32] See D. M. Armstrong, *A Materialistic Theory of the Mind* (London, 1968).

[33] *VI*, p. 139.

[34] *VI*, p. 146.

[35] *VI*, p. 250.

[36] *VI*, pp. 132–33.

[37] *VI*, p. 150.

[38] See Joseph Margolis, *Persons and Minds* (Boston, 1978).

[39] Richard Rorty, *Philosophy and the Mirror of Nature* (Princeton, 1979), pp. 3–13.

[40] Olafson, p. 175. Sallis talks about "the fundamental ambivalence attached to the phenomenological thought in its return to beginnings, the ambivalence of seeing to remain to the dimension of perceptual experience while, simultaneously, being obliged to maintain a distance from this dimension as they very condition of the possibility of the return" (*PP*, pp. 109–10). Lefeuvre writes: "Il faut remonter au delà et à travers la forme même du langage avec ses paradoxes, dévoiler un Etre opérant inaccessible, transcendent dont nous ne connaîtrons jamais la profondeur négative, heureux seulement d'en saisir les traces dans le renversement de sa négativité du monde qu'il est et aux oeuvres de cultures telles que les hommes les édifient" (*PP*, 371–77). Madison: "La question que Merleau-Ponty pose est en fin de compte cette même question éternelle qui, comme Aristote disait il y a plus de deux millénaires, était posée autrefois, est maintenant et toujours posée et toujours le sujet de notre embarras. L'ontologie de Merleau-Ponty est de la sorte la prise de conscience de l'Etre comme ce qu'est éternellement recherché et éternellement le sujet de notre étonnement, transcendance éternellement insaisissable dans l'immanence."

[41] See my "Merleau-Ponty ou la phénoménologie du sens," *Revue de Métaphysique et de Morale*, juillet-septembre 1979.

THÉRÈSE-ANNE DRUART

IMAGINATION AND THE SOUL–BODY PROBLEM IN ARABIC PHILOSOPHY

> The body, so to speak, is simply the riding-animal of the soul, and perishes while the soul endures. The soul should take care of the body, just as a pilgrim on his way to Mecca takes care of his camel; but if the pilgrim spends his whole time in feeding and adorning his camel, the caravan will leave him behind, and he will perish in the desert.
>
> Al-Ghazali, *The Alchemy of Happiness* [1]

Al-Ghazali's picturesque Middle-East analogy on the relation between soul and body would fit nicely with other analogies, famous in the history of Western philosophy, e.g., that (1) the soul is bound fast in a body as an oyster in its shell (Plato, *Phaedrus*);[2] and (2) the soul is not present to the body in the way a seaman is present to his ship (Descartes, *Meditations*, VI,[3] answering in some fashion Aristotle's query in *On the Soul*, II, 1).[4]

Yet this picturesque analogy, noted by Beatrice M. Zedler,[5] is not the proper starting point for our inquiry into the soul–body problem, since it comes from a classic of Muslim spirituality — not philosophy — whose author wrote a detailed critique of the Arabic philosophers, *The Incoherence of the Philosophers*.[6] Nevertheless, this analogy is revealing about what we will find in some texts. In *The Incoherence of the Philosophers* al-Ghazali denies that the philosophers have proved that the human soul, or intellect, is a spiritual substance, which exists in itself and is not impressed upon a body.[7] Al-Ghazali does not believe that the basic position of the philosophers is wrong but rather that there is no rational demonstration of it.[8] This at once gives us a hint that, according to al-Ghazali, Arabic philosophers have insisted on the independence of the human soul from the body. To understand this position and al-Ghazali's reaction to it, we must first examine its roots in Greek philosophy.

I. ORIGIN OF THE PROBLEM OF THE RELATION BETWEEN SOUL AND BODY AND OF ITS LINK TO THE RELATION BETWEEN INTELLECT AND IMAGINATION

Arabic philosophers often try to claim that Plato and Aristotle hold basically

327

A-T. Tymieniecka (ed.), Analecta Husserliana, Vol. XVI, 327–342.
Copyright © 1983 by D. Reidel Publishing Company.

the same opinion, so that the agreement between philosophers can be viewed as a sign of the absolute truth of philosophy. For instance, al-Farabi (870–950) wrote *The Harmony between the Two Sages,*[9] i.e., Plato and Aristotle, to reconcile their positions and therefore is led to reinterpret them. What do Arabic philosophers, like al-Farabi and Avicenna, know about Plato's and Aristotle's position on the relation between soul and body? Once we have determined this, we can better understand and evaluate their own contribution.

From Plato himself they know little, except the *Republic* and the *Laws* (this last work may be included only up to book IX).[10] The argument for the immortality of the soul in *Republic*, X,[11] implies a complete independence of the soul from the body. The evil of the body cannot affect the soul. The myth of Er, which follows, if taken literally implies that the same soul can take on different bodies in turn. Therefore the soul is considered as the sole aspect of human individuality. The soul is an eternal entity of its own that preexists the body and can use successive bodies. Since the connection between soul and body is purely accidental, transmigration is possible.

On the other hand, at the beginning of *On the Soul* or *De anima* Aristotle claims that there is an intimate connection between soul and body. This is why Aristotle thinks that all activities of the soul require the body, except perhaps intellection. For intellection not to require the body, it must not be a form of imagination or must exist apart from imagination.[12] Insisting on the intimate connection between soul and body, Aristotle goes on to show that the human being is a composite of body and soul. The whole human being, soul and body, is a substance and the soul is not an entity in itself but simply the form of the body.

And for this reason those have the right conception who believe that the soul does not exist without a body and yet is not itself a kind of body. For it is not a body, but something which belongs to a body, and for this reason exists in a body ....[13]

The soul is the actuality of the body: "Hence too we should not ask whether the soul and body are one, any more than whether the wax and the impression are one ...."[14] From this analogy Arabic philosophers develop the habit of speaking of the soul as "imprinted" or "impressed" upon the body.

Aristotle affirms the inseparability of body and soul. Yet in III, 5, after having distinguished an active and a passive aspect in the intellect, he speaks of the whole intellect, or only the active part of it, as immortal and eternal.[15] At the same time, contrary to his assertion at the beginning of the text, he also holds that the soul never thinks without an image.[16] Intellection requires imagination and also memory which is thought by Aristotle to be an aspect of imagination (phantasia). There seems therefore to be an inconsistency in Aristotle's position about imagination. Hence imagination appears to be a key concept to help one to understand the position of philosophers influenced by Aristotle. This is why I wish to focus my approach to the soul–body problem on the conception of imagination in the Arabic *De anima* tradition. As such the *De anima* text in III, 5, is rather cryptic, especially in what concerns the immortality and eternity of the intellect or part of it. The only thing said is that the active intellect acts "like light does; for in a way light too makes colours which are potential into actual colours . . . ."[17] To complete these meager indications and to avoid letting Aristotle contradict himself in affirming both that the human intellect, or part of it, is separate and that it requires images to think, Arabic philosophers following a Greek tradition maintain that the active intellect is not a part or an aspect of the human intellect but rather an immaterial entity, the lowest of the separate intellects in an emanation scheme. Furthermore, they have no qualms in developing Aristotle's analogy of light by borrowing from Plato some aspects of the analogy of the Form of the Good to the sun, since, as I indicated earlier, they insist on Plato's and Aristotle's agreement. This analogy is found in *Republic*, VI.[18] Yet these exegetical acrobatics to resolve the inconsistency of Aristotle's positions in what concerns imagination are not enough to suppress the importance of imagination for the human intellect nor to resolve the puzzle of how one may fit together Plato's conception of the substantiality of the soul and therefore its independence from the body and Aristotle's original insight about the interdependence of soul and body. To face these difficult issues, the best approach is to turn to Avicenna, considered by al-Ghazali as the main representative of the Arabic philosophers, and to al-Ghazali's critique of this position. In each case, we will see that the position concerning the degree of connection between soul and body entails a parallel degree of connection between intellection and imagination. Avicenna will lean in the direction of moderate Platonism in affirming the substantiality of the soul and therefore will also attenuate the connection between intellection and imagination. Al-Ghazali, who criticizes Avicenna's Platonism, will emphasize the necessary connection between soul and body and therefore will maximize the connection between intellection and imagination.

II. AVICENNA'S POSITION (980–1037)

In Avicenna the texts closest to the *De anima* tradition are those of his *Book of Healing*, in the sixth part of its natural treatises [19] and of its "digest," the *Book of Salvation*. [20] Though apparently following an Aristotelian point of view in the *Book of Healing* Avicenna indicates clearly that for him the human soul is a substance since it is a pure form, i.e., a form that is not *in* a substratum. [21] Therefore the human soul is not the form of the body. In a human being body and soul are both substances and the union between them is accidental. [22] Therefore corruption of the body does not entail corruption of the soul. [23] This recalls Plato's view in *Republic*, X. But then how does one explain the relationship between these two substances? On this point Avicenna rejects Plato's conception of preexistence of the soul and argues that the soul needs the body for its individuation just as the body needs the soul to be alive. Therefore "the soul comes into existence whenever a body does so fit to be used by it." [24]

Thus "the body which comes into being is its kingdom and tool." [25] The soul emanates from the first principles whenever some bodily matter suitable to become its body is available. The soul emanates for one particular body. From this it follows that for Avicenna the dualism between body and soul is lessened by the uniqueness of the relation of each soul to its body, since the body is necessary for the individuation of the soul. Therefore, though the human soul is a substance, it cannot preexist before the body, and there is a natural desire in the soul which inclines it to occupy itself with its own body, to use it, to care for it, and to be attracted to it. This natural desire binds the soul to its particular body and keeps it away from all other bodies. [26] The *Book of Salvation* adds that this is why the soul does not contact directly other bodies but does so always through its own body. [27] One's own body appears as the condition for access to other bodies. This natural desire of the soul for its own particular body not only distinguishes this body from all others but also explains why Avicenna rejects transmigration of souls. He seems to have considered transmigration as a serious point in Plato's position because he argues at some length against it, though without referring to Plato. [28] It seems, therefore, that Avicenna, though he defends the substantiality of the human soul, is trying to lessen its consequences in what concerns its relationship with the body. A natural desire links each soul to its own particular body which individuates it. The body is then presented as an instrument for the soul and as attracting the soul.

But is the body a necessary and indispensable tool for all activities of the soul? To clarify this question we have to deal with what we called imagination in Aristotle. Avicenna expands Aristotle's conception of imagination, because he posits a different faculty for each kind of object (i.e., here sensible qualities and intentions) and for each mode of relation to the object (simple apprehension or storing of it or even active combining and separating). Aristotle's imagination (phantasia) is expanded by Avicenna to reach five faculties at the animal level beyond the five external senses. These faculties are (1) the common sense; (2) the retentive faculty, i.e., a kind of memory, which simply stores sensible representations; (3) the faculty called sensitive imagination in animals and rational imagination in men, that combines and separates some of the representations stored in the retentive faculty so that one can form such things as winged horses for instance; (4) the estimative that perceives nonsensible intentions in particulars (for instance, it is by means of the estimative that the lamb "judges" that the wolf is to be avoided because it is inimical); and (5) the recollective, i.e., a kind of memory that retains the nonsensible intentions perceived by the estimative.[29] To simplify matters I shall refer to these five faculties (or inner senses) taken together as "imagination,"[30] since in all these faculties one finds images in a broad sense.

Following Aristotle Avicenna claims that all animal faculties are exercized by means of some organ and, therefore, that all five of the faculties which interest us require a physical organ.[31] This point must be important for Avicenna because he presents detailed arguments to justify this claim at the different levels of what we are calling "imagination." Therefore, it is clear that "imagination" requires the body.

But what about intellection? Intellection does not require the body, first, because what receives intelligibles must be immaterial. Therefore, the substance which receives intelligibles does not subsist in a body in such a way as to be in any sense a faculty residing in, or a form of, that body. The receiving substratum is the intellect or the human soul.[32] Avicenna presents some arguments for this position. These arguments are based on the nonsensible character of intelligibles. Second, intellection does not require the body because it does not require a physical organ,[33] since otherwise the intellect would not understand itself, nor understand that it understands.[34] The intellect understands directly through itself, whereas the senses use an organ and so cannot sense themselves. Avicenna adds that the retentive faculty because it requires a physical organ cannot perceive itself.[35] In other words, Avicenna emphasizes the contrast between sensation and intellection as well as the contrast between "imagination" and intellection. This leads him to claim that the human soul strictly requires neither "imagination" nor the body, since the intelligibles which are the objects of intellection are not abstracted from images.[36] It is why he has to break the progression in what can be called his theory of the degrees of abstraction.[37] The living being is looking for the form *qua* form, i.e., as separated from matter. Sensation does not effect a complete detachment from matter and this is why the presence of the external object is required. The retentive faculty separates the form from matter but not from material accidents, such as location, quantity, etc. But the highest form of "imagination," the estimative faculty, goes still farther and grasps "intentions" which in themselves are nonsensible but happen to be in particulars, i.e., the enmity of this wolf. These three degrees of abstraction are performed by the living being itself. But when one reaches the level of intellection, one sees that the abstraction of the form is perfectly performed but not any more by means of the human being. Strictly speaking the human soul does not abstract forms, i.e., intelligibles. The active intellect does this and communicates these intelligibles ready made to the human soul. Therefore, consideration of images does not help produce the intelligibles, but simply may prepare the soul to receive intelligibles which are given to it by the active intellect. We have to remember that the active intellect is the tenth and last pure intelligence and therefore exists outside of the human mind. The human soul is fundamentally passive. "Imagination" can help it to turn itself to the light of emanation, as a sunflower would turn itself toward the sun. Thus, even in the case of those intelligibles which are separated from matter but are not separate in themselves, images are not really required, since these intelligibles are not abstracted from images but handed over by the active intellect. Avicenna holds that there is a second kind of intelligibles,[38] intelligibles that are separate in themselves, such as God and the pure Intelligences, including the active intellect, and even oneself, i.e., one's soul. Since these intelligibles are in themselves separate from matter and without any relation to it, it follows that consideration of images cannot help one reach them. Therefore "imagination" is not necessary as such for intellection but is a usual preparation for it. Why? Because "imagination" helps in providing particulars. From these result four intellectual processes: (1) abstraction of universals; (2) negation and affirmation about them; (3) discovery of empirical premises; and (4) acceptance of informations because of a long tradition, such as historical and geographical facts.[39] These four processes require "imagination" to train the soul to leave the level of immediate sensation. "Imagination" and therefore the body are required to help the human soul acquire the basic starting points of conception and judgment. But once the human soul has acquired them the human soul returns to itself. Then if any of the lower faculties happens to occupy it, it diverts it from its proper higher activity.[40] Eventually the soul needs to go back to "imagination" from time to time but only to acquire a new starting point. This happens often at the beginning of intellectual life but rarely later on.

At a higher stage "imagination" is of no help at all; on the contrary, it distracts and hinders the mind.[41] Why? Because it seems that the soul can do only one thing at a time, though it has two activities, one in relation to the body, namely, to govern it, and another in relation to itself and its principles, namely, intellection.[42] Unfortunately, these two activities are so opposed and naturally work so much against one another that when the soul is busy with one, it turns away from the other because it cannot easily combine the two.[43] "Imagination" is one function of the soul, but a function in relation to the body since, as we have seen earlier, it requires a physical organ.[44] This inability of the soul to exercize its two activities at the same time explains why in cases of illness and old age the intellect appears to cease performing its proper activities. This apparent influence of the body on the soul is not a proof that the activity of the soul is incomplete without the help of the body.[45] For what happens is simply that in the case of illness the activity of the soul is only suspended. This suspension of activity explains why, once health is restored, the body often fully regains its understanding of all its previous objects.[46] For Avicenna there is no intellectual memory since memory understood as storage would imply a localization and therefore some relation to matter. We must remember that we are trained to turn to the active intellect so that we can receive the emanation of intelligibles at will. We do not store intelligibles; we simply become more and more apt to receive them at will all over again.[47] This again shows that the soul is not "imprinted" in the body, i.e., is not a form, nor subsisting in it. Since the soul in order to intelligize must have immediate contact with intelligibles, "imagination," which can initially help to trigger the process, must later on be rejected since it can only hinder the continuation and development of the process

The soul, therefore, is a substance which is in some way related to a particular body and thus is attracted to it; but in another way it is related to itself and higher beings. How must the human soul deal with these two relationships? First, the soul must use the body as a tool which at the beginning is useful to help the soul train itself to receive intelligibles emanating from the active intellect. Thus "imagination" can help initiate the process of emanation; but once this process is well established, "imagination" becomes a hindrance, a screen between the soul and itself, i.e., its self-awareness and its awareness of the pure intellects. It is clear, therefore, that for Avicenna, though the body is required for the individuation of the human soul and its coming into existence, it is nevertheless not part of the human self. Why? Because Avicenna claims that we do not need the body to reach self-aware-ness. Here he offers the famous "experiment" of the "Flying man" or "Man in the void." A man is suddenly created in the void with his eyes veiled and his limbs extended so that he can neither see, nor touch them. In addition, he cannot hear, since there is no sound. Such a man would receive no impres-sion from the senses, yet according to Avicenna he would know himself.[48] From this "experiment" both Verbeke[49] and Zedler,[50] among many others, conclude, that for Avicenna man is a spiritual substance, i.e., a soul, which is the self, and therefore man is spirit. Self-awareness for Avicenna is immediate and therefore an intermediary can only prevent it. Neither sensation nor "imagination" are required for self-awareness and therefore the body is not part of the self, and the soul must be a substance in its own right.

Now to sum up. For Avicenna man is made up of two substances, body and soul. His real self is the soul which needs the body to get its existence and individuation, but not to reach self-intellection. Therefore, though the soul has a natural attachment to its own body, it must deal cautiously with this natural attachment. At first this attachment, via its more immaterial form, "imagination," helps the mind prepare itself for reception of the intelligibles; but later this attachment, even in its highest form, "imagination," would hinder the soul's highest activities, which require immediacy. Avicenna, with Aristotle, maintains the specificity of intellection, but, against Aristotle's explicit statements, is led to admit not only that intellection is possible without "imagination," but even that "imagination" can hinder it. Thus Avicenna writes:

a man may need a riding-animal and some means to reach a certain destination. But once he has reached this destination and it happens that his mount and means prevent him from separating himself from them, then the very means of arrival become a hindrance.[51]

Since the soul is a substance it does not necessarily need the body nor "imagination." The two latter help the soul to reach awareness but once awareness turns to itself, then body and "imagination" are to be cast away, since for Avicenna awareness of self is immediate and therefore use of an intermediary would simply preclude it. Yet the "Flying man" did not need the body at all to reach this self-awareness. Here again we seem to be con-fronted with an inconsistency. For Avicenna to maintain the substantiality of the soul and of the self as soul limits the use of "imagination" to a necessary preparation for the first steps of intellection because the body is needed for individuation; but, on the other hand, the "Flying man" experiment seems to exclude the need for this first step, because for him self-awareness is immediacy and not self-reflection based on awareness of an object. This is why the "Flying man" has been compared to Descartes' approach to the *Cogito*[52] and why both Avicenna and Descartes have problems in dealing with the relationship between body and soul.

III. AL-GHAZALI'S POSITION (1058–1111)

Near the end of his life, Al-Ghazali wrote his intellectual biography in which he evaluates the respective merits of Islamic theology, philosophy, a form of Shi'ism, and finally mysticism. In this biography he devotes considerable space to dealing with philosophers, whom he considers particularly dangerous. Philosophers, according to him, disagree and therefore cannot claim, as al-Farabi and Avicenna do, to have reached *the* truth. Aristotle refuted Plato. He, al-Ghazali, refuted the main Muslim followers of Aristotle, that is, al-Farabi and Avicenna,[53] in showing that they are inconsistent. This general comment on philosophers, that they disagree, was already to be found in *The Incoherence of the Philosophers* in which al-Ghazali criticizes chiefly the positions of al-Farabi and Avicenna. There is no unity in philosophy, and al-Ghazali sets himself to the task of destroying the facade that philosophers had built to hide their lack of rigor. Aristotle had criticized Plato in the name of truth by saying, "Plato is dear to us. And truth is dear, too. Nay, truth is dearer than Plato."[54] In the name of truth al-Ghazali himself will begin to criticize the Muslim Aristotelians in showing the inconsistency of their positions. It is therefore interesting to see how he handles Avicenna's positions concerning the substantiality of the human soul and the relation of this soul to the body. Such questions are his topics in problems XVIII and XIX. Problem XVIII rejects any proof of the substantiality of the soul as not solid enough and problem XIX shows that a proof of the survival of the soul after the death of the body is therefore also impossible. Fundamentally though, as we have already seen, al-Ghazali agrees with the position of the philosophers; yet he claims that rational demonstration of this position is impossible.[55] To show the inconsistency of Avicenna whom he quotes at length, al-Ghazali will present an Aristotelian critique of Avicenna's Platonism. Avicenna's arguments for the substantiality of the soul are not cogent because they cannot prove that the human soul or intellect is *not* a bodily faculty, whereas the animal and vegetative souls are bodily faculties. To destroy the particular status granted to the human soul, al-Ghazali has to show that animal faculties are not fundamentally different from the intellect and that even the human soul requires the body. The points I want to examine are the following: according to al-Ghazali (1) some senses may perceive themselves, and even if they do not, the fact that the intellect does perceive itself does not necessarily make it radically different from the senses; (2) only what subsists in the senses come to subsist in the intellect; and "imagination," that links both senses and intellect, is akin in some fashion to grasping the universals; (3) the argument used to show that what receives the intelligibles must be immaterial applies to the estimative faculty as well; (4) the need of the soul for its own body for its individuation makes the connection between a soul and its body a necessary one; and (5) the self cannot be the soul, since self-awareness is always also awareness of the self as body.

We will first look at the continuity between the different faculties and then examine the consequences of this for the relation of the soul to the body, i.e., its own particular body.

First, intellection does not differ radically from sensation. Following Aristotle, Avicenna maintains that the senses do not perceive or sense themselves whereas the intellect intelligizes itself.[56] For the senses require a physical organ and intellection does not. Al-Ghazali objects to this position by contending, first, that some sense may sense itself, and second, that senses differ from each other in their functioning without their difference leading one not to consider them as bodily faculties. Thus, even if al-Ghazali's first claim, that some sense may sense itself, should prove to be false, one should argue that the fact that intellect exhibits some difference from the five senses does not prove that it cannot be considered a bodily faculty. Indeed, sight and touch, though both external senses, differ. One can have tactual perception only if there is contact between the object and the organ of touch;[57] but one cannot have a visual sensation if the object of sight is put on the organ because sight does also require a medium:

This difference between sight and touch does not make the two differ with respect to dependence upon the body. Why, then, is it improbable that that which is called the Intellect should be one of the bodily senses, differing from the other senses insofar as it perceives itself?[58]

Second, intellection does not differ radically from retention, i.e., a lower aspect of "imagination." Al-Ghazali rejects the position that claims that universals, i.e., intelligibles, do not come from the particulars, but are given directly by the active intellect. Everything which subsists in the intellect comes from the senses but in the sense it is "lumped together," whereas the intellect analyzes it. From the sight of one particular man, the intellect can discover the concept of man, i.e., the universal. Why? Because sensation itself in some fashion is already perception of the universal. Each time that one sees a particular, one does not always imprint a new image on the retentive faculty. For instance, if we see a certain quantity of water once in a certain vessel, an image will be imprinted in the retentive faculty; but if later on we see the same quantity of some other water in the same vessel, no new image will be imprinted in the retentive faculty, since water *qua* water, i.e., as universal, does not change. The same goes for the hand. After having once seen the same hand, the retentive faculty will not take in a new image. A new image will be imprinted only if one sees another hand, differing from the first one by its qualities but not *qua* hand. What is common to the two hands does not imprint a new image. This explains why such images have often been assumed to be universals.[59] Animals have some grasp of the class, therefore there is no great difference between the retentive faculty and the intellective. Some aspect of "imagination" is akin to intellection.

The similarity between "imagination" and intellection is also shown in the case of the estimative faculty, the highest aspect of "imagination." Avicenna indicates in his "theory of degrees of abstraction" that the estimative faculty is at the peak of the abstraction process as performed by the living being itself, since it receives "intentions," which, in themselves, are nonmaterial, although accidentally they happen to be in matter. Yet the estimative faculty can grasp these intentions only in connection with particulars, for instance, the lamb perceives the enmity in this wolf.[60] Al-Ghazali contends that, since intentions as such are immaterial, the argument for the immateriality of what receives intelligibles should be also applied to them and, therefore, the estimative faculty should not be considered as a bodily faculty. Therefore, one must either abandon the theory of the uniqueness of intellection or the explanation of the estimative faculty.[61]

On the three points we have just been examining al-Ghazali shows that Avicenna does not succeed in his attempt to make a good case for a radical difference between the intellect and animal faculties such as the senses and "imagination." Therefore, the intellect could be a bodily faculty and not be independent from the body. Let us then examine what al-Ghazali says about the relation between the soul and its body in the case of the human being.

First, al-Ghazali considers Avicenna's claims that the human soul needs a body for its individuation and for coming into existence but yet that the connection between the soul and its own body is accidental rather than essential.[62] Al-Ghazali argues that Avicenna's basic position on the necessity of a particular body for a soul to come into existence implies that the connection between a soul and its own body is essential, since it explains why twins are two different persons. Two sperm drops begetting twins may be present within the same womb at the same time. Conditions are perfectly identical; yet the twins are two persons, and the soul of *this* body cannot be the director of the soul of *that* body. Thus, if this particular connection between body and soul is required for the soul to come into existence, will it not be required also for its survival? Maybe the connection is necessary, for if the connection is accidental, why does the soul need a specific particular body, which is its own, and not simply any identical body? But if the connection is essential, as seems to be the case, then the soul cannot subsist without its body.[63] If the connection is necessary, there is no survival of the soul.

Second, if the connection between soul and body is essential, one should, according to al-Ghazali, reject the claims that the human soul can know itself directly and that the self is the soul. Al-Ghazali asserts that the human soul cannot know itself directly, because man is always aware of his body. This is why man affirms himself as body, to such an extent that he affirms himself as being in his clothes and in his house. Yet the soul as Avicenna defines it has no relation to houses or garments.[64] In other words, al-Ghazali notices that we speak of ourselves as wearing a garment or being in a house, but that souls are not wearing garments nor staying in houses since they are viewed as purely immaterial. Our way of speaking of ourselves implies that we are body. Some aspects of the body, however, seem indispensable for the self, while others do not. For a man may assume that he will subsist after amputation of his feet, but he cannot assume that he will subsist without his heart. Therefore, the claim that sometimes man is unaware of the body is false. It follows that pure self-awareness is also awareness of one's body and therefore that the body is an integral aspect of the self. Immediacy of the intellect to itself is impossible. Al-Ghazali does not share Avicenna's enthusiasm for the "Flying man" experiment.

To show the weaknesses of Avicenna's position, al-Ghazali follows Aristotle's claim that if there is a close connection between intellection and "imagination," there is also a close connection between the soul and body. He maintains that such connections exist and therefore that the substantiality of the human soul cannot be philosophically demonstrated. The imaginary experiment of the "Flying man" is countered by an analysis of human experience. A human being experiences himself not as pure soul but as body, too, and the close connection ascertained between the soul and its own body makes survival of the soul rationally unthinkable.

Avicenna rejects this connection between "imagination" and intellection, and claims that the soul is a substance and therefore that the body is simply a riding animal for the soul. As the pilgrim needs his camel to reach Mecca, but will never reach Mecca if he fusses too much about his mount, so Avicenna claims that the soul needs the body to reach the first step of its philosophical pilgrimage, the beginning of contact with the active intellect. But for this contact to develop fully, and therefore to pursue its pilgrimage up to its final step, the soul must rid itself of its riding animal which becomes only a hindrance. "Imagination" ceases to lead to intellection. The body does not help reach Mecca. It simply takes you as far as the first step on the way to Mecca.

Al-Ghazali, on the contrary, shows that a true Aristotelian approach — and for him such an approach is the best philosophically speaking — excludes such a conception of the body. The body is an integral aspect of the self and therefore the soul, as well as the self, are inseparable from the body. Yet al-Ghazali himself seems to accept Avicenna's basic claim about the substantiality of the soul. This is why in *The Alchemy of Happiness* he too speaks of the body as the riding animal of the soul. For him philosophy does not succeed at all in establishing this position, since such a demonstration is beyond the powers of human reason. Reason is linked to "imagination" and therefore to sensory experience and, therefore, rationally the body cannot be conceived only as the riding animal of the soul. Such a conception is only grasped by revelation. This is why al-Ghazali presents it only in a work of spirituality. Yet even then he seems to give more emphasis to the body than Avicenna since he claims that one needs the body to reach Mecca itself and not simply to reach the first stage of the pilgrimage. His awareness of the importance of the body for the self is too great to allow him to dismiss quickly the body. The soul needs the body to reach its final aim.

Maybe al-Ghazali and Avicenna are mistaken in claiming that the body should be considered as only the riding animal of the soul; but a look at their approach to this question shows clearly that Aristotle was correct in claiming that the connection between "imagination" and intellection goes with the connection between soul and body. By minimizing the connection between soul and its body Avicenna is obliged to minimize the connection between "imagination" and intellection. By emphasizing philosophically the connection between body and soul al-Ghazali is obliged to enhance the connection between "imagination" and intellection. Yet, like Aristotle, both are unable to remain faithful to and consistent with their philosophical insight. Perhaps deeper study of imagination could lead us to a more coherent position.

Georgetown University

NOTES

[1] Translated by Claud Field (London: John Murray, 1910), p. 44.

[2] 250 c.

[3] *Oeuvres de Descartes*, ed. Adam and Tannery, 8: 81.

[4] 413 a 8–9.

[5] 'The Prince of Physicians on the Nature of Man,' *Modern Schoolman* 55 (November 1977–May 1978): 176.

[6] Algazel, *Tahafot al-Falasifat*, Arabic text edited by Maurice Bouyges, S. J., Bibliotheca arabica scholasticorum. Série arabe, no. 2 (Beirut: Imprimerie Catholique, 1927); hereafter cited as Bouyges. English translation: Al-Ghazali, *Tahafut al-falasifah* (*Incoherence of the Philosophers*), trans. Sabih Ahmad Kamali, Pakistan Philosophical Congress Publications, no. 3 (Lahore: Pakistan Philosophical Congress, 1963); hereafter cited as Kamali.

[7] Bouyges, p. 297; Kamali, p. 197.

[8] Bouyges, p. 304; Kamali, p. 200.

[9] Arabic edition by Albert Nader (Beirut: Dar al-Mashreq, 1968). German translation by Friedrich Dieterici, *Alfarabi's Philosophische Abhandlungen* (Leiden: Brill, 1892), pp. 1–53.

[10] Cf. the end of al-Farabi's *Summary of Plato's 'Laws,'* in Alfarabius, *Compendium Legum Platonis*, ed. and Latin translation by F. Gabrieli, Plato Arabus, no. 3 (London: Warburg Institute, 1952); Arabic, p. 43 and Latin, p. 33.

[11] 608 d–611 a.

[12] I, 1, 403 a 3–10.

[13] II, 2, 414 a 19–22. D. W. Hamlyn's translation in *Aristotle's 'De Anima,' Books II, III*, Clarendon Aristotle Series (Oxford: Oxford University Press, 1968), p. 14.

[14] II, 1, 412 b 5–6; Hamlyn's translation, p. 9.

[15] 430 a 22–25.

[16] III, 7, 431 a 15–16 and 431 b 2; III, 8, 432 a 3–14.

[17] 430 a 15–17; Hamlyn's translation, p. 60.

[18] 506 d–509 c.

[19] *Avicenna's De Anima. Being the psychological part of 'Kitāb al-Shifā''*. Arabic text ed. F. Rahman, University of Durham Publications (London: Oxford University Press, 1959); hereafter cited as Rahman. Cf. the Arabic edition with a French translation: *Psychologie d'Ibn Sīnā (Avicenne) d'après son oeuvre aš-Šifā'*, ed. Ján Bakoš, 2 vols. (vol. 1, Arabic text; vol. 2, French translation) (Prague: Editions de l'Académie tchécoslovaque des sciences, 1956). The medieval Latin translation has been critically edited by S. Van Riet: *Avicenna Latinus. Liber de Anima seu sextus de naturalibus*,

2 vols. (Louvain: Peeters-Brill, 1968–1972). In the margin of this edition one can find the reference to the page in Rahman's edition.

[20] English translation by F. Rahman, *Avicenna's Psychology* (London: Oxford University Press, 1952); hereafter cited as *AP*; Arabic edition: *al-Najāt* (Cairo, 1938), pp. 157–93.

[21] *Healing*, I, 1 and 3; Rahman, pp. 4–16 and p. 29. (1) Commentary and (2) English translation of I, 1, by Lenn Evan Goodman, 'A Note on Avicenna's Theory of the Substantiality of the Soul,' *Philosophical Forum* 1 (Summer 1969): 547–54 and 'Text. On the Soul,' ibid., pp. 555–62.

[22] *Healing*, V, 4; Rahman, p. 227. *Salvation*, XIII; *AP*, p. 58.

[23] *Healing*, V, 4; Rahman, pp. 227–33. *Salvation*, XIII; *AP*, pp. 58–63.

[24] *Salvation*, XII; *AP*, p. 57. Cf. also pp. 56–58. *Healing*, V, 3; Rahman, pp. 223–27.

[25] *Healing*, V, 3; Rahman, p. 225. *Salvation*, XII; *AP*, p. 57.

[26] *Healing*, V, 3; Rahman, p. 225. *Salvation*, XII; *AP*, pp. 57–58.

[27] *Salvation*, XII; *AP*, pp. 57–58.

[28] *Healing*, V, 4; Rahman, pp. 233–34. *Salvation*, XIV; *AP*, pp. 63–64.

[29] *Healing*, IV, 1; Rahman, pp. 163–69. *Salvation*, III; *AP*, pp. 30–31.

[30] There are some controversies about the relation between imagination and common sense in Aristotle as well as about what exactly the "phantasia" is. On this topic and in function of our own purpose, Rahman's note in *AP*, p. 78 (on p. 31, 2) is very useful. Cf. pp. 79–83 (on p. 31, 19 and on p. 31, 23).

[31] *Healing*, IV, 3; Rahman, pp. 188–94. *Salvation*, VIII; *AP*, pp. 41–45.

[32] *Healing*, V, 2; Rahman, pp. 209–16. *Salvation*, IX; *AP*, pp. 46–50.

[33] *Healing*, V, 2; Rahman, pp. 216–21. *Salvation*, X; *AP*, pp. 50–54.

[34] *Healing*, V, 2; Rahman, pp. 216–17. *Salvation*, X; *AP*, pp. 50–51.

[35] *Healing*, V, 2; Rahman, p. 218. *Salvation*, X; *AP*, p. 52.

[36] *Healing*, V, 5; Rahman, p. 235.

[37] *Healing*, II, 2; Rahman, pp. 58–67. *Salvation*, VII; *AP*, pp. 38–40.

[38] *Healing*, V, 6; Rahman, p. 239.

[39] *Healing*, V, 3; Rahman, pp. 221–22. *Salvation*, XI; *AP*, pp. 54–56.

[40] *Healing*, V, 3; Rahman, p. 222. *Salvation*, XI; *AP*, pp. 55–56.

[41] *Healing*, V, 3; Rahman, p. 223. *Salvation*, XI; *AP*, p. 56.

[42] *Healing*, V, 2; Rahman, p. 220. *Salvation*, X; *AP*, p. 53.

[43] *Healing*, V, 2; Rahman, pp. 220–21. *Salvation*, X; *AP*, pp. 53–54.

[44] *Healing*, V, 2; Rahman, pp. 220–21. *Salvation*, X; *AP*, pp. 53–54.

[45] *Healing*, V, 2; Rahman, p. 219. *Salvation*, X; *AP*, p. 53.

[46] *Healing*, V, 2; Rahman, p. 220. *Salvation*, X; *AP*, p. 54.

[47] Thérèse-Anne Druart, 'Ibn Sina's conception of intellection in the *De anima* part of the *Shifa'* in its relation to Aristotle's positions in the *De anima*,' in a collection of essays edited by Parviz Morewedge (forthcoming). Cf. *Healing*, III, 6; Rahman, pp. 244–48 and H. A. Davidson, 'Alfarabi and Avicenna on the Active Intellect,' *Viator: Medieval and Renaissance Studies* 3 (1972): 162–63.

[48] *Healing*, I, 1; Rahman, pp. 15–16; English translation by L. E. Goodman, pp. 561–62 and *Healing*, V, 7; Rahman, pp. 255–57.

[49] Gérard Verbeke, in the introduction, "Le 'De Anima' d'Avicenne une conception spiritualiste de l'homme," in *Avicenna Latinus. Liber De Anima seu sextus de naturalibus*, ed. S. Van Riet, vol. 1 (for books IV–V), pp. 36*–46*.

[50] Zedler, p. 167.

[51] *Healing*, V, 3; Rahman, p. 223; the translation is mine. *Salvation*, XII; *AP*, p. 56.

[52] Cf. G. Furlani, 'Avicenna e il "Cogito, ergo sum" di Cartesio,' *Islamica* 3 (1927): 53–72 and Roger Arnaldez, 'Un précédent avicennien du Cogito cartésien?' *Annales Islamologiques* 11 (1972): 341–49.

[53] Richard Joseph McCarthy, S. J., *Freedom and Fulfillment*, an annotated translation of al-Ghazālī's 'al-Munqidh min al-Ḍalāl' and Other Relevant Works of al-Ghazālī (Boston: Twayne, 1980), p. 72, paragraphs 33–35; Arabic edition by Farid Jabre: Al-Ghazali, *Al-munqiḍ min aḍālal* (*Erreur et délivrance*), 2d ed. (Beirut: Commission libanaise pour la traduction de chefs-d'oeuvre, 1969), p. 20 of the Arabic text.

[54] Kamali's translation, p. 4; Bouyges' Arabic edition, p. 8.

[55] Kamali, p. 200; Bouyges, p. 304.

[56] *Healing*, V, 2; Rahman, pp. 216–21; *Salvation*, X; *AP*, pp. 50–54.

[57] In fact, because of that Aristotle wonders whether the skin is the organ of touch or only the medium and decides the flesh or skin is only the medium and the organ is beyond it, II, 11, 422 b 17–423 b 26. Avicenna, on the contrary, considers that there is medium in the case of touch, *Healing*, II, 3; Rahman, pp. 71–72.

[58] Kamali, p. 210; Bouyges, p. 317.

[59] Kamali, p. 218–19; Bouyges, pp. 330–32. Apparently Galen was known to have claimed that donkeys perceive universals, in *De methodo medendi*, II, pp. 138–39, in *Clauii Galeni Opera omnia*, ed. C. G. Kühn, vol. 10 (Leipzig, 1825). Ibn Bājjah (Avempace) indicates in a little treatise that animals look for water and not for some particular water and therefore are in some way in touch with the universal. Arabic text in Ibn Bājjah (Avempace), *Opera metaphysica*, ed. Majid Fakhry (Beirut: Dār al-Nahār, 1968), p. 108; English translation by M. Ṣaghīr al-Ma'ṣūmi, 'Ibn Bajjah on the Agent Intellect,' *Journal of Asiatic Society of Pakistan* 5 (1960): 32–33.

[60] *Healing*, II, 2; Rahman, pp. 60–61, and I, 5; Rahman, p. 45. *Salvation*, VII and III; *AP*, pp. 39–40 and p. 31.

[61] Kamali, pp. 202–4; Bouyges, pp. 306–8.

[62] *Healing*, V, 4; Rahman, pp. 227–31. *Salvation*, XIII; *AP*, pp. 58–61.

[63] Kamali, pp. 222–25; Bouyges, pp. 335–39.

[64] Kamali, p. 212; Bouyges, pp. 320–21.

PART VII

HUSSERL AND THE HISTORY OF PHILOSOPHY

WOLFGANG SCHIRMACHER

MONISM IN SPINOZA'S AND HUSSERL'S THOUGHT

The Ontological Background of the Body-Soul Problem

I. ACTUALITY OF THE PROBLEM AS THE QUESTION OF MAN'S SELF-UNDERSTANDING

The classic body-soul problem formulates unaltered a key philosophical problem. The experience of being a citizen of two worlds, existing in the corporeal world while at the same time living a life of the mind ("Leben des Geistes"),[1] persistently disturbs our self-understanding. Who is man, what relation has his thinking to corporeality? Is "soul" merely an archaic name for "mind," or does the soul possess a reality of its own? Can the assumption of an immaterial world be reconciled with scientific, technical reality? Can present-day brain research offer us solutions to these questions[2] whereby man proves himself to be an electrochemically controlled mechanism? Or is psychology the science which has to probe and examine our emotional and cognitive faculties? These are questions which affect our existence. But scientists do not determine what science means;[3] and whether body, soul, mind are even accessible to scientific investigation becomes ever more questionable. The question of man's being remains in its anthropological constraint without significant answer.

The experience of the otherness of mind, played out in the traditional hierarchy of world, man, and God, which tempted the modern age into subjectivity,[4] is at the same time the experience of the powerlessness of the mind, of its inherent dependence on the senses: without body there is no mind, without world, no body.[5] In order to know how man is we must learn what constitutes his world. The anthropological body-soul problem refers compellingly to ontology. The philosophical treatment of this problem is thus characterized in the past and present by an evident proximity to fundamental ontological determinations. Whether the world is perceived as monistic or dualistic, whether the unity of all beings is stressed or the difference intoned, has immediate effect on the body-soul problem. There is however no obligation. The monist can argue materialistically or idealistically; the advocates of dualism are locked in hopeless dispute.[6]

The ontological background is, however, by no means without implications. On the contrary, it decisively determines the manner in which man

345

A-T. Tymieniecka (ed.), Analecta Husserliana, Vol. XVI, 345–352.

understands himself and his behavior in the world. A dualistic division of the
world is the necessary condition for human dominance, justification of the
exploitation of nature and fellow-man. Even if the ontological dualism vaguely
acknowledges a prior unity of being,[7] it expresses a human claim to order
which divides body and soul. Man is no longer one with the cosmos, but
forms a center of his own, understanding himself at best in interaction with
the nonhuman.[8] Then the foreignness of things begins, and their misuse as insignificant objects at our disposal. Modern instrumental science would be
inconceivable without Descartes' dualistic ontology, his division of thought
and corporeal world. But this dualism which made man proud and free is
responsible for our fatal incapacity to understand the present destruction of
the world as human self-destruction. That which the body experiences daily
and which our mind declares not relevant to itself, can only then be carnally
present and become comprehensible when the unity of the world reveals itself
as horizon of the phenomena. We live and die in one single world.

II. ONTOLOGICAL AND METHODOLOGICAL MONISM: HUSSERL VERSUS SPINOZA

The body-soul problem is the problem of our self-cognition. It is a matter of
the quality of our insight into the universal coherence of the cosmos. Does
this prescribe an ontological monism? Certainly not in the sense of a mono-
causal interpretation — ideologic in character[9] — of the world, but perhaps as
a concentrated "seeing" (*Blick*) into the essential being which endures and
which first of all renders particularity possible. This meditative thinking does
not seek ideas which differ from the matter (*Sache*), but rather follows the
sense of being. If it loses sight of beings at first, it is only to recover their
unity and proportion within the context of beings (*Sachverhalt*).[10] Our point
of departure must be our self — the only phenomenon about which we can
make an authentic statement. The cosmic unity proves itself through the
human way of existence, in body and soul. Spinoza's *amor Dei intellectualis*
and Husserl's epoché are tried and tested methods of approach to this monistic
evidence (*Aufweis*). They accomplish the same task. Nevertheless there are
differences not only in the two thinkers' terminology and in their descriptions
of the body-soul relationship. Their respective concepts of monism are
essentially different.

Husserl turns away from Spinoza's ontological understanding of unity. He
does not discuss the "what" of the world, decisive for us is the "how." The
shaking up of traditional ontology takes effect in Husserl's thought, leads

to a human self-restriction which Kant anticipated.[11] The unity of method springs from the unity of being. But Husserl's methodological monism [12] could not be sustained. The phenomenologists who followed him are characterized by a remarkable hunger for world. Heidegger, Merleau-Ponty, and Sartre fill out Husserl's transcendental structure with *Dasein*, bodiliness, society. Husserl's conception of *Lebenswelt*, emergent in his later works, unintentionally gave this development yet another name pregnant with meaning. The world in and around us returns powerful in the determination of man's being. The ontological element which Husserl's phenomenological reduction excludes becomes anew the determining element. But upon closer inspection this phenomenological ontology shows itself to be neither a return to Spinoza nor a dissent from Husserl. Instead it signifies a necessary passage through to an operative knowledge in which truth and method form no antithesis. The dualistic predisposition of modern metaphysics, the separation of body and soul which occurs epistemologically as subject-object division, is what the critics of the present so fatally dominant thought want to overcome. Spinoza, Husserl, and the phenomenological movement deny themselves in this respect the all two simple solution of taking sides: either to name everything mind or matter. The experience of the difference must not be allowed to be explained away by the necessity of comprehending the world as unity.

III. THE BODY–SOUL PROBLEM AGAINST ITS ONTOLOGICAL BACKGROUND

1. *Spinoza's Certainty of Being*

Spinoza, Descartes' contemporary, student and exemplary critic, designated thought and extension as the two modi of the undivided cosmos which we human beings alone are able to comprehend through our body and our mind.[13] Spinoza's presentation of the relationship of body and soul in his *Ethics* carefully traces the phenomenological experience without metaphysically distorting such through interpretation. Everything the mind knows it knows through the body. The mind has, however, no possibility of causally effecting the body — a statement made by Spinoza[14] which Husserl as well accepted.[15] Yet body and soul always cooperate. The well-being of the body heightens the thinking power of the mind and vice versa. This is more than merely a claim of psychophysical unity,[16] for man develops a more adequate attitude. Only the desire of knowledge, of the *amor Dei intellectualis* accords with the most realized bodiliness. Of course, the mind comprehends, not the body, but the mind's delight in the adequate idea encourages the body toward attaining its own appropriate bodiliness. Spinoza's *Ethics* can be read as the phenomenology of such cooperation between body and soul which brings forth humanness. Psychophysical solidarity which constitutes human life is, however, founded alone in the one substance which is both nature and God. Without this ontological background, Spinoza's presentation of the body-soul relationship would remain arbitrary and the contended cooperation, coincidental. Spinoza's ontology of substance is nevertheless not to be understood as dogmatic positing, but rather, as the *Ethics* progresses, it verifies itself as indispensable. The principle of universalization makes the particular case possible, admits of the individuality of body and soul. For neither one could ever fall out of the substance and become irresponsible towards the whole. This would then, according to Spinoza, contradict self-preservation common to all beings in which cosmic unity most evidently expresses itself.[17] In other words: an irresponsible attitude in theory and praxis condemns to death without fail. Humanity repudiates the ontological judgment but will have to endure it nonetheless.

2. *Husserl's Critique of Belief in the World*

Spinoza's serene certainty of being which he contrasted with the Cartesian humanization of the world, was to Husserl a rationalism whose system had in the meantime collapsed.[18] Spinoza's apparently immutable substance was shaken by the modern experience of historicity; in any case, that is how we see it. The world alters itself with our interpretation: how then can adequate knowledge — certainty in thought and action — assert itself?[19]

Descartes' universal doubt is the modern principle. Husserl tests his conclusions on Descartes, not on Spinoza. His phenomenological reduction suffices this doubt, but avoids Descartes' two-substance doctrine with its weighty consequences and the resulting subject-object division as well.[20] Husserl reacts to the historical shaking up of ontology by entirely pulling the ground out from under it. The mundane attitude, without which a relevant discourse on being[21] is impossible, is restricted. The question of "what" proves to be empty, despite the excess of answers; exact knowledge can be found only in "how." Phenomenologically man experiences himself as intentional being. The undivided structure of intentionality is his transcendental ego to which each empirical subject responds.[22] Body and soul show themselves to be differing attitudes whose constitutions are precisely definable.

Husserl differentiates between body, soul (psyche), and spirit. In the relevant passages of the second book of his *Ideen*[23] he determines at the outset the traditional "Doppelrealität" (p. 342) of man, his "induktiv-reale psychophysische Einheit" (p. 353). Body and soul are distinct from and yet dependent upon one another in their unity. The "Einheit des seelischen Seins und Lebens," according to Husserl, is "verknüpft . . . mit dem Leib als Einheit des leiblichen Seinsstromes, der seinerseits Glied der Natur ist" (p. 343). Occurrences experienced by the soul belong to the "sinnlichen Sphäre" (p. 339). The empirical ego is "das Ich der leibseelischen Natur . . . . Es ist nicht selbst leibseelische Einheit, sondern lebt in ihr" (p. 339). This naturalist attitude must be supplemented by the personalist (cf. pp. 281 ff.). The soul reveals itself as "Untergrund des Geistes" (p. 334). This spirit comprises "Ich-Mensch" and culture as objectified spirit. "Zum reinen Wesen der Seele gehört die Ich-Polarisierung" as well as the "Notwendigkeit einer Entwicklung, in der sich das Ich zur Person und als Person entwickelt" (p. 350). According to Husserl, only as mind-person does man comprehend himself in complete concretion, "als Subjekt der Intentionalität" (p. 355). Mind sustains itself as body and soul in the structure of intentionality; this becomes self-manifest to the mind above all. Husserl's universal intentionality succeeds to the place of Spinoza's substance, and yet the historicity which acts destructively upon substance is integrated as the hermeneutics of *Dasein* — at least by Heidegger. Body, soul, and mind accomplish assignments which can be authenticated. Husserl's transcendental idealism need not yet be objectionable, rather it is to be understood more as the delimitation of an area of investigation than as ideology.

IV. THE ONTOLOGICAL THEME IN THE PHENOMENOLOGY MOVEMENT AFTER HUSSERL

Yet Husserl's evidencing (*Aufweis*) of the monistic structure of intentionality and of the precedence of the unity-promoting mind he emphatically advocated, in no way ended the body-soul discussion. Husserl started a movement whose phenomenological interpretation of man happens upon ontological questions. The further development, revision, or falsification — construed according to standpoint — of the Husserlian phenomenology begins with his determination of man. It was too well-known that Husserl's radical point of departure passed over certain elements of human life not to be relinquished. Are we then primarily spirit-being, consciousness, intentionality? Can the excluding of that which after all constitutes our world be allowed? Should our knowledge merely view coldly and foreignly that which we live through in love and hate? The world in which we live is not a stage play, for in this world we also die. A spectator does not live, for living means changing. Such objections, however, apply to Husserl only to a certain extent. His extensive investigations are in no way intended to hinder man from life, but rather to allow him to comprehend better how he lives his daily as well as scientific life. The full correlation ego-*cogito-cogitatum* appears in just such an interpretation for the first time. Erroneous paths, subjectivistic as well as objectivistic can now be avoided. Husserl does not construct a world-for-us in the manner of a theorist, but rather analytically descriptively evidences its constitution. That which has been set by "I" is then no longer product, but correlate; the intentional object plays the role of the transcendental guide.[24] Absolute cognition of the world can perhaps be realized as strict science in the phenomenological attitude of the viewer.[25] But Husserl's radically theoretical attitude does have certain consequences which subsequent philosophers, also phenomenologically oriented, were not willing to accept: (1) the thorough transformation of sense perception into categorical perception; (2) the presenting of a consciousness which lives entirely in the present; (3) a misapprehending of historical existence which occurs only as past and not as finiteness.[26]

The phenomenological *Blick* of man's being-in-the-world ("In-der-Welt-sein") shows, according to Heidegger, Merleau-Ponty, and Sartre, that theory itself must be comprehended as descendent attitude. In contrast, our knowledge must be understood in an existentially, bodily, socially involved manner. As "geworfener Entwurf" (Heidegger) man becomes transformer of the world and establisher of unity in one. *Dasein* holds together what belongs together, being and time.[27] Husserl's intentionality completely fills itself to a way of life. *Dasein* itself *is* intentional achievement (Heidegger), as body I *have* world (Merleau-Ponty) and *bring* world to totality in a dialectic of "I" and "We" (Sartre). Those elements of being — coming to ripeness, ambiguity, limit, negation — absolutely excluded by Husserl, return. The event of which man partakes is finite.[28] The inconceivable concealment of emerging (*entbergenden*) being reveals itself conclusively in event technology, which can keep us alive, but whose distortion in the form of science-technology leads the human race instead toward certain destruction.[29]

V. SUMMARY: TRUTH AS TECHNOLOGY

Husserl's phenomenological *Blick*, which intended to clarify even up to the absolute streaming, has been broken by the experience of its invalidity

(*Nichtigkeit*). Nevertheless, the evidencing of nothingness (*Nichts*) only strengthened the unity of being. Indeed, the insight into the finiteness of man hinders us from traditional ontology, but intentionally, existentially, bodily, and socially we constantly manifest a unity which accords with that of the cosmos. Precisely in failure and loss is this unity most evident. Monism is not an intellectual construction which must be secured in dogmata. It expresses the simple experience of aliveness – the fundamental experience of body and soul. Through interpretation it is merely misplaced (*verstellt*). Husserl's decisive step forward would be forgotten if one attempted to define the monistic experience with regard to its content. In its enduring continuity universal unity shows itself to be of diverse, particular elements. We constantly lay claim to them.[30] But it can be neither theoretically – as Husserl assumed – nor practically – as Sartre hoped – guaranteed that we affect the unity of body and soul. Praxis acknowledges only power, and theory wants to be without responsibility. Both forms of life are anthropocentrically false and have been for some time. But we refuse to comprehend truth as technology, which gives reliable information through failed or successful operation. Thus the dying human rase sees sentence passed daily on its technology of life. The instrumental technology which dominates our planet, together with its science, realizes with great efficiency our fall from unity of being. In accord with the remaining spirit is the poisoned body and the decayed soul. This constitutes the body-soul problem of our time.

Translated by Virginia Cutrufelli

NOTES

[1] Cf. H. Arendt, *The Life of the Mind*, 2 vols. (New York, 1978).

[2] Cf. J. C. Eccles and K. R. Popper, *The Self and its Brain* (New York, 1977).

[3] Cf. M. Heidegger, 'Science and Reflection,' in *The Question Concerning Technology and Other Essays*, by Heidegger, trans. W. Lovitt (New York, 1977), pp. 155–82.

[4] Cf. W. Schulz, *Ich und Welt: Philosophie der Subjektivität* (Pfullingen, 1979).

[5] Cf. R. M. Zaner, *The Problem of Embodiment* (The Hague, 1964).

[6] Cf. J. Seifert, *Das Leib-Seele-Problem in der gegenwärtigen philosophischen Diskussion* (Darmstadt, 1979), pp. 126–30.

[7] Ibid., p. 127.

[8] Cf. K. Löwith, *Aufsätze und Vorträge* (Stuttgart, 1971).

[9] Cf. Seifert's critical notes, pp. 54ff.

[10] Cf. M. Heidegger, *On Time and Being*, trans. J. Stambaugh (New York, 1972), pp. 4ff.

[11] Cf. E. Fink, 'Die Idee der Transzendentalphilosophie bei Kant und in der Phänomenologie,' in *Nähe und Distanz*, by Fink, ed. F. Schwarz (Freiburg, 1976), pp. 7–44.

[12] Cf. E. Husserl, *Cartesianische Meditationen und Pariser Vorträge, Husserliana*, vol. 1 (The Hague, 1963), pp. 15f.

[13] Cf. Spinoza, *Ethics*, pt. 2, scholia 23.

[14] Ibid., pt. 3, scholia 2.

[15] Cf. E. Husserl, *Ideen zu einer reinen Phänomenologie und phänomenologischen Philosophie*, bk. 2, ed. M. Biemen, *Husserliana*, vol. 4 (The Hague, 1952), p. 283.

[16] Ibid., pp. 355ff.

[17] Spinoza, *Ethics*, pt. 5, scholia 10.

[18] Cf. E. Husserl, *Die Krisis der europäischen Wissenschaften und die transzendentale Phänomenologie*, ed. E. Ströker (Hamburg, 1977), pp. 70ff.

[19] Cf. K. Hammacher, 'Spinoza,' in *Grundprobleme der grossen Philosophen: Philosophie der Neuzeit I*, ed. J. Speck (Göttingen, 1979), pp. 101–38.

[20] Cf. E. Husserl, *Krisis*, secs. 10ff.

[21] Cf. the articles of A. Lingis, U. Claesges, and J. N. Mohanty in *Analecta Husserliana*, vol. 1, ed. A-T. Tymieniecka (Dordrecht, 1971).

[22] Cf. P. Aubenque, *Le Problème de L'Etre chez Aristote* (Paris, 1966).

[23] E. Husserl, *Ideen*, Beilagen I–XIV.

[24] Cf. L. Landgrebe, *Der Weg der Phänomenologie* (Gütersloh, 1967), pp. 9–39.

[25] Cf. E. Tugendhat, *Der Wahrheitsbegriff bei Husserl und Heidegger* (Berlin, 1967), secs. 8ff.

[26] Cf. H. G. Gadamer, 'Die phänomenologische Bewegung,' in *Kleine Schriften*, vol. 3 (Tübingen, 1972).

[27] Cf. M. Heidegger, *On Time and Being*, p. 4.

[28] Ibid., p. 54.

[29] Cf. W. Schirmacher, *Ereignis Technik: Heidegger und die Frage nach der Technik* (Hamburg, 1980).

[30] Cf. Y. Nitta's article in this volume.

A. Z. BAR-ON

HUSSERL'S BERKELEY [1]

I

What are the sources of phenomenology? I have attempted elsewhere[2] to trace phenomenology to its bedrock by distinguishing five sources: (1) the ancient skeptics; (2) Descartes; (3) Locke, Berkeley and Hume; (4) Kant; and (5) Franz Brentano.

This was a working hypothesis rather than a closely documented, historiographical thesis. Even so, my suggestion has some merits. First, explicit references can be found in Husserl's writings to each of these areas of the recent and remote philosophical past. Husserl names some of the above mentioned thinkers as those whose thought contained the first "phenomenological impulses" or the philosophical challenges which paved the way for phenomenological method and theory. Apart from these references, I find in these philosophical domains of the past seeds of our philosophical present as embodied in phenomenology: conceptions, considerations, arguments, which have been suggested in the past, but achieved their elaboration in one way or another in the phenomenology of Husserl and his followers. Indeed, I am convinced that the "sources" mentioned in my "working hypothesis" deserve a closer examination than previously made concerning their impact on both Husserl and other phenomenologists.

I will not venture to grade these sources as to their significance. All of them are, I think, important. If I chose the third of them, or rather one branch of it, for closer scrutiny, I do it more for subjective reasons than because of some principle. Berkeley's philosophy appeals to me more than that of the others, and I find it particularly rewarding to trace the history of his ideas in recent philosophy. Also, should I succeed in substantiating my conjecture about the strong impact of Berkeley's philosophy on Husserl, it will turn out that the former happens to have two heirs in contemporary philosophy who are independent of, and even antagonistic to, each other: phenomenalism on the one hand and phenomenology on the other. I might then cherish the hope that the uncovering of such connections could have a therapeutic effect upon the philosophical struggles of our day.

What I said in my working hypothesis about Berkeley's influence on

353

A-T. Tymieniecka (ed.), Analecta Husserliana, Vol. XVI, 353–363.
Copyright © 1983 *by D. Reidel Publishing Company.*

Husserl was based mainly on the *Logical Investigations* and on the first volume of Husserl's *Erste Philosophie*. I knew, however, that this evidence cannot possibly be claimed as sufficient to justify my thesis. In these texts Husserl deals with Berkeley's positions in the context of specific problems of a restricted scope. He does not refer directly to Berkeley's doctrine in its entirety.

In my intial investigations, I looked for evidence to substantiate my thesis in the Husserl archives at Louvain. There, among some still unpublished Husserl manuscripts, I found one[3] which fills at least part of the evidential gap. This manuscript consists of an introduction to philosophy, and seems to be a nearly complete text of a semester course read by Husserl twice at the University of Freiburg: the first time when he was offered the philosophy chair in 1916 and the second time in 1918. The first chapter of this introduction (about 8,000 words) deals almost exclusively with the principles of Berkeley's philosophy. The other chapters are entitled 'Leibnitz − Formal and Material Ontology,' 'Criticism of Kant,' and 'A Final Look at post-Kantian Philosophy.'

This is quite a peculiar selection of topics for an introductory general course in philosophy. The pattern apparently reflects Husserl's particular conception of what his students needed − then and there − to cross the threshold of phenomenology or to begin philosophizing (which was more or less the same thing for Husserl). Two aspects are particularly significant for my thesis: (1) that Husserl chose to open his introductory course with a whole chapter on Berkeley, and (2) that we are given in this chapter a direct and extensive analysis of Berkeley's basic philosophical principles.

II

Two strata of content may be distinguished in the text under consideration: a historical one and a systematic or critical one. At the historical plane Husserl attempts to present Berkeley's doctrine as an intermediary phase in the history of the classical British empiricism, i.e., as the link between Locke and Hume. This development Husserl interprets as follows: Locke was a dogmatic dualist who argued that reality consists of two kinds of substances, material and spiritual. It is, however, in Locke's own system that the notion of substance suffers erosion. Still, as Husserl rightly points out, Locke was unable to offer a world picture not founded on a dualistic metaphysics.

Berkeley reduced the number of substances to one, thereby replacing the dualistic conception by monism. His arguments were intended to under-mine the notion of material substance and enable him to present reality as spiritual through and through.

Hume went even further than Berkeley by attempting to do away with spiritual substance as well. He eventually ended in the arms of skepticism.

So far the historical layer of the text. On the systematic level Husserl seems to disconnect his analysis of Berkeley's theory from its historical context. He deals with it directly, examines the content of its propositions, their coherence and truth, paying particular attention to the problems which though raised were not solved by Berkeley himself and which ought to concern us "here and now." In what follows I shall concentrate on this systematic aspect only.

Husserl opens the discussion with what may be considered Berkeley's central thesis: "As to the phenomena of things — their *esse* is *percipii*. Only spirits have absolute existence" (Dingerscheinungen — ihre *esse* ist *percipii*. Absolute Existenz haben nur Geister, p. 2a).

In this formulation the Berkeleyan thesis clearly presupposes one onto-logical *distinction* and contains two ontological *statements*. The distinction is between relative and absolute existence, while the statements are: (1) absolute existence is to be assigned to spirits (Husserl later speaks in the same sense of the spiritual subject, the knowing self or the percipient); and (2) to the physical objects we may attribute only relative existence, i.e., the existence in the mind of a subject.

In view of this pattern of Berkeley's central thesis the discussion divides naturally into two stages. Husserl deals first with physical bodies and their mode of existence, with Berkeley's explication of the concept of a physical thing and of *the world* of physical things (including organisms), i.e., the concept of nature. He stresses the originality of Berkeley's analysis of these notions, but also criticizes them. He tries to tell us how are these concepts to be amended in order to make them phenomenological.

The same applies to the *second* stage of the discussion, where the statement about mind and spiritual subject is considered. Husserl presents Berkeley's conception of the mind and the self, as Berkeley had meant it, to be entirely spiritual. Here too the smallest alterations are sought, which when introduced into these concepts will make them acceptable phenomenologically.

To put it otherwise, Husserl does not think that there is any need to overthrow these two conceptions of Berkeley's in order to arrive at the correct rendering of nature and mind. We only have to overcome their limitations.

Let us now see how Husserl deals with Berkeley's explanation of the first part of his central thesis and defense of it, i.e., how he explicated the ontological status of the physical object and of nature in general.

A subtle distinction is detectable here between a direct and positive justification and an indirect one by negation. Although Husserl called Berkeley's theory "paradoxical immaterialism," he did not forget to mention that Berkeley himself had denied explicitly the paradoxical character of his theory; it was for him but an adequate expression of the attitude of the plain man, of common sense, of the natural world-outlook, a theory which would appear self-evident to anyone who did not allow the philosophers to maneuver him into a blind alley.

At this junction Husserl mentions a number of possible and actual objections to Berkeley's theory, which in one form or another can be found in Berkeley's texts themselves, like the objection contained in Locke's so called "representational theory" of perception in particular and knowledge in general; the objection based on the distinction between primary and secondary qualities, which according to Husserl lies at the roots of the natural sciences; possible considerations as to the knowability of material substances, on the assumption that they exist; etc. In these dialectical arguments, Husserl argues, Berkeley employed a concept, or a set of concepts, which deserve a philosopher's closest attention.

Concerning the constructive side of the argument. Berkeley argues that in order to see whether his theory reflects faithfully the common-sense outlook, it suffices to examine how the plain man applies the word "existence" or "exist" to what is called "things." Husserl writes:

When I say "this writing table exists," it means that I see it or feel it; it is true that I also say that it exists when I am not in my study; then however only in the sense that I will see it if I enter the study.

Clearly, this "enter the study" reduces for Berkeley to sequences of groups of sense-data (*Gruppenfolgen von Sinnesdaten*), which are connected empirically with the re-appearing of the ideas of the writing table in an actual impression. But the question may be asked, how would it be if I were not around at all, if I died; would the table, according to the natural attitude, continue to exist? Certainly, replies Berkeley. That the table exists does not mean that it is being perceived just by myself, or that it can be perceived by myself under certain experiential circumstances. It means instead that either I or somebody else, man or God perceives it. (pp. 3a–4a)

It may seem that Husserl is simply paraphrasing here a fragment of the third paragraph of Berkeley's *Principles of Human Knowledge*. A closer look reveals, however, that Husserl inserted into this fragment some important exegetical hints which make the quotation suitable for our discussion.

Husserl added to the fragment his own explanation of the phrase "enter the study," thereby complicating Berkeley's example, in which only one physical object is included: "my writing table." Another physical object is now added to it, at least by implication: "my living body." It is not against Berkeley's contention to consider the living human body, mine, yours, anybody's living body, as an example of a physical object. However, even a brief reflection will show that such a classification yields considerable difficulties. We will return to the problem of the status of the living body later. At this stage let us restrict our argument to the realm of physical objects exclusive of my living body. (For the same reason we will postpone to a later stage the problem of *the other*, man or God — it is "other spirit" in Berkeley's text, but this makes no difference for our present argument.)

It is not easy to see exactly how Berkeley defends the first part of his thesis, unless we take into account a number of distinctions presupposed in his reasoning. The first distinction is between simple ideas, e.g., a smell, a tinge of a color, a sound, a feeling of warmth or cold, and complex ideas or complexes of ideas, like a table, a house, or the body of an animal. Berkeley did state his position with reference both to the simple and the complex ideas. Some of his conceptions, however, hinge upon a theory according to which a certain relation holds between the former and the latter. It is that the simple ideas are also the more elementary, the primary ones, or that the simple ideas are those which are *given* to each of us in the fundamental sense of givenness, while the complex ideas are products of our experience, the outcome of an experiential process in which our mental faculties, such as association and habit, are involved. Indeed, it is impossible to exaggerate the importance of this assumption, however vague and unsystematic its formulation in the text may be.

The other distinction to be taken into account in this context is a more subtle one; it is the distinction between the *truth criteria* of a statement of the form "*x* exists" and the *meaning* of such a statement, or between the meaning of the concept of existence and the correctness of its application to particular cases. Berkeley denies the need of holding these two aspects of existential statements apart, but some of the difficulties in his position, indicated by Husserl, will not be understood properly if this distinction is overlooked.

Against the background of these distinctions the following dialogue can be imagined, as a result and conclusion of Husserl's critical analysis of Berkeley's position.

(1) *The materialist or epistemological realist*: It is clear to me a priori that

there are material things which endure in their identity and possess a mani-fold of properties. Under certain conditions I perceive these things in their identity, endurance, and at least some of their properties.

(2) *Berkeley, speaking on behalf of the plain man*: Nonsense! I do not know anything a priori about material things and have no reason to assume that they exist. I know what my eyes see, my hands touch, and generally, what I experience by my senses. This table exists for me if I see it, can touch it, move it, or even break it. To begin with, there are single, simple sensations. Experience connects one sensation with another, one datum with another. It transforms the separate elements of sensation into complexes of sensations of relative endurance. For easier identification in practical life these complexes are given names. But all this is a product of my experience by means of association, habit, perhaps also anticipation and memory. There is nothing more. I do not presuppose identity and endurance; I have therefore no right and no need to assume the existence of material things outside of mind and experience.

(3) *Husserl*: Canceling the a priori assumption of the existence of things perceivable but not actually perceived by anybody is one of the most impor-tant decisions in the whole history of philosophy. This is the beginning of an immanent, empirical philosophy of which my phenomenology is but a logical continuation. There is no reason, however, to ascribe it to the plain man, the prephilosophical attitude. In our prephilosophical behavior and beliefs we indeed assume that this table exists when nobody perceives it. Moreover, we assume that this table, as much as any other physical thing, has a unity of properties and endures in its existence. This assumption *is* a part of the natural attitude, an attitude which should be taken into careful consideration in our philosophical reflection; we have to include it into our description and analysis of consciousness, even if it is entirely clear to us that the assumption of things not mind-dependent is groundless.

III

We may now consider the second part of Berkeley's central thesis, as pre-sented and criticized by Husserl. But before we do so a remark ought to be made on the status of my living body. Husserl characterizes Berkeley's position on this matter as follows:

my living body,[4] like any other material thing, is no more than an economy-of-thought complex of sensual ideas; hence, my living body is perceived content, it is not a perceiving

subject . . . . The connection between my mental experience and that complex of ideas
which I call my living body is given to me at first empirically . . . . (p. 7a)

Berkeley's position concerning the body of the human subject seems to be
accurately presented. It appears that even here Berkeley remained faithful to
his basic ontological conception and did not hesitate to define his own body
as a complex (or a series of complexes) of sense perceptions. Indeed, he did
not ignore the connection between himself as a spiritual subject and his body.
However, the connection was conceived by him merely as an empirical fact
and thus the ontological scheme remained as it were intact.

Husserl does not, however, draw attention to the fact that Berkeley had
overlooked the peculiarity of that connection, which is that I not only sense
my body or some of its organs, but also *by means* of it I sense *it* as well as
other bodies. Or more precisely, my living body is sensing and sensed at the
same time, and there is no way here to separate the subjective from the
objective aspect of my experience.

We are all aware of the difficult problem involved in this peculiar fusion
of objectivity and subjectivity in the subject's body. By now there should
of course be no question about Husserl's interest in this problem. There is,
however, no trace of it in the text under discussion. Husserl missed here
a rare opportunity to delineate one of the grave limitations of Berkeley's
philosophy, its failure to give a philosophical account of the human living
body as the body of the subject.

IV

Up till now we have dealt with the immanent realm of knowledge, the knowledge the object of which allegedly exists in consciousness, in the mind. Is
there also a transcendent knowledge, the objects of which are independent
beings whose existence is absolute? Berkeley's answer to this question is
of course affirmative. Such knowledge is a fact and we may define at least
three kinds or modes of it: (1) my knowledge of myself as a spirit, as an
independent subject; (2) my knowledge of the other, the fellowman as a
spirit, as an independent subject; (3) my knowledge of God as a spirit and an
independent subject in the highest degree.

In the text under consideration Husserl shows that Berkeley produces
a special argument for each of these three kinds of epistemological transcendence. In the first instance it is a direct but absolutely certain intuition;
in the second it is analogical inference and empathy, while in the third

it is an etiological argument, i.e., an argument based on the principle of causality.

Husserl writes on the first kind of transcendence: "And what about knowing the mind? At first we have knowledge of this kind directly in the form of reflection . . . . " And he continues:

The self, the spiritual subject, is not given in the way that a complex of ideas is given; it has nothing to do with the senses at all. It is not perceived by sensation but by reflection. Moreover, it is so perceived intuitively in full certainty and indubitability; and it is so perceived as an active principle, as a subject of a manifold of activities, as a spiritual substance. (p. 7b)

At this junction Husserl underlines Berkeley's position concerning the concept of substance in general:

Thus, substance in general is not an empty word or contradictory notion, according to Berkeley; only the material substance is something of this kind. The concept of substance is justified and necessary, and as such it is the concept of a spiritual subject. To be substance means to be active or to live as a self, to operate as a self, to activate causal connections and experience them. But it also involves having, as a self, sense ideas and to act on them as we do in our sense imagination. (p. 7b)

While Husserl appreciates Berkeley's efforts to disclose the structure as well as the dynamics of the mind, he criticizes the final outcome of these efforts. First of all, Husserl finds the notion of *givenness* to be problematic. Berkeley opens his analysis with an assumption that the self is given to each of us intuitively in reflection. However, givenness is precisely what is characteristic of the idea, the relationship between the object of sense perception and the subject experiencing that object. No wonder, therefore, Husserl maintains, that Berkeley had to qualify his statement immediately.

The subject *is not a sheer datum*. His being is active, it is activity, and this you do not come across as you do with a sense datum. Therefore Berkeley refrains altogether from applying the word "idea" to the spirit. He favours a precise and thereby restricted concept for the term "idea," such that denotes (umgrenzt) only *passive givenness*. (p. 8a; my italics)

We may generalize and say that Berkeley attempted to explicate the difference between relative and absolute existence, between being perceived and perceiving, between object and subject in the fundamental sense of these notions. At their outset these attempts deserve recognition as the beginning of the phenomenological method in philosophy. However, if one thinks of them as more than just a beginning, they must be viewed as a failure.

Berkeley failed because he stopped short of discovering *intentionality*, which characterizes more than any other feature the structure of consciousness and the mode of being of the spirit.

This criticism will not be properly understood if we fail to take into account the train of intricate reasoning, the aim of which in Berkeley's doctrine is to demonstrate the transcendence involved in the knowledge of God. The argument in a somewhat free formulation is as follows. Let us examine the range of relationship holding between consciousness and its natural objects, the ideas. Among various distinctions which can be made in this field one stands out as self-evident and extremely important: on the one hand, there are ideas and complexes of ideas which seem to be forced upon me, and my perceiving them does not depend upon my will; on the other hand, there are ideas and complexes of ideas which I entertain in consciousness at will and on my own initiative. When I look at my writing table I cannot help seeing its shape and color; when I touch it I have no choice but to feel its smoothness or roughness; when I try to lift it I have to feel its weight; etc. A whole manifold of complexes of ideas with its various applications is being forced on me and there is nothing I can change in this respect. I may indeed try to stop the activity of my senses. I may close my eyes, block my ears, avoid touching. However hard I try to do it, I can hope for only a partial success. Be it as it may, whenever I enable my senses to act, the contents of my sense perceptions and experiences are not under my control. They are given to me, forced upon me.

It is different when we deal with imagination. Here I am free to entertain images and to connect them into various complexes at will. I can introduce any changes I like into them. Nothing can stop me from transforming in my imagination this room into a tennis court where a match between Borg and myself is taking place, or from inviting Immanuel Kant here and have him lecture on Berkeley's philosophy, for example. Everything is now under my control.

This clear-cut and fairly obvious distinction gives rise to some puzzling problems. One of them, which had baffled Berkeley, is as follows. From my inspection of these inner dynamics of my consciousness I can learn at least two things. The first one is the notion of a cause. The second one is the fact that I myself function sometimes as a cause, that I possess causal efficacy. I can learn both of these things from the act of directing my attention to a certain object or aim, for example, an act I so often experience.

Picking up the various threads of reasoning that Husserl ascribes to Berkeley it becomes clear that while I possess causal efficacy in the domain of imagination, I am deprived of it in the domain of sense perception. Here the question arises, What causes all these phenomena, all the ideas which are forced on me in the way they are forced on me? And according to Berkeley's tacit assumption, a cause they must have, since nothing happens without a cause.

If *I* am not the cause of nature and of the specific order which prevails among its phenomena, their cause must be outside of my mind. It cannot be a material cause, since there are no material things outside of the mind. Hence the cause must be spiritual. I cannot however locate it in the alter ego, in other people. Empathy and inference by analogy tell me that they are subjects of a similar nature to mine, i.e., they are also deprived of causal efficacy regarding that which reveals itself in my experience as the world of physical objects.

Berkeley's conclusion is, as we all know, that we have to assume the existence of an infinite spirit or mind, a subject whose causal efficacy applies not only to imagined objects and states of affairs, but also to physical nature; in other words we have to assume the existence of a deity.

And again, although Husserl considers this analysis of the structure and dynamics of mind a good beginning of phenomenological procedure, at the last resort he regards it as a gross simplification. The source of this simplification or even distortion is again Berkeley's failure to discover the basic feature of the structure of consciousness which we now call constitutive intentionality. The constitutive intentionality as a basic and entirely primordial characteristic common to both modes of relationship between consciousness and its objects is prior to them. Were Berkeley aware of this point, he would not be trapped into the theoretical impasse, from which he had tried to extricate himself by means of the theological hypothesis.

This was Husserl's criticism of the second half of Berkeley's main thesis. The argument did not receive a detailed treatment in the text under discussion, but this could hardly be expected. After all, the phenomenology of Husserl may be regarded as an elaboration and development of this critical argument.

The Hebrew University of Jerusalem

NOTES

[1] An earlier, Hebrew version of this paper was published in the *Iyyun: A Hebrew Philosophical Quarterly* **29** (1980): 44–54.

2 In the introduction to my Hebrew translation of Husserl's *Cartesian Meditations* (Jerusalem: Magnes Press, 1972), p. 21.

3 Husserl papers, Louvain, F.I. 30; hereafter cited by page in the text. I am immensely indebted to the director of the Husserl Archives, Professor S. IJsseling and to his collaborators in the Archives, Professor Karl Schumann, Dr. Edy Marbach, and G. Van Kerckhoven, for their assistance in locating the manuscript and transcribing it for me. The quotations from it are published with the permission of the Husserl Archives; the translation from the German original is mine.

4 *Leib*: the German language has a very wise distinction between body in general, any body – *Körper*, and the living human body, *Leib*. The difficulties of rendering this distinction in English haunt all of Husserl's translators (I had the same problem translating Husserl into Hebrew). In translating *Leib* as "living body" I followed Dorion Gairn, the translator of the *Cartesian Meditations* into English. I am not very happy with this rendering, but cannot think of a better one.

ANNEX

ANNA-TERESA TYMIENIECKA

THE OPENING ADDRESS OF THE SALZBURG CONFERENCE

Eure Exzellenz!
Herr Präsident!
Liebe Mitarbeiter des Weltinstitutes für Phänomenologie!
Verehrte Gäste!

Im Namen des Weltinstitutes und als Programmdirektor des achten Kongresses für Phänomenologie möchte ich vor allem unseren beiden hochverehrten Gastgebern, seiner Exzellenz dem Herrn Erzbischof von Salzburg, Dr. Karl Berg, und Professor Georg Pfligersdorffer, Präsident des Internationalen Forschungszentrums für die Grundlagen der Wissenschaften, einen aufrichtigen Dank aussprechen für die so schönen und freundlichen Worte des Willkommens mit welchen sie uns gewürdigt haben. Es ist uns eine besondere Freude zu sehen, dass sie unserem philosophischen Bemühen ein so tiefes Verständnis für die Rolle der Philosophie in unserer dürftigen Zeit entgegenbringen.

Ich möchte jetzt ein paar Worte der Begrüssung an Sie richten und auch etwas über den Kongress selbst und über die Arbeit des Weltinstitutes hinzufügen.

Es genügt einen Blick in den Saal zu werfen, um zu sehen, dass wir hier die Zusammenkunft einer geistig-philosophischen Gemeinschaft vor uns haben. Oder noch besser wir haben vor uns eine Weltfamilie die das Weltinstitut für Phänomenologie bildet.

Unsere alten und neuen geschätzten Mitarbeiter sind hier aus vier Kontinenten und 12 Ländern zusammengekommen, um die weitreichenden philosophischen Diskussionen, in welchen viele von ihnen seit mehreren Jahren stehen, gemeinsam weiter zu führen.

Es ziemt sich denn, das ich zu-erst die Repräsentanten der dem Institut angegliederten Zentren und Gesellschaften begrüsse.

Dappertutto mi rallegro tanto di rivedere gli nostri amici italiani con la professoressa Angela Ales Bello, directrice del Centro Italiano delle richerche fenomenologiche di Roma. Puis quelle joie d'avoir avec nous Madame Maria da Penha Petit, directrice de la section française de notre Institut à Paris. Combien de Congrès avons nous déjà organisé ensemble entre les italiens et les français! It is with greatest pride that we welcome also Professor Luciana

367

A-T. Tymieniecka (ed.), Analecta Husserliana, Vol. XVI, 367–370.
Copyright © 1983 by D. Reidel Publishing Company.

O'Dwyer, who represents the Australasian Phenomenology Society, recently established and affiliated with us.

Heartiest welcome to an old friend Professor Zvi Bar-On who represents our Israeli Phenomenological Society. Professor Hans Rudnick, der sich der praktischen Leitung des jetzigen Kongresses gütigst gewidmet hat, repräsentiert unsere Internationale Gesellschaft für Phänomenologie und Literatur.

Mit einem besonderen Vergnügen möchte ich unsere japanischen Phenome-nologen, vor allem den werten und lieben Mitarbeiter Professor Yoshihiro Nitta und Hirotaka Tatematsu aus Tokyo und Nagoya begrüssen, die schon seit langem mit uns in einer äusserst produktiven Mitarbeit stehen.

Alle Honoratioren, alle Teilnehmer, und alle öffentlichen Gäste, seien Sie uns alle herzlich willkommen!

Nun darf ich vielleicht einige Worte über unsere Anliegen sagen. Was für ein Ziel verfolgt unser Kongress? Was für ein Ziel verfolgt das Weltinstitut für Phänomenologie? Was für eine Rolle trauen wir der Philosophie überhaupt zu?

Wie angekündigt ist es der achte Internationale Phänomenologie Kongress des Institutes. Das Institut selbst aber und die Internationale Husserl Gesell-schaft, die ihn vorbereitet hat, haben zu Angang des Jahres 1979 die erste Dekade ihrer Existenz und ihrer internationalen phänomenologischen Arbeit gefeiert. Und mit grosser Genugtuung sehe ich Dr. Erling Eng aus Lexington, Kentucky, hier vor mir. Als ich im April 1969 auf unserem ersten Phänomeno-logie Kongress in Waterloo, Ontario, die Internationale Husserl Gesellschaft gestiftet habe, war Dr. Eng der Mitbegründer und hat die Entwicklung der Gesellschaft unterstützt und gefördert. Was das Institut anbelangt, so es ist offiziel und rechtlich als eine akademische Forschungs- und Lehrinstitution gegründet und registriert. Es ist vollkommen unabhängig und wird autonom durch uns selbst verwaltet. Es veranstaltet ein einheitliches Forschungspro-gramm über die Grundlagen und die weitere Ausarbeitung der Phänomeno-logie, erstens als Philosophie, die alle philosophischen Disziplinen würdigt, wie: Epistemologie, Anthropologie, Ethik, Esthetik, Ontologie, Kosmologie und Metaphysik. Sie wird aber in unseren Forschungsplänen, die wir Schritt für Schritt in unseren Symposien, Seminarien und Kongressen ausführen, besonders als die Grundlage der Interdisziplinären Forschung ausgearbeitet. D.h. wir folgen dem Traum von Aristoteles, Leibniz und Husserl selbst, in der Philosophie — phänomenologischen Philosophie — eine *mathesis Universalis* zu entfalten.

Dieses ist eine grosse und äussert wichtige Aufgabe. Die Kommunikation zwischen den im aktuellen Denken isolierten Wissenschaften wiederherzustel-len, ist bestimmt eine grosse Aufgabe, die nur die Philosophie ausführen kann.

Eine Aufgabe, die nicht nur für den Fortschritt der Wissenschaften selbst, sondern auch für ihr Bewerten im einfachen menschlichen Leben, für dieses Leben selbst, und unsere Kultur, eine dringend erforderte Aufgabe ist.

Husserl selbst hat das Vorhaben der Phänomenologie auf eine ähnliche Weise definiert. Das Vorhaben unserer Forschung hat sich aber, auf breiter und tiefer angelegte Ansätze stützend, eine besondere Wendung genommen: die *Phänomenologie des Menschen-in-seiner-Kondition* ist die ursprüngliche Grundlage dieses interdisziplinären Dialoges, in welchen wir in unseren Seminarien, Symposien und Kongressen eintreten. Auf den Menschen kommt es doch immer an!

Aber nicht nur die Wissenschaften, sondern auch der einzelne Forscher, Gelehrte, Philosoph ist in unserer Welt isoliert. In dieser schnellebigen Welt gibt es kulturelle Kreise, aber keine 'Schulen' mehr. Der Philosoph ist von einer Aspiration auf Weisheit zu einer beruflichen Tätigkeit herabgesetzt.

Auf den üblichen Kongressen anderer Gesellschaften werden normalerweise die beruflichen Interessen weit mehr verfolgt als das Suchen nach wissenschaftlicher Kommunikation, der Austausch der Forschungsresultate und das gegenseitige Anspornen. Dieser Institut steht in einer radikalen Reaktion dazu. In einer Welt, wo die intellektuelle und geistige Kommunikation der Gelehrten zugrunde geht, geht auch der Mensch zugrunde. Es ist unser Anliegen, die Menschen-Gemeinschaft der Philosophen aller Kulturen, Rassen, Sprachen und Konfessionen in einem gemeinsamen philosophischen Forschen herzustellen.

Was würde sich dazu besser eignen, als die Phänomenologische Philosophie? Wie Sie aus dem Programm ersehen können, hat Husserl's Gedankengut, das langwierig in der deutschen Sprache erzeugt wurde, sich in allen Hauptsprachen der Welt heimisch gemacht. z.B. die zahlreichsten phänomenologischen Veröffentlichungen hat man das letzte Jahr in der japanischen Sprache verfasst! Und obwohl man phänomenologische Untersuchungen auf französisch, holländisch, polnisch, spanisch, hebräisch, koreanisch, usw. ausführt, behalten sie doch immer einen gemeinsamen geistigen Kern. Die phänomenologische Reflektion ist offensichtlich tiefer in der menschlichen Natur angelegt als die kulturellen Besonderheiten.

So viel über unsere Arbeit. Ich möchte noch meinen tiefen Dank an unsere würdigen Gastgeber, Seine Exzellenz den Herrn Erzbischof und den werten Kollegen Herrn Präsident Pfligersdorffer für ihre grosszügige moralische, und praktische Hilfe aussprechen, ohne welche dieser Kongress nicht zustandegekommen wäre. Dabei darf ich auch nicht vergessen, einen schönen Dank an die werten Kollegen von der Österreichischen Phänomenologie

Gesellschaft, Professoren Karl Ulmer und Peter Kampits von der Universität Wien und an den Universität-Dozent, Dr. Hans Köchler von Innsbruck für ihre praktische Hilfe in den Vorbereitungen des Kongresses ausrichten.

Eure Exzellenz, Herr Präsident, verehrte und liebe Kollegen, ich erkläre den Kongress offiziel eröffnet.

Das Wort hat jetzt Professor Karl Ulmer, der als Präsident der Osterreichischen Gesellschaft für Phänomenologie sie auch begrüssen wird.

INDEX OF NAMES

ANALECTA HUSSERLIANA

The Yearbook of Phenomenological Research

Editor:
ANNA-TERESA TYMIENIECKA
The World Institute for Advanced Phenomenological Research and Learning
Belmont, Massachusetts